DISINFECTION

WATER AND WASTEWATER

edited by

J. DONALD JOHNSON

Professor of Environmental Chemistry
Department of Environmental Sciences and Engineering
School of Public Health
University of North Carolina
Chapel Hill, N.C.

P. O. Box 1425, Ann Arbor, Michigan 48106

Library of Congress Catalog Card No. 74-14425
ISBN 0-250-40042-1

Manufactured in the United States of America

PREFACE

This book describes the chemical and microbiological basis of the treatment of drinking, swimming and wastewater with chlorine, bromine, iodine and ozone. The main concern of water disinfection practice long has been with chlorine and its effectiveness in killing bacteria. Today, toxic compounds formation as a result of chlorination, and the questionable efficiency of chlorine as a bactericide and also as a cyst- and virucide, prompt the need for critical analysis of the alternatives to water disinfection by chlorine. The primary objective is better control of water chlorination and/or disinfection with other chemicals—to produce a minimal impact on man and our environment. Control of water disinfection practice is presented by discussing analytical methods available to assure that the minimum concentration of disinfectant has been added, consistent with the combined objectives of microbiological safety and minimum environmental impact.

Chemistry of water chlorination is discussed because the analytical methodology measures the effective disinfectant forms of free available chlorine, hypochlorous acid, as well as many relatively poor disinfectants such as hypochloride ion and the chloramines. Judging the effectiveness of microbiological disinfection depends on the quantitative assessment of the germicidal efficiency of the disinfectant. The concentration of the effective chemical species of disinfectant, multiplied by the time that this concentration has been in contact with the microorganism to be disinfected, equals the total dose. However, this total dose response to disinfection processes is not simple, but requires careful treatment and analysis.

This volume answers the question of how to assess efficiency of disinfection so that the process can be controlled with an analytical method of required selectivity, at the minimum necessary and desirable concentration (avoiding cost and toxicity problems caused when excess disinfectant is used).

Unfortunately, many problems are associated with the safety of handling gas chlorine, such as toxicity, tastes and odor, and also the ineffectiveness of chloramine residual. This makes it necessary to continue the search for better disinfectants and application methods in water and wastewater treatment. Bromine, bromine chloride, iodine and ozone have recently enjoyed considerable renewed interest as possible alternates to chlorination. The chemistry and microbiological efficiency of these disinfectants is discussed and compared (among themselves, and especially with chlorine) in their effectiveness against the more difficult to disinfect microbiological systems: viruses and cysts.

Sanitary engineers, chemists and biologists, as well as water and wastewater treatment plant personnel will find practical data on how to do a better job in treating their water and wastewater. Tools are also presented for developing a more critical understanding of the disinfection process, for the classical old as well as the new, soon to be tried water disinfectants.

Much of this volume evolved from conferences sponsored by the division of environmental chemistry of the American Chemical Society. Considerable additional material then was required to make a complete and useful book.

J. Donald Johnson
Chapel Hill, June 1975

CONTENTS

Chapter

1 Aspects of the Quantitative Assessment of Germicidal Efficiency 1
J. Carrell Morris

2 A Multi-Poisson Distribution Model for Treating Disinfection Data 11
Julie H. Wei, Shih L. Chang

3 Recycle–What Disinfectant For Safe Water Then? . . 49
Mark A. McClanahan

4 Water Disinfection–Chemical Aspects and Analytical Control 67
A. T. Palin

5 Selection of a Field Method for Free Available Chlorine 91
Charles Sorber, William Cooper, Eugene Meier

6 Interhalogens and Halogen Mixtures as Disinfectants . . 113
Jack F. Mills

7 The Comparative Mode of Action of Chlorine, Bromine, and Iodine on f_2 Bacterial Virus 145
V. P. Olivieri, C. W. Krusé, Y. C. Hsu, A. C. Griffiths, K. Kawata

8 Bromine Disinfection of Wastewater Effluents 163
F. W. Sollo, H. F. Mueller, T. E. Larson, J. Donald Johnson

9 Bromine Disinfection of Wastewater 179
J. Donald Johnson, William Sun

10 Comparison of Bromine, Chlorine and Iodine as Disinfectants for Amoebic Cysts 193
Richard P. Stringer, William N. Cramer, Cornelius W. Krusé

11 Disinfection of Water and Wastewater Using Ozone . . 211
E. W. J. Diaper

12 Economical Wastewater Disinfection with Ozone . . . 233
Harvey M. Rosen, Frank E. Lowther, Richard G. Clark

13 Chlorine and Oxychlorine Species Reactivity with Organic Substances 249
David H. Rosenblatt

14 The Chemistry of Aqueous Nitrogen Trichloride . . . 277
Jose Luis S. Saguinsin and J. Carrell Morris

15 Kinetics of Tribromamine Decomposition 301
Thomas F. LaPointe, Guy Inman, J. Donald Johnson

16 Dechlorination by Activated Carbon and Other Reducing Agents 339
Vernon L. Snoeyink, Makram T. Suidan

17 Comparative Inactivation of Bacteria and Viruses in Tertiary-Treated Wastewater by Chlorination 359
Peter P. Ludovici, Robert A. Phillips, Wayburn S. Jeter

18 Disinfection of Wastewater Effluents for Virus Inactivation 391
John T. Cookson, Jr., C. Michael Robson

CHAPTER 1

ASPECTS OF THE QUANTITATIVE ASSESSMENT OF GERMICIDAL EFFICIENCY

J. Carrell Morris

Division of Engineering and Applied Physics
Harvard University
Cambridge, Massachusetts

Although the bases for the quantitative expression of the effectiveness of germicidal agents have been known for more than 60 years, there has been relatively little application to the systematic tabulation of the relative potencies of disinfectants. Only a very small fraction of the total published literature on germicidal action is sufficiently complete or in form suitable for satisfactory quantitative analysis. Moreover, there is no consensus on which method of tabulation is most convenient. Probably the most common technique is to list the concentrations required to give a fixed percentage of kill with a given time of contact, but there is no unanimity with regard to either the percentage or the time. Other workers prefer to use the Chick's Law constant as a measure of relative effectiveness.

In recent years interest in germicidal research has been stimulated by concern for disinfection of wastewaters and inactivation of pathogenic viruses. A growing fraction of these studies give full attention to the dynamics of the germicidal process, providing the sort of information that can lead to fundamental tabulations.

It is time to try to achieve some general acceptance of terminology and forms for presentation of germicidal data. Proposals will be made in this paper in the hope that any discussion initiated will lead to a rational, systematic, and broadly accepted concordance of disinfectant efficiency for aqueous solutions.

THE SPECIFIC LETHALITY COEFFICIENT

Let us begin with a presentation of hypothetical but typical germicidal data in which the logarithms of the surviving populations of organisms, N, are plotted as a function of t, the time of contact of germicidal agent with the organisms, as shown in Figure 1-1. Often the data points for such a study are made to relate by a smooth curve of some sort, such as that of the dashed trace in Figure 1-1. Any such smoothed curve represents some degree of conceptualization, however; the soundest way to relate such data is empirically to connect the adjacent points by straight lines, as shown by the solid traces in Figure 1-1.

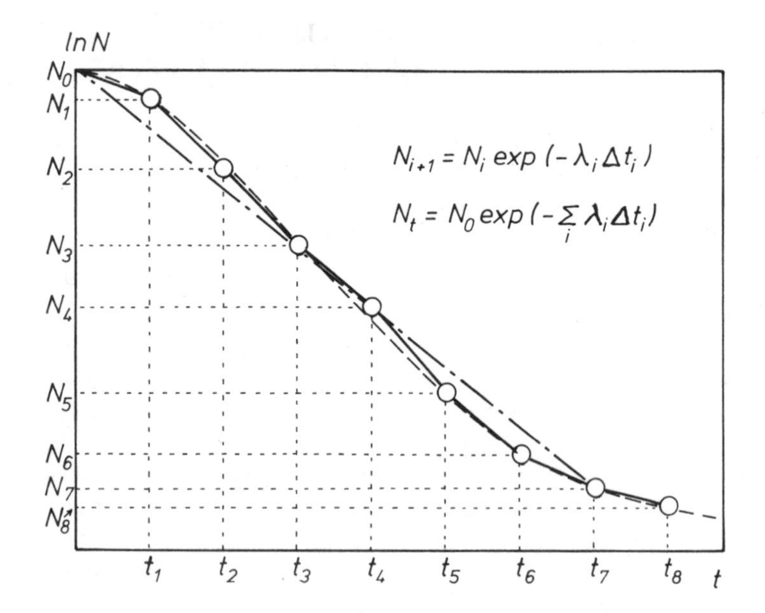

Figure 1-1. Graphical representation of model germicidal data.
N = numbers of viable organisms per standard unit sample
t = time of action of germicidal agent.

Each line segment connecting adjacent points may be represented by an equation

$$\ln(N_{i+1}/N_i) = -\lambda_i (t_{i+1} - t_i)$$

$$= -\lambda_i \Delta t_i \tag{1}$$

where $-\lambda_i$ is the slope of the line. These equations may also be written

$$N_{i+1} = N_i \exp(-\lambda_i \Delta t_i) \tag{2}$$

Moreover, the total decrease in N over the total t intervals of time is given by the expression

$$N_t = N_o \exp \left(-\sum_{i=1}^{t} \lambda_i \Delta t_i\right) \qquad (3)$$

There is no requirement that the successive λ_i values be the same. When they are, the data conform to Chick's Law, but in the general case the λ_i may vary with the course of the germicidal action as a result of changing response of the organisms, of changing concentration or form of germicide, of changing temperature, or possibly other factors. Depending on the situation λ may be treated as a function of the time or as a function of the surviving fraction of organisms, N_t/N_o.

It should be noted also that, in common with other first-order processes, the summation of equation (3) gives the same result as would be obtained if all the intermediate points were ignored and equation (2) were written simply for N_t and N_o.

The parameter, λ, is a reflection of the ease or rapidity with which the organisms are destroyed or inactivated. It may, therefore, be termed, appropriately, the susceptibility coefficient. It contains, as some kind of factor, the concentration of the germicidal agent, the dependence being generally expressed in the form

$$\lambda_i = \Lambda_i C_i^n \qquad (4)$$

where n is called the coefficient of dilution. Substitution of this relation into the term $\Sigma\lambda_i\Delta t_i$ of equation (3) then gives the expression

$$\sum_i \lambda_i \Delta t_i = \sum_i \Lambda_i C_i^n \Delta t_i \qquad (5)$$

When C_i is unity, the right side of this equation reduces to $\sum_i \Lambda_i \Delta t_i$.

Values of Λ_i relate to the germicidal process in two ways: one, they indicate the sensitivities or resistances of organisms to inactivation; two, they give the relative potencies of disinfectants at unit concentration. Accordingly, Λ may have two designations, one the *specific susceptibility coefficient* when organisms are compared, the other the *specific lethality coefficient* when germicides are compared.

Since there is no dependence on concentration of germicide in Λ, any variation in this parameter at constant temperature should involve the properties of the organisms being inactivated. Indeed, the most likely explanation for the variations in the logarithmic rate of inactivation at constant temperature and concentration of germicidal agent that are commonly observed in laboratory studies of disinfection appears to be a clumping or aggregation of the organisms. Yet it is not at all clear that this behavior is a fundamental property of the organisms, for the methods employed to culture and purify highly concentrated populations of microorganisms so that large N_o values may be realized may serve also to induce clumping.

If this is the case, then natural populations of the microorganisms should exhibit Λ values more nearly constant throughout the disinfection process, with values characteristic of isolated rather than clumped organisms. As a xeroth approximation and to facilitate additional considerations, it will be assumed throughout the remainder of this paper that Λ is a constant for a given reagent and organism at a specified temperature. The objective of quantitative germicidal studies then becomes the evaluation of values for Λ–specific lethality coefficients.

A problem that remains is the establishment of a single parameter equivalent to Λ for germicidal reactions in which λ is found to vary significantly during the course of the action. Methods are available for particular forms of variation; some of these appear in articles in *Disinfection*."[1] Discussion of such individual evaluations is beyond the scope of this chapter.

THE GERMICIDAL DOSE

If Λ is constant throughout the course of germicidal action, then equation (3) becomes

$$\ell n(N_o/N_t) = \Lambda \sum_i C_i^n \Delta t_i \tag{6}$$

The latter part of this expression, $\sum_i C_i^n \Delta t_i$, which can also be written as $\int C^n dt$ for situations in which the concentration of agent changes continuously as a function of time, may be termed the germicidal dose. Then

$$\text{dose} = D = \sum_i C_i^n \Delta t_i = \int_o^t C^n dt \tag{7}$$

E. L. Hall has fruitfully utilized this concept of dose to assess the effectiveness of chlorination in waters with considerable demand.[2] In his presentation, however, the additional simplification was made that n = 1.0, so that equation (7) was reduced to

$$D = \sum_i C_i \Delta t_i = \int C dt$$

Hall justified this simplification at some length for the germicidal species and organisms with which he was concerned. The simplification appears broadly applicable to other systems, at least in approximate fashion and provided the data are used for the same order of magnitude of concentrations as those from which they were obtained. Most of the evidence for n values much different from unity with potent aqueous germicides is based on scanty or unreliable data. It is difficult to carry out experiments over a wide range of concentrations to establish n reliably without encountering serious problems of bias in the measurement of time or of concentration.

TABULATION OF SPECIFIC LETHALITY COEFFICIENTS

Whether the simplification that n = 1.0 is used, the fundamental germicidal equation with Λ constant can be written in the form

$$N_t = N_o \exp(-\Lambda D)$$

where D is the dose as defined previously. For further discussion the simplified formulation $D = \int C dt$ will be used, however.

The particular dose, $D_L = \Lambda^{-1}$, can now be called a *lethal unit* or a *lethe* of the germicide under consideration. It is the dose required to reduce N_o to N_o/e. So, it is equal to the constant concentration needed to reduce N_o to N_o/e within unit time. The specific lethality coefficient, Λ, then is also the number of lethes provided by unit concentration for unit time.

There is a question of units for parameters like Λ and D_L. For practical reasons of size, intelligiblity and ease of application, I favor the use of mg/l and minutes as the standard concentration and time units respectively. Molar units of some sort are needed for theoretical comparisons, but such units are not likely to have broad engineering utility.

The use of specific lethality coefficients or some similar parameter for comparison of the potencies of germicides has the major advantage that the number is greater the more potent the agent, whereas tabulations of concentrations yielding a specified fraction of inactivation in a certain time exhibit the inverse relation. So, also, do tabulations of times required for a specified fraction of kill at a standard concentration.

Values of Λ can, however, be obtained readily from the other types of data. For example, this author presented in 1967 a tabulation of concentrations giving 99% inactivation of organisms within 10 minutes.[3] The equation

$$N_t/N_o = \exp(-\Lambda Ct)$$

then becomes

$$0.01 = \exp(-10\,\Lambda\, C_{99:10})$$

from which

$$10\,\Lambda\, C_{99:10} = 4.6$$

$$\Lambda = 0.46/C_{99:10}$$

Values of Λ so computed from the 1967 tabulation are shown in Table 1-1.

Table 1-1. Values of Λ at 5°C

Λ in (mg per ℓ)$^{-1}$ (min)$^{-1}$

Agent	Enteric Bacteria	Amoebic Cysts	Viruses	Spores
O_3	500	0.5	5	2
$HOCl$ as Cl_2	20	0.05	1.0 up	0.05
OCl^- as Cl_2	0.2	0.0005	<0.02	<0.0005
NH_2Cl as Cl_2	0.1	0.02	0.005	0.001

THE KILLING TIME

One of the problems encountered in simplifying the results of germicidal studies is the lack of an adequate concept of complete kill or inactivation, either practically or theoretically. Theoretically the percentage of inactivation according to standard equations is always something less than 100. Practically, the number of surviving organisms is reported as less than one per so many milliliters because of limitations of test methods and sample sizes.

There is a way around this dilemma, making use of the idea of a "whole-life" developed by H. A. Thomas for radioactive decay processes,[4] but equally applicable to any other first-order process with constant λ. The presentation that follows is essentially his, with some embellishments and emendations.

Assume a population of N_o particles–organisms in this instance–with a die-away parameter, λ, such that

$$-d\,\ell n\,N/dt = \lambda$$

The probability that a given organism will survive for a time period of zero to t is $\exp(-\lambda t)$. The probability that it will die or be inactivated is then equal to $1 - \exp(-\lambda t)$. With an initial N_o organisms at $t = 0$, the probability that exactly x of these will have died or been inactivated for the interval zero to t is the product of the probabilities for each of x organisms to die and each of $(N_o - x)$ to survive. This is given by

$$P(x) = \binom{N_o}{x} (1 - e^{-\lambda t})^x \, e^{-\lambda t(N_o - x)} \tag{9}$$

The probability that *all* the organisms will have been inactivated in time t, is

$$P(N_o) = \binom{N_o}{N_o} (1 - e^{-\lambda t})^{N_o} e^{-\lambda t(N_o - N_o)} = (1 - e^{-\lambda t})^{N_o} \qquad (10)$$

Set $P(N_o) \equiv \alpha$. For example, α might be 0.5, in which case there would be equal probability for complete disinfection and for incomplete disinfection. The previous inability to describe a complete killing time is shifted to a less than unit probability of complete kill, but this latter concept is easier to handle for it is used all the time intuitively in daily living.

The rationale for this approach is, in part, the fact that the mathematical equation relates to a continuum whereas the particles decaying or organisms being devitalized are a collection of discrete unitary individuals. When the numbers involved are small and particularly when the mathematical equations yield numbers indicating survival of a fraction of an organism, then the numbers must be interpreted on a statistical basis. Results can then be expressed either as a most probable number, as is done with standard tube methods, or as the probability of finding one or more survivors. The accuracy of data or computations is not decreased by expressing them in either of these ways. The true significance and variability are made explicit.

Equation (10) may be solved for t in terms of the chosen value for α, and known or specified values for λ and N_o. The resulting time, designated $\tilde{t}$ or t_{100}, may be considered the "killing time" (Thomas's "whole life") for the population, N_o.

From equation (10) and $\alpha \equiv P(N_o)$

$$\alpha^{(1/N_o)} = 1 - \exp(-\lambda\tilde{t}) \qquad (11a)$$

$$\lambda\tilde{t} = -\ln[1 - \alpha^{(1/N_o)}] \qquad (11b)$$

$$\tilde{t} = \frac{1}{\lambda} \ln[1 - \alpha^{(1/N_o)}]^{-1} \qquad (11c)$$

Because of the relation between $\tilde{t}$ and λ, it is also possible to express $\tilde{t}$ or t_{100} in terms of other commonly used kinetic parameters. Thus, the relation between $\tilde{t}$ and t_{90} is given by

$$\tilde{t} = t_{100} = t_{90} \log_{10}[1 - \alpha^{(1/N_o)}]^{-1} \qquad (12)$$

Define

$$\epsilon \equiv 1 - \alpha^{(1/N_o)} \qquad (13)$$

Then from equation (11b)

$$\lambda\tilde{t} = -\ln \epsilon \qquad (14)$$

From equation (13)

$$\alpha^{1/N_o} = 1 - \epsilon$$

$$(1/N_o)\ \ell n\ \alpha = \ell n(1 - \epsilon) \tag{15}$$

If the Stirling approximation, $\ell n(1 - \epsilon) \cong -\epsilon$, is made, then

$$(1/N_o)\ \ell n\ \alpha \cong -\epsilon \tag{16}$$

This approximation is valid within 5% for $\epsilon \leqslant 0.1$, corresponding to $N_o \geqslant 10$ to 100, depending on the value assigned to α. In most instances values of N_o will be much greater in germicidal studies.

Substitution of the approximate value of ϵ from (16) into equation (14) gives

$$\tilde{t} = -(1/\lambda)\ \ell n[(1/N_o)\ \ell n\ (1/\alpha)]$$

$$= (1/\lambda)\ell n[N_o(\ell n\ \alpha^{-1})^{-1}] \tag{17}$$

This equation provides a simplified relationship between $\tilde{t}$ and N_o for the selected α and established λ.

The choice of α is an arbitrary one, depending on the certainty of complete kill that is required. Three natural or simplifying choices are possibilities for an agreed-upon standard:

(a) Choice of $\alpha = 0.5$ is an instinctive one, representing a 50% chance that complete kill has been attained. For this choice of α and with the substitution $\lambda^{-1} = t_{90}$ log e, there results

$$\tilde{t} = t_{100} = t_{90} \log (N_o/0.693)$$

$$= t_{90}(0.16 + \log N_o) \tag{18}$$

(b) If somewhat less certainty of complete kill than $\alpha = 0.5$ can be accepted, then a choice of $\alpha = e^{-1} \cong 0.37$ is useful, for it leads to additional simplification. With $\alpha = e^{-1}$ equation (15) simplifies to

$$t_{100} = t_{90} \log N_o \tag{19}$$

and the total decay or reaction time is directly proportional to log N_o.

(c) If considerable certainty of complete kill is wanted, then a value of α near 0.9 seems logical. This would mean that 9 out of 10 samples would give negative findings. Such a result conforms approximately to the coliform standard that permits 10% of 10-ml MPN tubes to be positive. A convenient specific choice is $\ell n\ \alpha^{-1} = 0.100$, corresponding to $\alpha = 0.903$, for then

$$t_{100} = t_{90} \log (N_0/0.100) = t_{90} \log 10\, N_0 \tag{20}$$

The last of these possible choices provides the most advantages for use in working with disinfectants for water systems and is suggested as a standard definition of "killing time" or "complete kill."

Each of the equations (18), (19) and (20) predicts a linear variation of t_{100} with $\log N_0$ regardless of the particular value of α. That this variation does occur can be shown, for example, from the data of Chang on the thermal destruction of cysts of *E. histolytica.*[5] In Figure 1-2, the minimum times to achieve apparent complete kill are shown plotted against the logarithms of the initial inocula of organisms, N_0. The linear relationship is clearly shown.

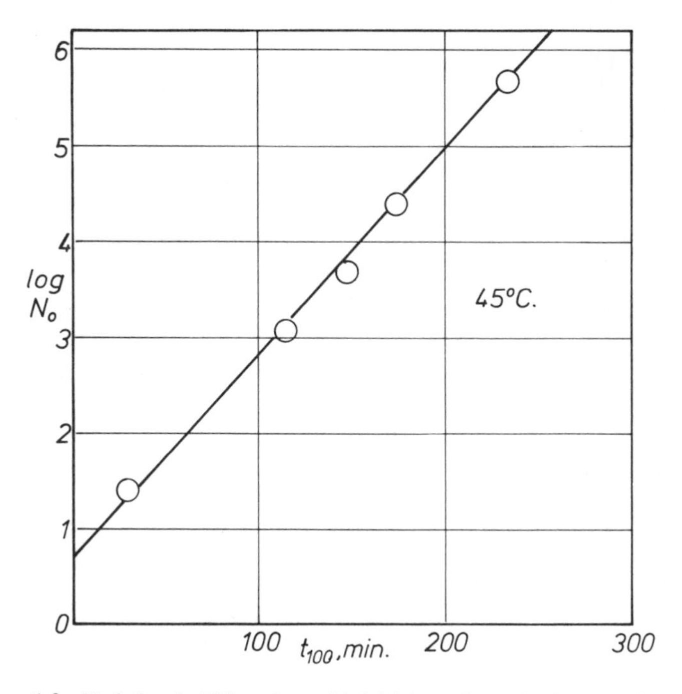

Figure 1-2. Variation in killing time with initial numbers of microorganisms. Thermal destruction of cysts of *E. histolytica* at 45°C.[5]

As pointed out by Chang, the data also suggest that he needed about six viable cysts to get a positive culture. The value of N_0 at the intercept $t_{100} = 0$ for the linear plot in Figure 1-2 may be regarded as a threshold N_0, a number of organisms required for a positive test result. It is clear that, according to the plot, values of N_0 equal to or less than the threshold value will show complete inactivation even when no biocide is present. With an intercept of log N_o = 0.78, the corresponding threshold N_o is 6.

There remains some question about the volume of water to which N_o refers. This must be the sample size or volume in which it is expected that zero active organisms will be found. So for example if, as in the standard MPN tests, it is the portions of 10 ml in which negative results are expected 90% of the time, then N_o should be the initial population of organisms per 10 ml. The quantity, 10 N_o, is then the organisms per 100 ml, a unit already in widespread use in water and wastewater standards.

The time for complete kill is related to the specific lethality coefficient, Λ, by the relation, $\Lambda C = \ell n\, 10/t_{90}$. This yields

$$t_{100} = \frac{1}{\Lambda C}\, \ell n(10\, N_o) \tag{21}$$

with equation (20). From this equation can be plotted either times required for complete kill at a given concentration(s) for complete kill in a given time, or doses for complete kill, all as a function of the initial population. Any one of these is a rational guide to disinfection practice, for it is clear that the degree of treatment needed for adequate disinfection will vary with the initial degree of contamination of the water.

REFERENCES

1. *Proc. Natl. Specialty Conf. on Disinfection* (New York: American Society of Civil Engineers, 1970).
2. Hall, E. L. "Quantitative Assessment of Disinfection Interferences," *Water Treat. Exam.* **22**, 153-174 (1973).
3. Morris, J. C. "The Future of Chlorination," *J. Amer. Water Works Assoc.* **58**, 1475-1482 (1967).
4. Thomas, H. A. Personal communication.
5. Chang, S. L. "Kinetics in the Thermodestruction of Cysts of *Endamoeba histolytica* in Water," *Amer. J. Hyg.* **52**, 82-90 (1950).

CHAPTER 2

A MULTI-POISSON DISTRIBUTION MODEL FOR TREATING DISINFECTION DATA

Julie H. Wei

Department of Civil Engineering
California State Polytechnic University
Pomona, California

Shih L. Chang

The National Environmental Research Center
U.S. Environmental Protection Agency
Cincinnati, Ohio

INTRODUCTION

As the chemistry of disinfectants in water and their interactions with interfering substances become better known, the state-of-the-art becomes more a subject of the effect of the microbiological physical state on the rate process. In analyzing disinfection data according to first-order kinetics, investigators have become increasingly puzzled by aberrant survival curves. Ironically, Chick, whose celebrated experimental studies laid the foundation of "Chick's law," observed exponential survival only in the destruction of anthrax spores; the survival curves obtained for the enteric bacteria and staphylococci were nonexponential.[1,2]

Chang pointed out that disinfection is kinetically a first-order rate process;[3] but in data treatment one must consider factors such as the cytostructure and the physical state of organisms, which may cause departure from this first-order exponential survival and confuse the investigator. In an earlier report Chang indicated that budding in young cultures of *Saccharomyces cerevicae* apparently produced survival data on destruction by X-radiation,[4] as if two Poisson variates existed in the yeast population.[5]

He also indicated that the existence of *Endamoeba histolytica*, only in the discrete state, was responsible for the consistent conformity of cysticidal data to first order kinetics.[6-9]

Berg and his associates[10] observed a persistent shoulder formation in survival curves in their study of the kinetics of enterovirus destruction by I_2 and analyzed their virus survival data by a model we call multi-hit. This model is similar to the multi-target model of the radiation literature.[11] Berg computed the values for k, the inactivation rate constant, from the linear section of the survival curves and obtained good results in their analysis. The values for n, the size of clumps, computed by this equation, may be regarded as the average size of all clumps; hence, a value of less than unity for n, which they obtained for Coxsackievirus A9,[10] seemed unreasonable.

Since clumps of organisms are most unlikely to be confined to one size, the multi-hit equation was expanded to a multi-multi-hit model expressed as

$$P_t = \sum_{i=1}^{N} A_i \, [1-(1-e^{-kc_i t})^{n_i}]$$

where P is the probability of survival after time t, N is the number of groups of clumps by size, n is the size of clump, A_i is the initial concentration (100%) of plaque- or colony-forming units, k is the destruction rate constant, and c is the destruction coefficient depending on the size and geometry of clumps.[4] Computation becomes too complex and time-consuming when the value for N is larger than 2.

In a more empirical approach Clark and Niehaus[12] developed a double summation equation for fitting aberrant survival curves. The summation equation was based on a number of simultaneous differential equations, each of which prescribes a state of destruction in which one plaque- or colony-forming unit is rendered nonviable. The equation is expressed:

$$S_i = \sum_{j=1}^{i} \sum_{n=1}^{i} \frac{(k_j \cdot k_{j+1} \ldots k_{i-1})(e^{-k_n t})(S_j^0)}{(k_i - k_n)(k_{i-1} - k_n)\ldots(k_j - k_n)}$$

where S is the nonsurvival, k is the destruction rate constant, and i and j are subscripts. For instance, S_i^0 is the initial state at t = 0 with all units surviving; hence, $S_i^0 = 0$. When $j = n$, $k_j - k_n = 1$.

This model assumes that the destruction rate constant is different at different states of clump destruction; it emphasizes the size of clumps, which dictates the number of the differential equations and must be assumed intuitively. The model is aimed at fitting a survival curve and finding the frequency distribution of clumps by size. The k in the last state of the destruction chain is, then, the rate constant for the destruction of discrete organisms.

The drawback of multi-hit model is that it is based on a simultaneous-hit concept. Hence an unusually large clump size is required to produce even a mild shoulder (Figure 2-1). In the destruction of a clump of organisms, nonsurvival of the clump as a plaque- or colony-forming unit is a result of cumulative "hits" during contact time. This clearly indicates the unsuitability of the multi-hit model for analyzing disinfection data.

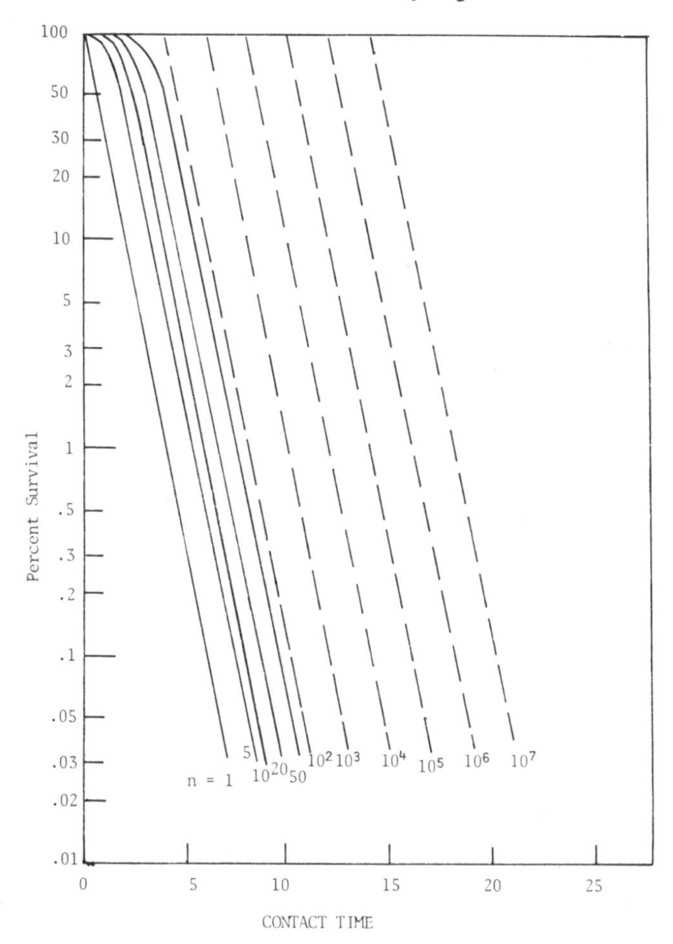

Figure 2.1. Predicted survival curves generated by multi-hit equation with clump size up to ten organisms per clump.

The double summation model is more applicable than the multi-hit in handling aberrant survival curves. Because of their empirical nature, the values for k and their subscripts must be intuitively selected. Hence, in the hands of the inexperienced, irrational numbers may occur. In another report Clark presented a procedure for estimating the parameters.[13] There are still operating difficulties, however, due to inexperienced handling of the model.

Since deaths of organisms in a disinfection process result from random collision between molecules of a disinfectant and microorganisms, which are present in extremely small numbers in comparison with those of the former, the collision rate can be expressed as a Poisson probability. Because clumps of organisms may exist in various sizes, it is apparent that the probability of collision involves more than one Poisson population. It is on these premises that the multi-Poisson distribution model is proposed to analyze disinfection data.

MATHEMATICS OF THE MULTI–POISSON DISTRIBUTION MODEL

Let us start with the first-order reaction:

$$\frac{dX}{dt} = -kX \tag{1}$$

where X is the concentration of organisms as plaque- or colony-forming units, t is the contact time and k is the destruction rate constant for single organisms. In the destruction of organisms in a clump, the following series first-order steps are envisioned:

$$X_n \xrightarrow{k} X_{n-1} \xrightarrow{k} X_{n-2} \xrightarrow{k} X_{n-3} \xrightarrow{k} \dots X_{n-n} \; (X_{n-n} = X_0) \tag{2}$$

where X_n is the concentration of organisms in clumps of size n, X_{n-1} is the concentration of clumps of size n less 1 giving n-1 survivors until X_{n-n} is the concentration of clumps with the last survivor destroyed; hence, we have X_0. The rate constant, k, is for the destruction of single organisms.

The destruction process of a clump can, therefore, be expressed by the following sequence of equations, assuming that r number of organisms–singles from clumps of size n–are destroyed:

Differential forms	**Integrated forms**
$\frac{dX_n}{dt} = -kX_n$	$X_n = X_n^o e^{-kt}$
$\frac{dX_{n-1}}{dt} = k(X_n - X_{n-1})$	$X_{n-1} = (kt)X_n^o e^{-kt}$
$\frac{dX_{n-2}}{dt} = k(X_{n-1} - X_{n-2})$	$X_{n-2} = \frac{(kt)^2}{2!} X_n^o e^{-kt}$
. . .	. . .

. . . . . .

$$\frac{dX_{n-r}}{dt} = k(X_{n-r+1} - X_{n-r}) \qquad X_{n-r} = \frac{(kt)^r}{r!} X_n^o e^{-kt} \qquad (3)$$

Down to clumps with single survivors:

$$\frac{dX_1}{dt} = k(X_2 - X_1) \qquad X_1 = \frac{(kt)^{n-1}}{(n-1)!} X_n^o e^{-kt}$$

The rate of formation of disinfected clumps is:

$$\frac{dX_o}{dt} = kX_1$$

When the number of disinfected clumps at any time X_o is

$$X_o = X_n^o - (X_n + X_{n-1} + \ldots + X_2 + X_1) \qquad (4)$$

The number of survivors S_n is:

$$S_n = X_n^o - X_o = \sum_{i=1}^{n} X_i \qquad (5)$$

Therefore, for a single Poisson population, we have

$$S_n = X_n^o \sum_{r=C}^{n-1} \frac{e^{-kt}(kt)^r}{r!} \qquad (6)$$

where X_n^o is the initial concentration of clumps of size n and S_n is the fraction of X_n^o surviving after time t.

When r = n-1 which means all clumps contain single survivors, equation (6) is simplified, and the survival curve becomes exponential. When more than one clump size is present in the destruction system—that is, there are sizes $n_1, n_2, n_3 \ldots n_i$ at an initial concentration of $X_{n_1}^o, X_{n_2}^o, X_{n_3}^o, \ldots X_{n_i}^o$, the survival curve becomes a composite of curves, the number of which corresponds to the number of clump sizes. The equation expressing the composite survival becomes:

$$S_t = S_{n_1} + S_{n_2} + S_{n_3} \ldots + S_{n_i} = \sum_{i=n_1}^{n_i} [X_i^o e^{-kt} \sum_{r=0}^{i-1} \frac{(kt)^r}{r!}] \qquad (7)$$

For a mixture of clumps of sizes from 1 to n, equation (7) is rewritten as:

$$S_t = \sum_{i=1}^{n} [X_i^o e^{-kt} \sum_{r=0}^{i-1} \frac{(kt)^r}{r!}] \qquad (8)$$

which is the multi-Poisson distribution equation.

Equation (8) can also be expressed

$$S_t = [c_0 + c_1(kt) + c_2 \frac{(kt)^2}{2!} + \ldots + c_{n-1} \frac{(kt)^{n-1}}{(n-1)!}]\ e^{-kt} \tag{9}$$

and

$$S_t = \sum_{j=0}^{n-1} c_j \frac{(kt)^j}{j!} e^{-kt} \tag{10}$$

where

$$c_j = \sum_{i=j+1}^{n} X_i^o$$

If we substitute y for $\log_e S_t$, equation (9) becomes

$$y = \log_e \left[c_0 + c_1(kt) + c_2 \frac{(kt)^2}{2!} + \ldots + c_{n-1} \frac{(kt)^{n-1}}{(n-1)!} \right] - kt \tag{11}$$

The slope of survival curves has, then, the following expression:

$$s = \frac{dy}{dt} = \left[\frac{c_1 k + c_2 k^2 t + \ldots + c_{n-1} k^{n-1} \frac{t^{n-1}}{(n-2)!}}{c_0 + c_1(kt) + c_2 \frac{(kt)^2}{2!} + \ldots + c_{n-1} \frac{(kt)^{n-1}}{(n-1)!}} \right] - k \tag{12}$$

When t = 0, we have

$$s_0 = \frac{dy}{dt} = \left[\frac{c_1}{c_0} - 1 \right] k \tag{13}$$

and as t approaches infinity, we have

$$s_\infty = \frac{dy}{dt} = -k \tag{14}$$

Equation (13) enables us to compute the percentage of discrete organisms, which is the expression

$$\frac{c_0 - c_1}{c_0} = \frac{-s_0}{k}$$

Equation (14) shows that the slope of $\log_e$ survival curves eventually approaches -k.

FITTING SURVIVAL CURVES BY THE MULTI-POISSON MODEL

Theoretical survival curves can be generated with ease with equation (6) in the case of single-Poisson populations by assuming certain values for n, k and t. For mixed clump sizes (n_1, n_2, n_3, ...) theoretical curves can be generated if the number of clump sizes is not more than 3. Because the per cent distribution of each clump size may range from 1 to 99 and the values for k/min normally encountered in disinfection may vary from 0.1 to 2.0, the number of curves that can be generated will be enormous. Even if such curves were generated, it would be very time-consuming to search for the one that would best fit the observed survival curve. Furthermore, it is quite possible that more than one generated curve would fit the observed equally well, and one would have to be chosen for use.

Hence, while the mathematical theory of the model is sound, the analysis of aberrant curves by this model must be intuitive and individualistic due to the nonuniformity of aberrance. In the case of exponential survival, the model is naturally reduced to a linear $\log_e S$ first-order equation. Survival curves, $\log_e S$, exhibiting a relatively flat shoulder and a linear tail, similar to those reported by Berg and his associates in the destruction of enteroviruses by I_2[10] can be analyzed with the single Poisson equation (6).

Curves exhibiting an exponential survival at the start, followed by a retardant die-away, clearly indicate high percentages of discrete organisms with small numbers of clumps, whose per cent and size distribution can be estimated by the location and length of the retardant section. If a retardant die-away segment is located below the 5% survival level, the value for k computed from the linear segment can be taken as that in the destruction of singles. If, on the other hand, the retardant segment is located above the 5% survival level–especially above the 10% level–the survival of the clumps will significantly influence the slope of the linear segment of curve and cause the latter to concave gently before the retardant segment is reached. A stepladder-shaped survival curve indicates the existence of more than one widely separated clump size.

To analyze the more complex aberrant survival curves, the latter must be partitioned into arbitrary segments on the time scale. These segments can be analyzed statistically for the per cent distribution and size of n and for values of k. In such an analysis the value for k will differ for each segment, but that for the segment showing survival of singles or clumps having *the last survivors* will be the rate constant sought. This method is elaborate, hence time-consuming; but the values for n_1, n_2, n_3, ... so ascertained should be quite dependable. The manipulation would be even more time-consuming if the number of clump sizes is larger than 4.

Another method of approach is the graphic "peeling off" originally used by Feurzeig and Taylor in analyzing nonlinear regression curves.[14] Their regression curves were expressed by the following equation:

$$y = \sum_{i=1}^{n} y_1 = \sum_{i=1}^{n} A_i e^{-k_i t} \qquad (15)$$

Equation (15) differs from the multi-Poisson equation (8) in that the partitioned segments are linear, and therefore A_i is a constant. Partitioned segments by the Poisson equation are nonlinear, except the segment for the survival of singles or clumps having the last survivors. The Poisson curves are dependent upon kt. Instead of peeling off one linear segment at a time by equation (15), segments of a Poisson curve are peeled off one at a time in accordance to the value of kt.

To illustrate let us assume four survival curves like A, B, C and D shown in Figure 2-2a. To facilitate analysis by the multi-Poisson equation (8), the four curves are reproduced separately (Figures 2-2b, c, d and e). Prior to analysis we must decide how many segments to peel off from the survival curve. The greater the number of segments, the more complex the analysis. As stated previously, the analysis becomes too complex and time-consuming when the number of clump size groups is more than three.

How important is the knowledge of clump size distribution? In any disinfection study the most important parameter sought–in either theoretical investigation of kinetics, practical examination of disinfectants for relative efficiencies or effects by environmental factors, such as pH and temperature–is the destruction or inactivation rate constant k. This is the

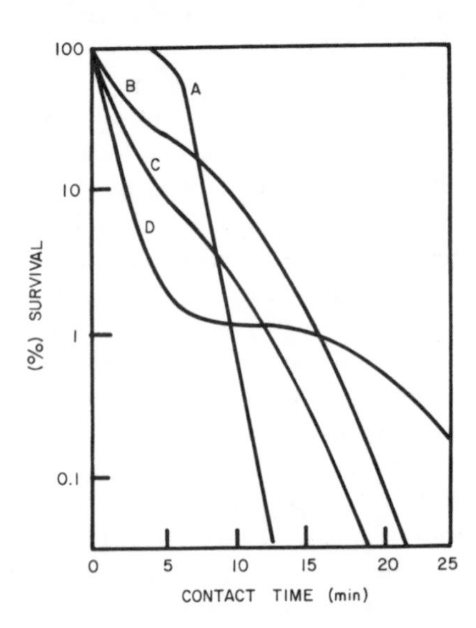

Figure 2-2a

Figure 2-2. Four hypothetical survival curves to be analyzed by the multi-Poisson method. [Each of the four curves will be presented separately in Figures 2-2b, 2-2c, 2-2d and 2-2d to demonstrate the partitioning process by equation (8).]

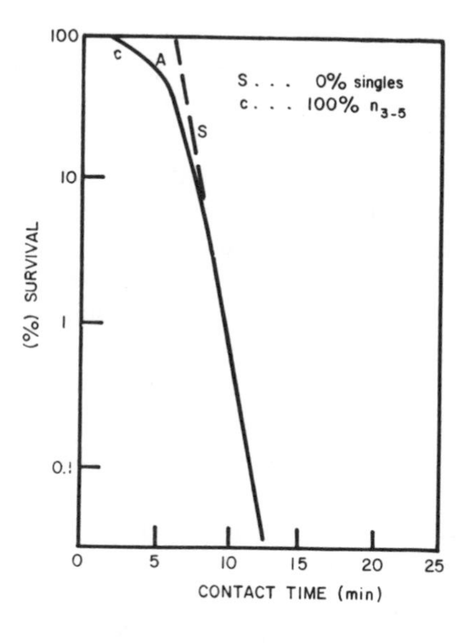

Figure 2-2b. Partitioning curve A.

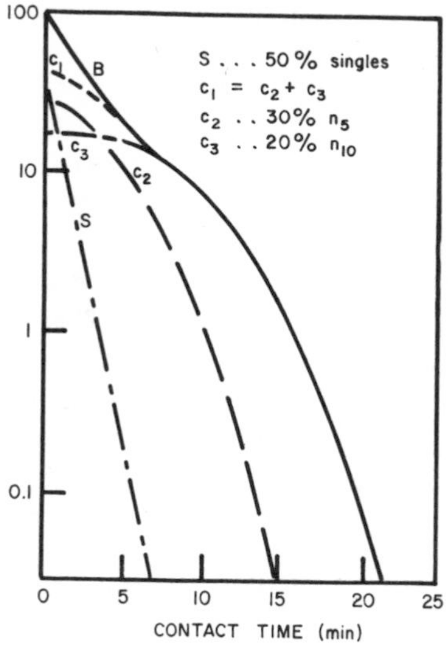

Figure 2-2c. Partitioning curve B.

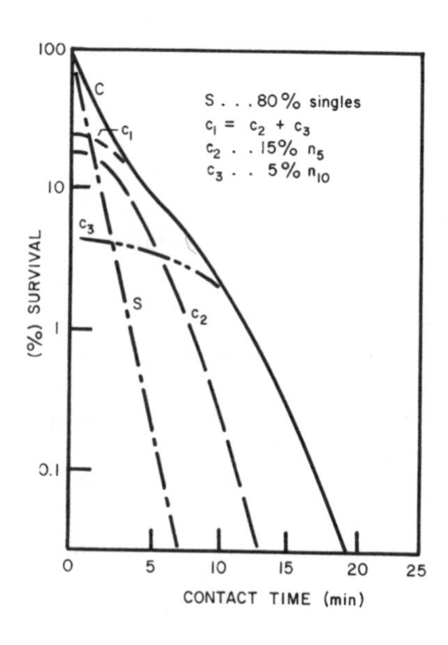

Figure 2-2d. Partitioning curve C.

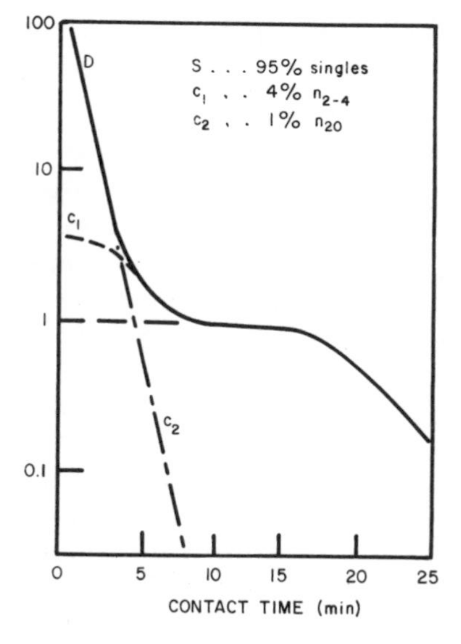

Figure 2-2e. Partitioning curve D.

only true constant in the destruction of single organisms, and the clump size distribution is important only as it locates the segment of the survival curve that represents the survival of single organisms of the clumps having the last survivors.

By employing equation (8) each of the four curves is peeled off segment by segment, with the clump size groups (n) limited to three. The segment representing the singles and those representing the survival of clumps of different sizes, and their per cent distribution are computed and shown in each figure (2-2b, c, d, e). This approach is still time-consuming requiring the service of a computer, hence, not suitable for ordinary use.

To make the model practical for normal use, the graphic "peeling off" method is recommended. This method requires a set of preprepared Poisson curves with the per cent survival plotted logarithmically against values of kt from 0-25 for values of n from 1-25. Such curves (Figure 2-3) were prepared with the aid of an IBM 360 at the University of Cincinnati. When the Poisson curves are plotted on tracing paper and an observed survival curve is plotted on the same $\log_e$ per cent survival and time scale, the latter curve can be easily partitioned and each segment analyzed graphically for the size and per cent of n. The value for k can be read from the group. Since the model has only the k for destruction of singles or clumps having the last survivors, the confusion derived from using different rate constants for different segments is eliminated.

VALIDATING THE MULTI-POISSON DISTRIBUTION MODEL

To validate the multi-Poisson model, we tested how well the calculated values for n_i agreed with the observed frequency distribution of clumps by size, and how well the values for k computed from aberrant survival curves agreed with those from exponential curves.

A unicellular microorganism well suited for a test organism is the free-living amoeba *Naegleria gruberi*. This amoeba forms cysts in clumps of various sizes on a buffered sucrose trypose agar.[15] Suspensions of cysts can be prepared by washing plate cultures with distilled water and purified by repeated washing and centrifuging. The cysts are relatively uniform in size–about 15μ in diameter (Figure 2-4a). When a suspension of single cysts is desired, the above-prepared suspension is sonicated at 20 kilocycles sec^{-1} for 5 minutes under refrigeration, and the sonicated preparation is filtered through a 25μ nylon strainer. To avoid reclumping, the cysts were washed and resuspended in HCl/NaAc solution buffered at pH 3.0. The final suspension is diluted in distilled water to give a cyst concentration of 1,000 ml^{-1} and used as test water.

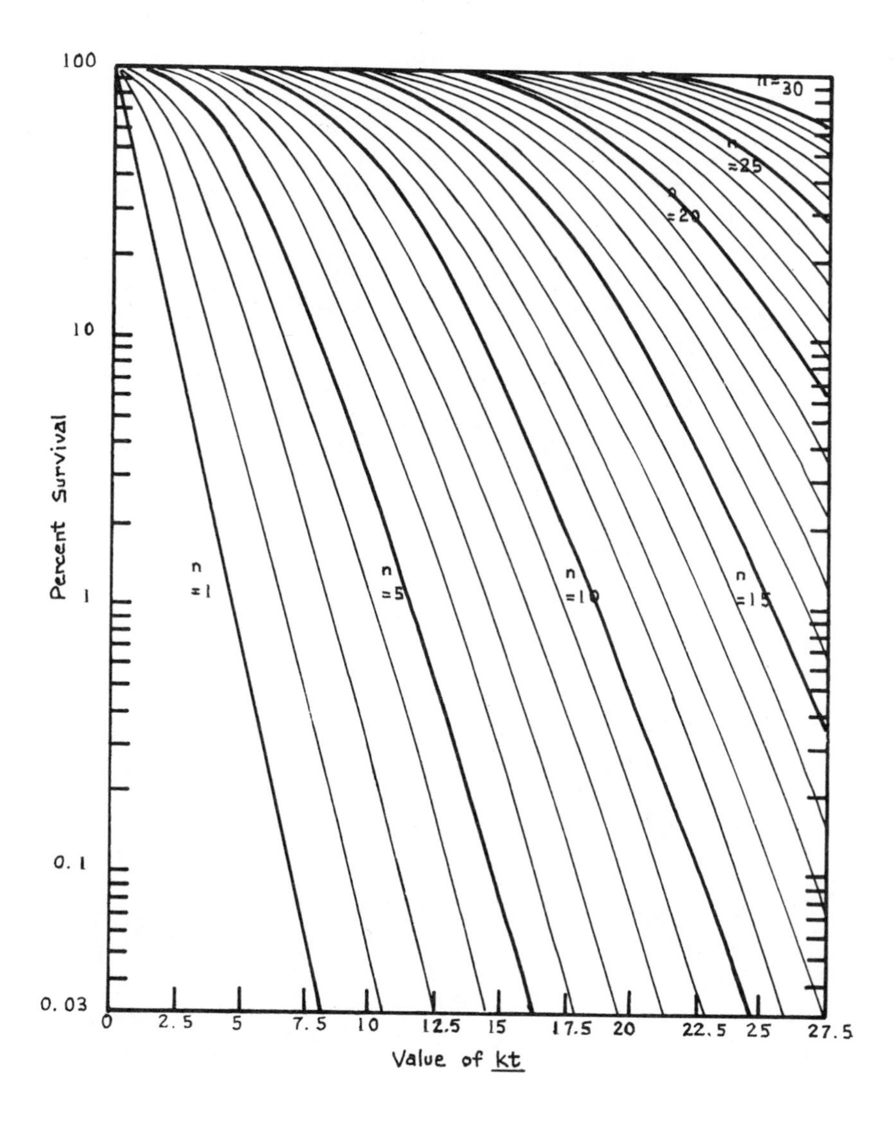

Figure 2-3. Predicted survival curves generated by multi-Poisson distribution model and plotted as per cent survival vs. kt for different values of n.

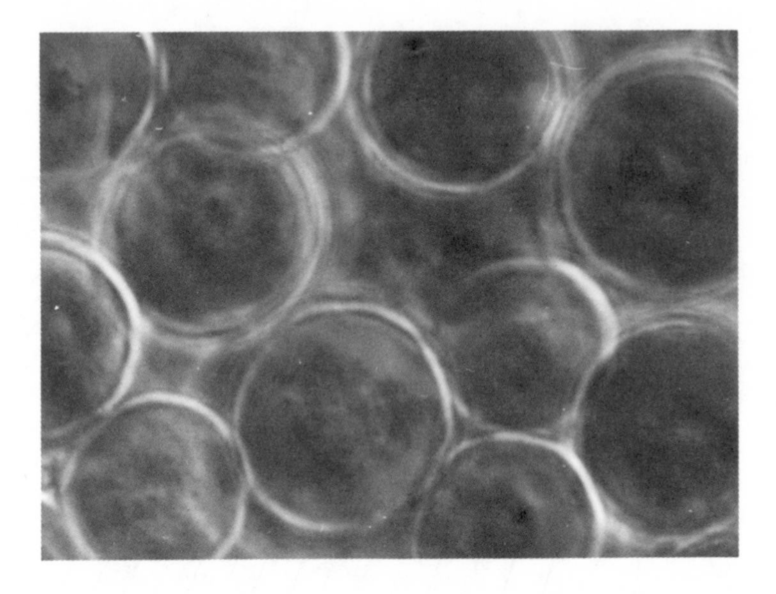

Figure 2-4a. A photomicrograph of cysts of *N. gruberi* in a clump (sheet formation). Notice the area of contact between cysts. 1500X.

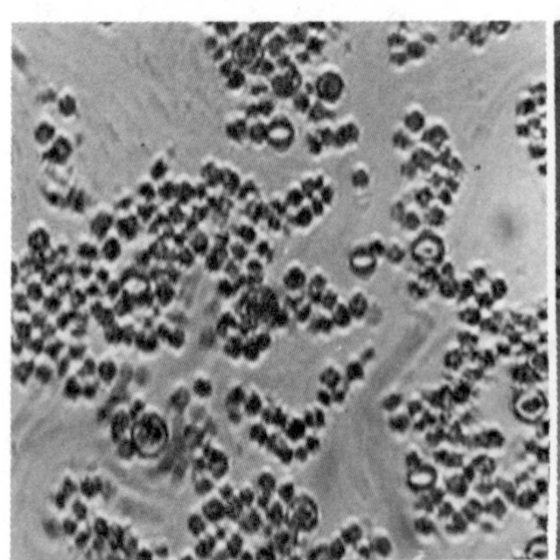

Figure 2-4b. A photomicrograph of *N. gruberi* cysts in a crude suspension. Notice the clumps of various sizes. 100X.

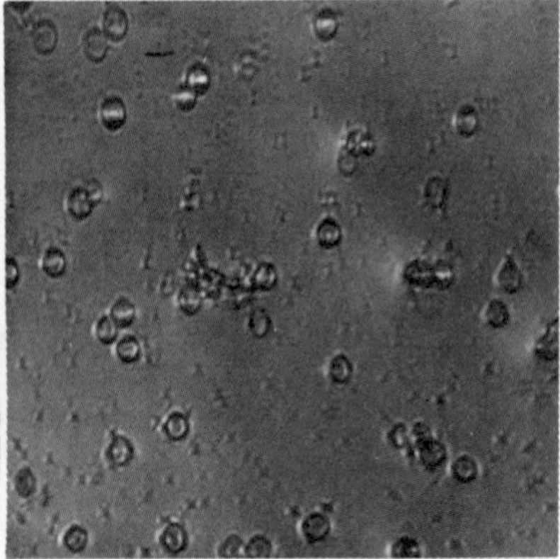

Figure 2-4c. A similar photomicrograph taken of a washed suspension after a 5-min sonication. 150X.

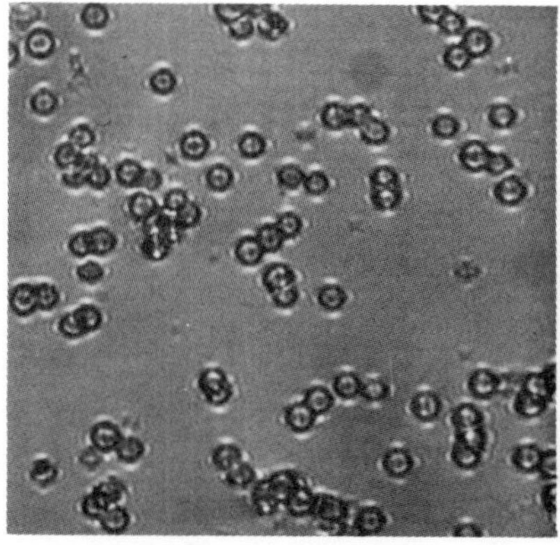

Figure 2-4d. A similar photomicrograph taken of a cyst suspension after 3 min sonication. Notice the presence of more clumps than 1c. 150X.

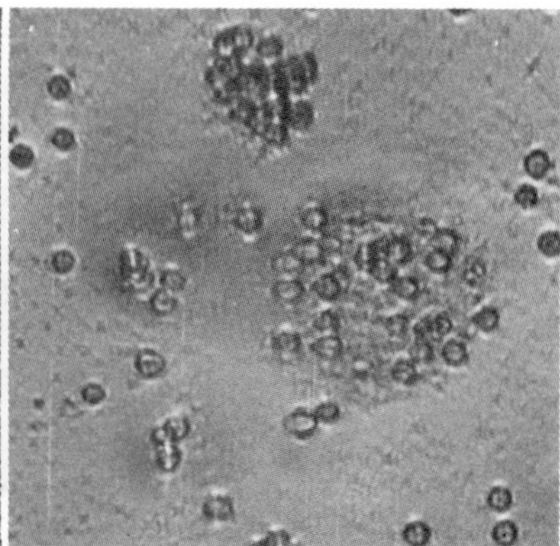

Figure 2-4e. A photomicrograph of a mixed suspension containing 95% soniciated-filtered cysts and 5% untreated cysts.

Suspensions containing various cyst clump size distribution patterns can be prepared by sonicating a stock suspension for 1-4 minutes and diluting each sonicated suspension with enough HCl/NaAc buffer to give a concentration of 1000 units (singles and clumps) ml^{-1}. These suspensions are then ready for use as test waters.

The cyst clump size and per cent distribution can be ascertained by direct enumeration under 100-150X magnification in three replicate drops of the dilute suspension so prepared. The examination is repeated after a test to ensure that there is no significant change in clump size and per cent distribution during a test.

Survival of cysts can be ascertained by plaque-count technique similar to that employed in bacteriophage concentration determination. The technique has been described in detail.[16]

With a strain of *N. gruberi* isolated from a swimming pool in Cincinnati and maintained on an agar medium,[16] suspensions containing discrete cysts and mixtures of singles and clumps of various size and per cent distributions were prepared by the procedure described previously. Three beakers containing 400 ml of a cyst suspension were dosed with enough I_2 solution to give a residual of about 6.0 (5.9-6.2) mg/l. Five-ml samples were taken after scheduled contact times and placed in test tubes containing dried Na_2SO_3 (I_2 neutralizing agent). The neutralized samples were plaqued to determine cyst survival. When the per cent destruction of cysts became high toward the end of the test, the size of samples taken was increased 10-fold, and cysts in each sample were concentrated by a 10μ millipore membrane and resuspended in 5 ml of distilled water for plaquing. The results are expressed as plaque forming units (PFU).

In one series of tests to obtain survival curves of the retardant die-away type, a suspension of discrete cysts was mixed separately with three amounts of an untreated suspension to achieve 1, 5 and 20% of the mixed. These mixed suspensions were also treated with iodine to produce a residual of 6.5 mg/l, and the cyst survival in each mixture was ascertained as above.

The cyst clump size distribution in an untreated suspension is typified by the results shown in Table 2-1. A glance reveals that clump size distributions were essentially the same at 5°, 15° and 25°C.

Table 2-1. Distribution of *N. gruberi* Cysts in Suspensions After Sonication and Filtration through a 25μ Microstrainer

Cyst clump size (n)[a]	% Distribution in Suspension		
	5°	15°	25°
singles	93.4	94.5	95.0
2s	5.0	4.3	3.6
3s	1.5	1.0	1.4
4s	0.1	0.2	0.0
5s	0.0	0.0	0.0

[a] Each number is an average of triplicate counts.

Preliminary experiments showed insignificant change in clump size distribution after an exposure to either 2 or 5 mg/l of I_2. Experiments were conducted with both suspensions containing discrete cysts and those subjected to 0, 1, 2 and 3 minutes of sonication, and exposed to 5.9-6.1 mg/l of I_2. Tests made with sonicated-filtered and with untreated cyst suspensions were conducted at 5°, 15° and 25°C, and those with suspensions were subjected to 1, 2 and 3 minutes of sonication at 25°C only. The per cent distribution of cyst clumps by size in the untreated suspension and those with 1-, 2- and 3-minute sonication observed in triple microscopic counting are shown in Table 2-2.

To observe the shapes of survival curves of cysts as singles and in clumps of different size distribution, results are presented in Figures 2-5 and 2-6. Figure 2-5 demonstrates clearly that clumping of cysts is the only factor responsible for deviation of survival curves from exponentiality, since linear curves were obtained with discrete cysts and curves with a shoulder were obtained with untreated cysts. These survival curves are relatively simple

Table 2-2. Per Cent Distribution of *N. gruberi* Cysts by Clump Size Ascertained by Direct Microscopic Counting in Suspensions After Varying Periods of Sonication

Clump Size (n)	% Distribution[a] After Stated Sonication Time			
	0	1 min	2 min	3 min
1	0.02±0.01	3.8±0.9	13.1±2.1	38.2±7.5
2	0.05±0.01	4.3±1.0	17.2±2.9	28.5±6.0
3	1.13±0.3	5.2±0.8	16.5±2.4	18.1±3.9
4	2.50±0.6	6.1±1.1	20.3±3.3	11.4±2.0
5	4.00±0.9	7.5±0.8	17.6±2.8	3.0±1.2
6	4.80±0.8	8.2±1.3	12.1±2.1	0.8±0.2
7	4.40±1.3	10.3±1.5	3.2±0.8	0.0
8	6.10±0.9	14.2±2.8	0.0	
9	6.20±2.1	16.2±3.6		
10	10.50±1.9	9.3±2.0		
11	13.40±2.1	7.2±1.0		
12-15	22.3 ±3.5	5.5±1.1		
15-20	18.6 ±3.1	2.2±0.4		
>20	2.0 ±0.4	0.0		

[a]Average of triple plaque counts.

to analyze—the linear curves by equation (1) and those with a shoulder by equation (6). The values for k and n so computed are tabulated as follows:

Temperature °C	Sonicated and Filtered Cysts		Untreated Cysts	
	k	n	k	n
5	0.53	1	0.53	18
15	0.77	1	0.75	18
25	1.16	1	1.14	18

It should be noted that the values of 1 and 18 for n, especially the latter, must be considered averages. As shown in Table 2.1, the per cent distribution of clump size observed in direct count in a sonicated-filtered cyst suspension was: singles, 93.4-95%; 2s, 3.6-4.0%; 3s, 1.0-1.5%; and 4s, 0.0-0.1%. But the presence of such small percentages of small clumps is too insignificant to influence the linear survival of the overwhelmingly large percentage of singles. The per cent distribution of clumps by size in the

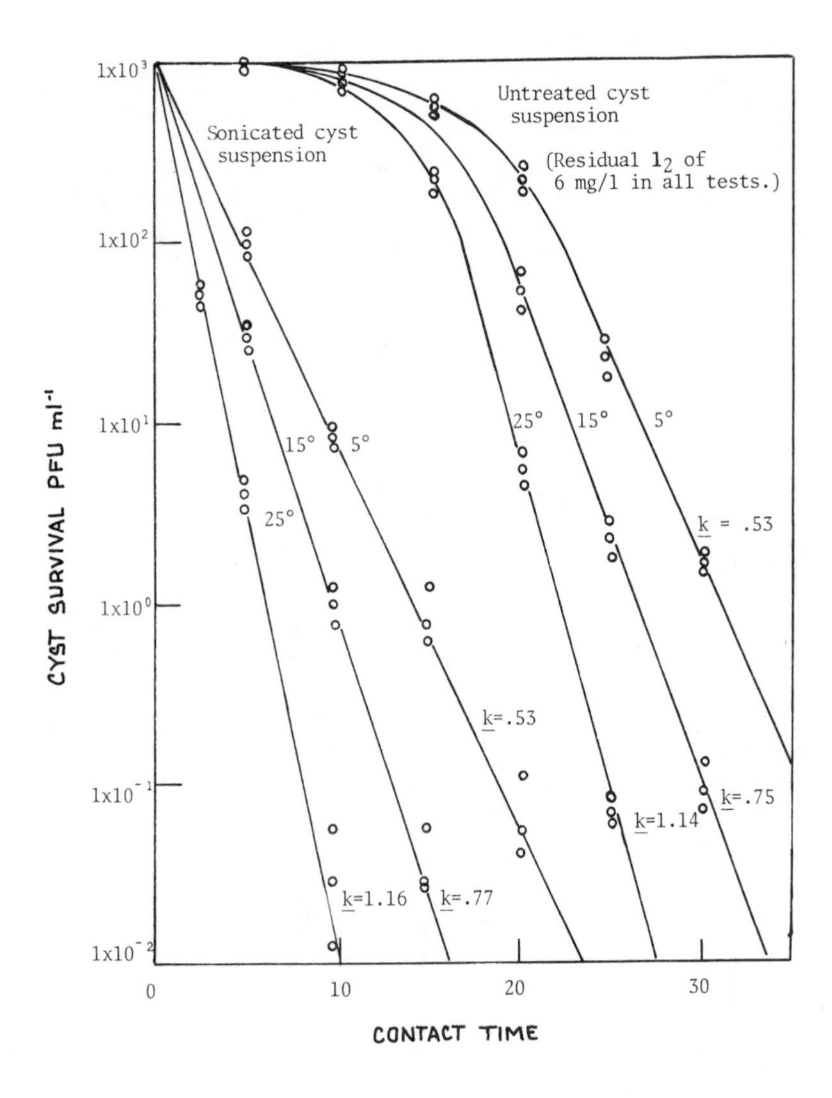

Figure 2-5. Destruction of cysts of *N. gruberi* in water by I_2 at varying temperatures and pH 6.0-6.2.

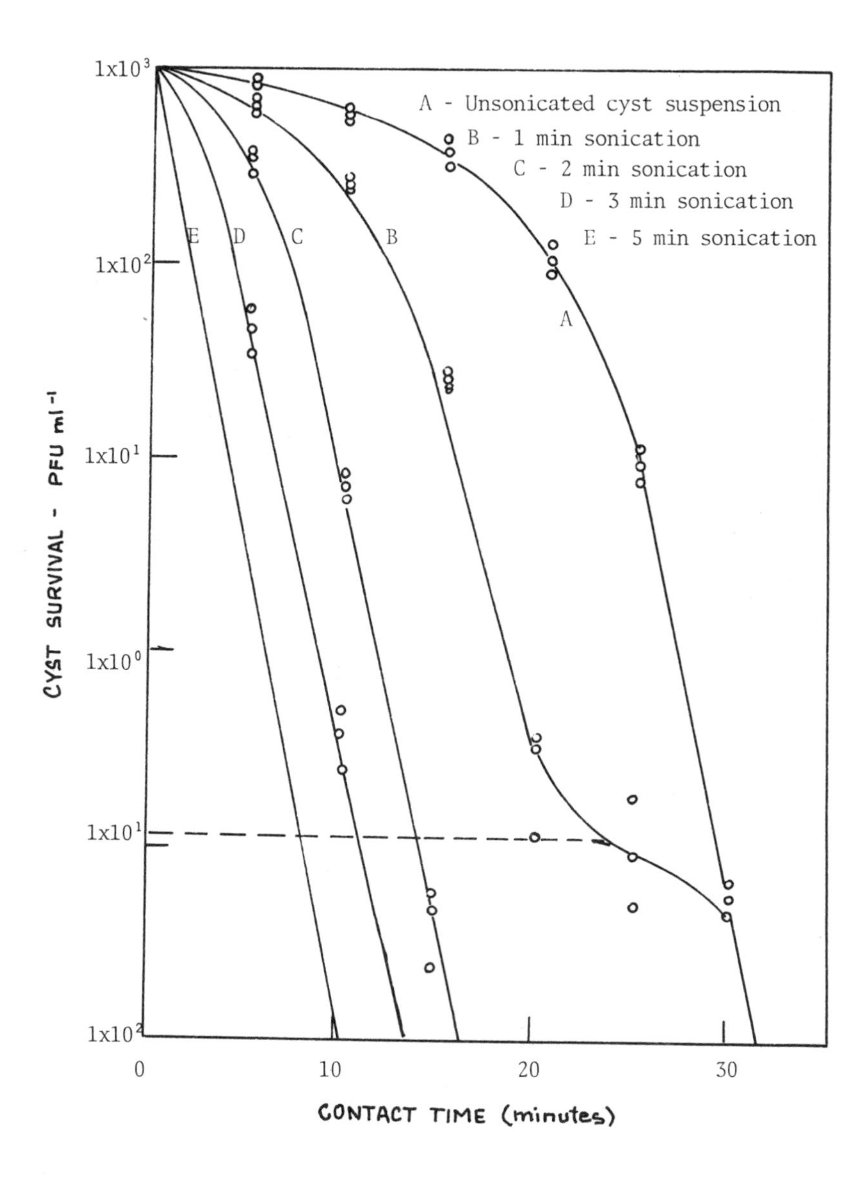

Figure 2-6. Effect of size and per cent distribution of clumps of *N. gruberi* cysts on survival in iodinated water at 25°C and pH 6.0-6.2 with a residual I_2 of 6.0 mg/l.

untreated suspension was similar to that shown in Table 2-2 except that 99% of clumps were in the 8 to >20 size range, with <0.1% of 1s and 2s, and 0.8-0.9% of 3s to 7s.

The values for k for these 2 sets of survival curves were similar at three temperatures. One might question the effect of the geometry of clump formation on the value of k. In plate cultures, cysts clumps are formed only in sheets, although the cysts in a sheet may not be exactly in the same plane (Figure 2-3a). The maximum number of contact areas on a single cyst is 4, but often there are only 3 (Plate 2-3a). We estimated that no more than 5% of the total surface area of a cyst is not exposed to a disinfectant because of contact, and concluded that the effect of contact of cysts in a clump is not significant enough to be considered in computing the values for k.

In Figure 2-6 points of the survival curve for sonicated-filtered cysts were not shown because the curve was reproduced from Figure 2-5. The shoulder of the curve for the untreated cysts which was shorter than that in Figure 2-5 at 25°C, was apparently due to a higher percentage of clumps in the small size range. The curve for the cysts receiving a 3-minute sonication exhibited a second shoulder at the 0.01% survival level. The close correlation between the mildness of the shoulder and the degree of sonication is significant. When these survival curves are examined in the light of clump counting data shown in Table 2-2, it is quite clear that the higher the percentage and the larger the size of clumps, the more prominent is the shoulder.

A visual concept of the clumping phenomenon is believed helpful in interpreting cyst survival data and their treatment. Photomicrographs of cysts in suspensions after varying degrees of treatment, are shown in Figure 2-3. Figure 2-3b shows cysts in clumps in a crude suspension prior to washing. Figure 2-3c shows cysts mostly as singles in a suspension after a 5-minute sonication; the small clumps were removed by filtration. Figure 2-3d–a suspension after a 3-minute sonication–shows 1/3 single cysts and 2/3 clumps, most of which are 2s and 3s. Since every single cyst can excyst and multiply in the trophozoite form, it is readily seen how the cyst clumps cause aberrance in survival curves.

It is important to note the value in Table 2-3 of 1.15 for k/min, and the closeness of this value to that obtained from the survival curves at 25°C in Figure 2-5. The values for k at the three different temperatures appeared to be reliable, since the values obtained from linear survival curves and those obtained from curves with shoulders coincided well. Hence, these values are plotted on a log scale against the reciprocals of the respective absolute temperatures in Figure 2-7. The slope of the linear relationship gives a value of 7,700 calories per degree for the energy of activation. This agrees with

Table 2-3. Values for k[a], Per Cents of Single and Clumped Cysts, and Size of n_m[b] in Sonicated Suspensions Exposed to I_2 at 25°C

Parameters	Values for Suspensions Having Stated Sonication Time							
	0 min		1 min		2 min		3 min	
	Observed	Calculated	Observed	Calculated	Observed	Calculated	Observed	Calculated
% singles	0	0	3.8	5	13.1	12	38.2	35
% clumps	100	100	96.2	95	86.9	88	61.8	65
n_m	20	18	15-20	10	7	5	5	3
k	1.15		1.15		1.15		1.15	

[a] k, destruction rate constant min^{-1} for discrete cysts

[b] n_m, maximum size of cyst clumps

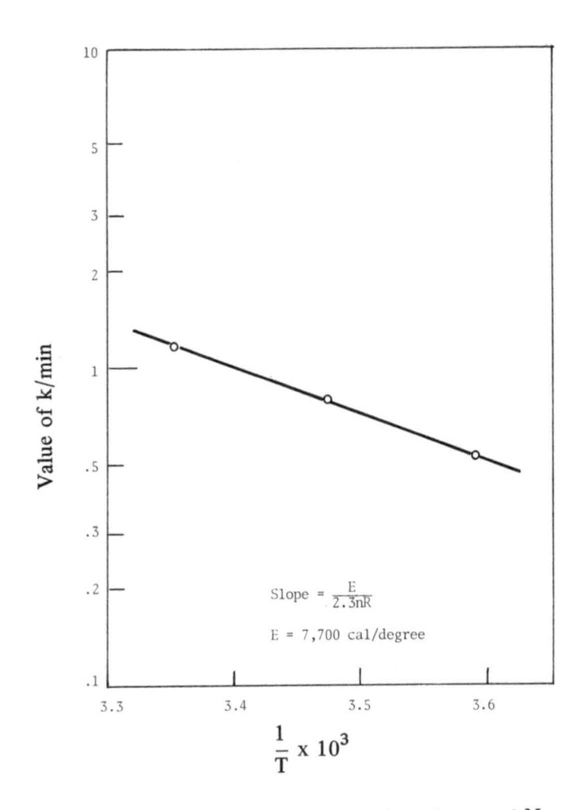

Figure 2-7. Effect of temperature on destruction of cysts of *N. gruberi* in water by I_2 (6.0 mg/l) at pH 6.0-6.2.

the energy of activation reported earlier in the destruction of cysts of *Endamoeba histolytica* by I_2.[17]

To produce survival curves of the retardant die-away type, mixed suspensions containing 80, 95 and 99% sonicated-filtered cysts and 20, 5 and 1% untreated cysts were treated with 6.5 mg/l of I_2 at 25°C. The procedure for taking samples and plaquing to ascertain cyst survival was the same as that used in preceding experiments. Results of these experiments are summarized in Table 2-4. The per cent distributions of cysts by clump size in the three mixed suspensions are presented in Table 2-5 and the per cent in the 95%-5% suspension in Figure 2-3a.

For the analysis, the 20 and 5% data shown in Table 2-4 are plotted as survival curves in Figures 2-8a and 2-8b. From cyst clump distribution data, it is clear that the retardant die-away is, indeed, a manifestation of

Table 2-4. Survival Data on Destruction of *N. gruberi* Cysts by I_2[a] in Mixtures of Sonicated[b] and Nonsonicated Suspensions at 25°C

Contact Time (min)	Cyst Survival in Mixtures with Stated % of Nonsonicated Suspensions					
	20%		5%		1%	
	PFU/10 ml	%	PFU/10 ml	%	PFU/10 ml	%
0	11000±2300	100	11700±2300	100	10900±2100	100
	320±42		60±8		23±4	
5	260±37	2.480	55±8	0.494	18±2	0.170
	240±30		48±6		15±2	
10	15±3		9±2		3±1	
10	13±2	0.118	7±2	0.067	2±1	0.022
	11±2		6±1		2±1	
	2±1		4±1		1.5±1	
15	1.5±1	0.015	3±1	0.030	1.5±1	0.012
	1.5±1		3±1		1±1	
	3/4±2/4		1.5±1		1/4±1/4	
20	2/4±1/4	0.005	1.5±1		1/4±1/4	0.0015
	2/4±1/4		1±1		0±	
	2/4±1/4		1/4±1/4			
25	1/4±1/4	0.002	0±	0.0008		
	0		0			

[a]6.5 mg/l residual I_2

[b]Sonicated for 15 minutes and filtered through a 25μ microstrainer

Table 2-5. Per Cent Distribution by Clump Size of *N. gruberi* Cysts in Mixtures of Sonicated and Nonsonicated Suspensions Ascertained by Direct Microscopic Enumeration

Cyst Clump Size	% Distribution[a] in Stated % of Nonsonicated Suspensions		
	20%	5%	1%
Singles	75.2 19.2	90.0 20	97.1 19.2
2s	7.0 4.0	2.5 2.8	1.1 0.8
3s	4.1 1.9	3.5 1.0	1.5 0.4
4s	5.1 0.4	3.1 0.3	0.05 0.03
5s-10s	8.48 0.52	0.595 0.16	0.04 0.03
>10	0.12 0.09	0.306 0.03	0.21 0.03

[a]Average of triple counts

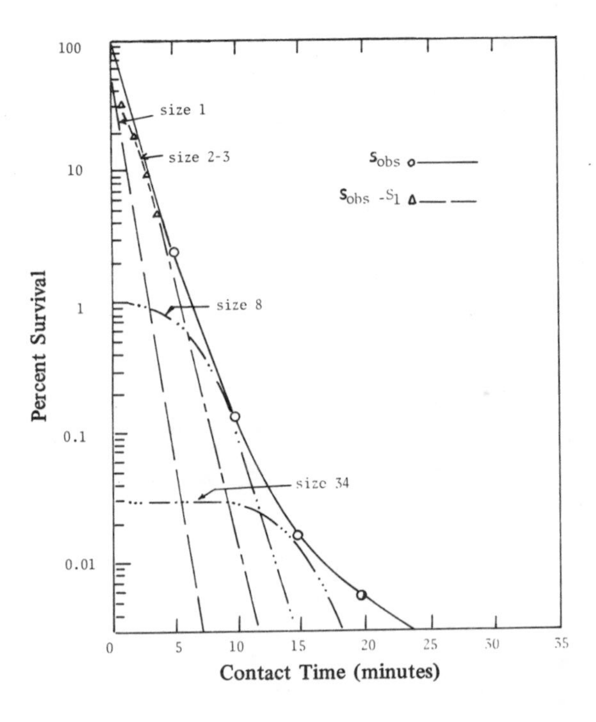

Figure 2-8a. Survival curve for *N. gruberi* cysts in suspension containing sonicated and 20% untreated cysts, exposed to 6.5 mg/l I_2 at 25°C.

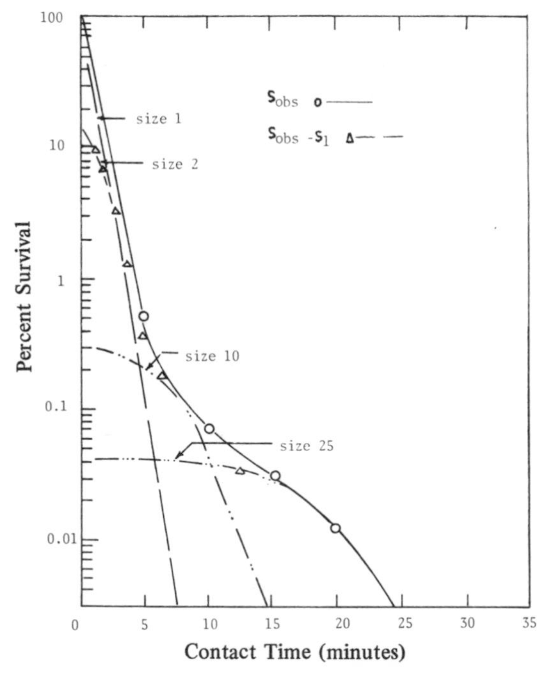

Figure 2-8b. Survival curve for cysts in suspension containing sonicated and 5% untreated *N. gruberi* cysts, exposed to 6.5 mg/l I_2 at 25°C.

the survival of high percentages of discrete organisms, and the higher the per cent of singles the more marked the initial exponential drop in survival.

The ease with which the retardant die-away is analyzed by the multi-Poisson equation depends on how apparent the exponential segment of the survival curve is and how accurate and elaborate a clump size and per cent are sought. With the values for kt known, equation (11) can be employed to compute the values for n_i. The process should not be too complex and time-consuming with a computer if we limit to three the number of clump sizes.

The linear segments in curves in Figure 2-8b are so apparent that the values for k can be easily computed. The curve in Figure 2-8a presents some degree of uncertainty as to the section of the linear segment, due to the very slight curving of the initial part of the curve. However, careful measurement of the linearity of the curve showed that the segment above the 20% survival was quite linear, and this per cent was taken as the cutoff point.

With the values for k ascertained, the curves were partitioned and analyzed with equation (12) for s and sizes of n_i. The Poisson curves for the partitioned segments are plotted in the respective figures. From the location of the partitioned curves, the percentages of clumps by size are read from the graphs. The values for per cent and size of clumps calculated from the survival curves and those obtained in direct microscopic examination, together with the values for k are presented in Table 2-6.

Table 2-6. Values for Clump Size and Per Cent, k Computed from Data on Destruction of *N. gruberi* Cysts by I_2[a] in Mixed Suspensions at 25°C

	Values for Mixtures Having Stated % of Nonsonicated Suspensions					
	20%		5%		1%	
Parameter	Calculated	Observed	Calculated	Observed	Calculated	Observed
% singles	79.47	75.20	96.65	90.00	99.03	97.10
% of n_{2-4}	14.00	16.20	3.20	9.10	0.50	2.65
% of n_{5-10}	6.53	8.48	0.34	0.60	0.44	0.04
% of n_i 101	0.03 (n_{34})	–	0.04 (n_{25})	–	0.03 (n_{14})	–
k/min	1.18		1.17		1.18	

[a]Residual I_2 6.2 mg/l

Table 2-6 discloses the fairly close agreements between the calculated and observed per cent of single cysts and also the calculated and observed per cent of each of the 2-clump size groups. The values of 1.18, 1.17 and 1.18 for k/min are almost identical for the three survival curves, and not too different from the values of 1.15 obtained for the survival curves at 25°C in the preceding experiments. These observations support the belief expressed earlier, that clumping is the major, if not the only, factor responsible for aberrant survival curves in carefully controlled disinfection studies. They also substantiate that the effect is manifested only by the number of surviving units in a clump, and that the surface area of each surviving unit in a clump neutralized by contact with other units is not enough to significantly affect the destruction rate.

To substantiate that the tail of a retardant die-away curve is caused by clumps of unusually large size, the test made with a mixed suspension containing 1% untreated cysts was repeated. After a 10-minute contact when a 99.9% destruction was expected, a 200-ml sample was removed and neutralized with Na_2SO_3. The test was terminated after the 25-minute contact was reached. The 200-ml neutralized sample was centrifuged and resuspended in 10 ml of distilled water and sonicated for 5 minutes. The entire sonicated suspension was plaqued for surviving cysts. The results thus obtained are shown as follows:

Contact Time (min)	PFU/10 ml Sample	
	Unsonicated	Sonicated
0	9,500	
5	21	
10	3	32
15	1	
20	1	
25	1	

The sonicated sample was more than 10-fold that of the unsonicated. This increase in the number of surviving cysts is too large to be accounted for by experimental error and must be explained as the increase in surviving units caused by breaking up the large clumps by sonication.

APPLICATION OF MULTI–POISSON MODEL TO ANALYSIS OF REPORTED DATA

Application of the multi-Poisson model was extended to reported studies, in which bactericidal and virucidal data were obtained in carefully controlled experimental conditions. We selected the data reported by Butterfield and his associates[18] on the destruction of *Escherichia coli* by free chlorine and those by Berg and his associates[10] on the inactivation of enteroviruses by I_2 as the most suitable for the analysts.

Treatment of Data on Destruction of *E. coli* by Free Chlorine

Butterfield *et al.* reported their very carefully conducted study on destruction of *E. coli* by free-chlorine,[17] but plotted the survival curves in a manner that obscured a very interesting phenomenon in survival curves of the bacteria. To reveal this phenomenon these data were replotted on a semilog scale; their results obtained at 2°-5°C are shown in Figures 2-9a, b, c and d, and those at 20°-25°C in Figures 2-10a through 2-11.

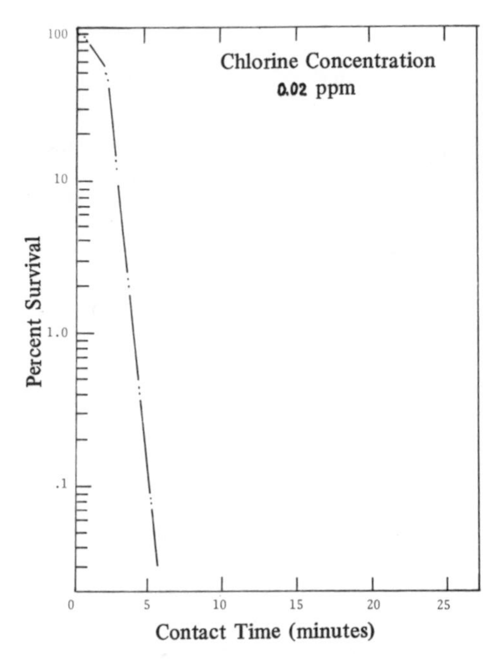

Figure 2-9a. Survival of *E. coli* in destruction by free chlorine at 2-5°C, pH 7.0

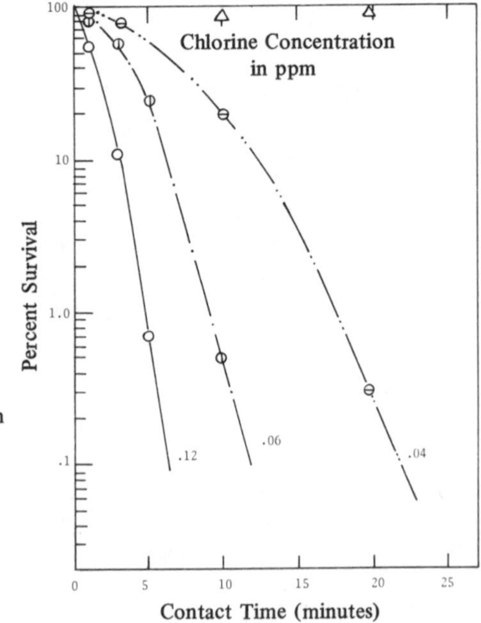

Figure 2-9b. Survival of *E. coli* in destruction by free chlorine at 2-5°C, pH 8.5

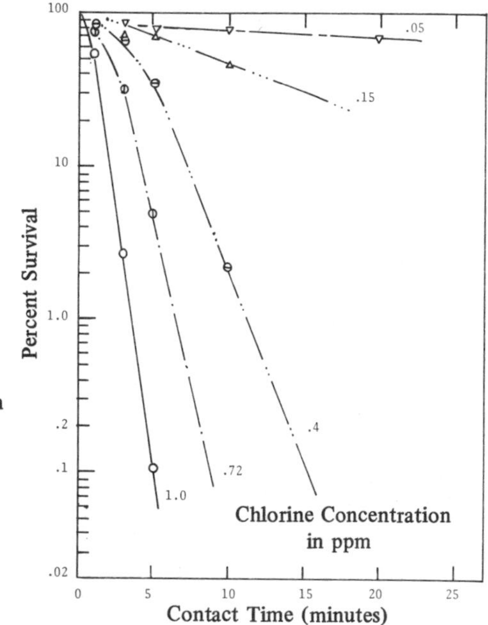

Figure 2-9c. Survival of *E. coli* in destruction by free chlorine at 2-5°C, pH 9.8

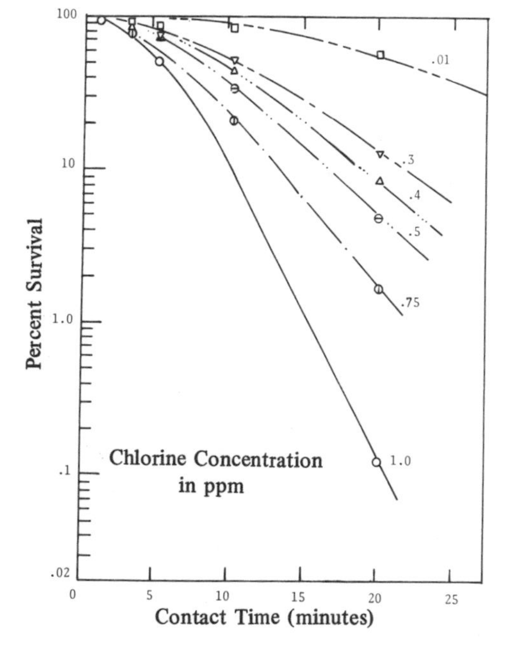

Figure 2-9d. Survival of *E. coli* in destruction by free chlorine at 2-5°C, pH 10.7

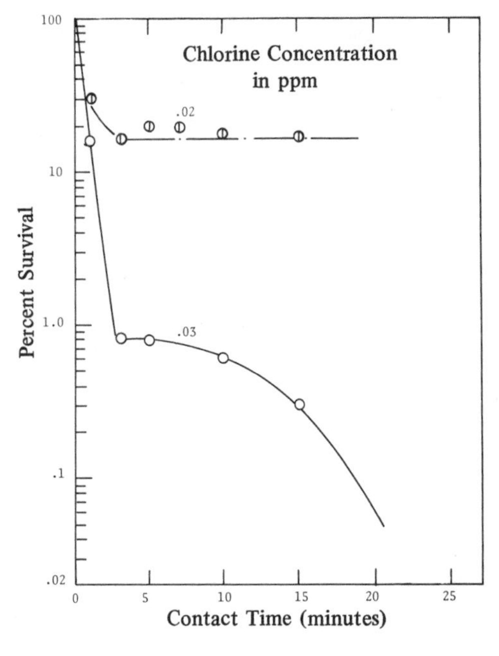

Figure 2-10a. Survival of *E. coli* in destruction by free chlorine at 20-25°C, pH 7.0

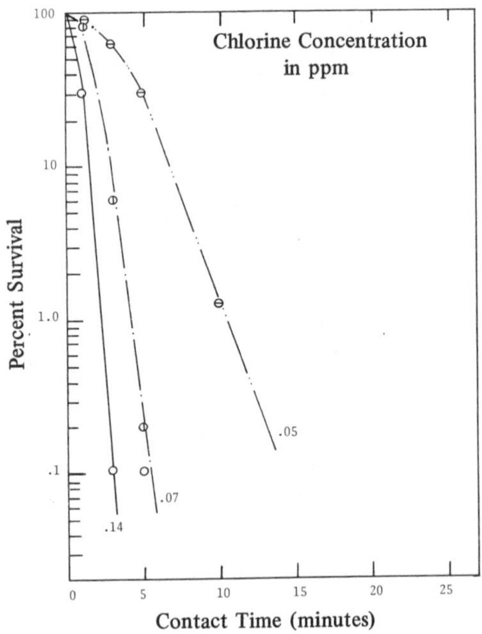

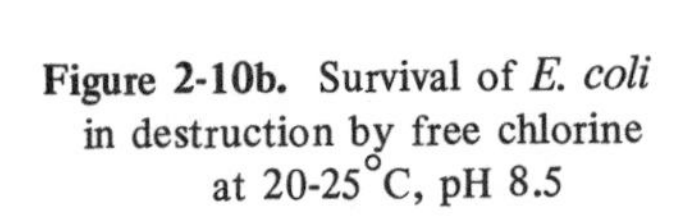

Figure 2-10b. Survival of *E. coli* in destruction by free chlorine at 20-25°C, pH 8.5

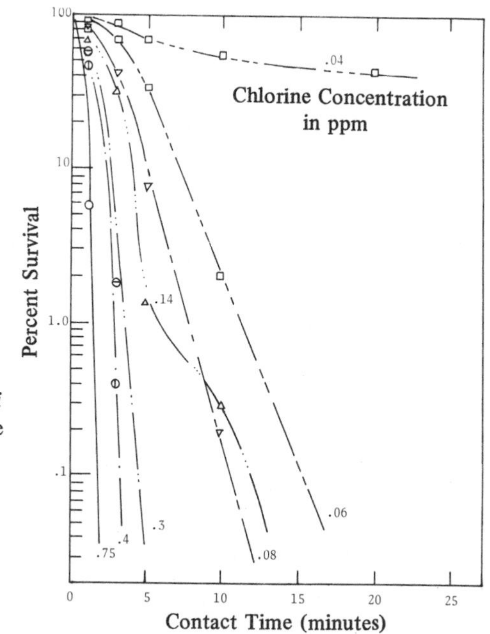

Figure 2-10c. Survival of *E. coli* in destruction by free chlorine at 20-25°C, pH 9.8

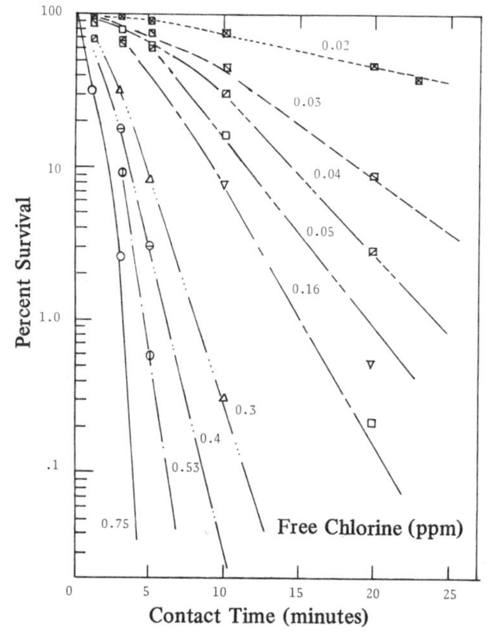

Figure 2-10d. Survival of *E. coli* in destruction by free chlorine at 20-25°C, pH 10.7

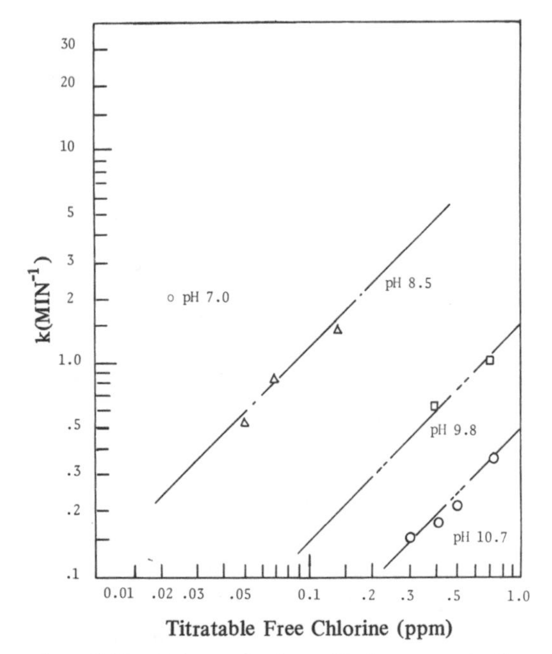

Figure 2-11. Rate constant vs. chlorine concentration in destruction of *E. coli* at 2-5°C

Examination of these survival curves shows the formation of a shoulder in all curves except those in Figure 2-10, in which two retardant die-away type curves were obtained. The smaller the chlorine dosage, the lower the temperature or the higher the pH value; *i.e.,* the slower the killing, the more marked were the shoulders. In most of these survival curves, the linear tails were so apparent that the values for the rate of disinfection, k, can be ascertained without difficulty. With equation (6) the values for the clump size, n, have been computed and came out to be between 2 and 5, mostly 4.

These values of n, clump size, are most interesting when they are interpreted in relation to an electron micrograph of *E. coli* prepared with a 16-hr old culture. It is clear that most of the *E. coli* cells were in a state of 2-about-to-divide-into-4 at this age. Since these authors used young cultures of *E. coli* in their study, it is not surprising that they obtained a shoulder in most of their survival curves; but the interesting point is that the computed values of n are in such good agreement with the electron micrographic findings observed with *E. coli.*

The values for k computed from each survival curve are presented in Table 2-7, with their respective temperatures, pH values and chlorine concentrations. The values of k obtained for different pH values at 2-5°C are plotted against their respective residual chlorine on a log-log scale in Figure 2-11 and those at temperature 20-25°C in Figure 2-12. With the exception of the points for pH 10.7 at 20-25°C, which were quite irregularly located, all other points fitted reasonably well the lines whose slopes have a value of 1.0. There was only one observation at pH 7.0 at each temperature, for which no lines could be drawn. It may be presumed that the same slope applied to k-chlorine residual relationship for pH 7.0. It appears then, that doubling the free chlorine residual doubles the destruction rate. This concentration-destruction rate relationship is consistent with the first-order kinetics.

The relative bactericidal efficiency of HOCl and OCl^- is demonstrated in Figure 2-13, in which the free chlorine residuals required to yield a value of 1.0 for k/min are plotted on a log scale against the respective pH values. Examination of both curves reveals a ratio of about 100:1 which means that HOCl is about 100 times as bactericidal as the OCl^- ion. From the values for the size and parameters in the analysis obtained with the multi-Poisson equation, it is concluded that this model is applicable to the analysis of bactericidal data.

Treatment of Data on
Inactivation of Enteroviruses by I_2

Since Berg and his associates[10] obtained survival curves that all exhibited a shoulder and computed the values for k from the linear tails, these k values should be identical with those computed by the multi-Poisson model. The values for the clump size (n) were computed with the multi-hit equation and, therefore, must be quite different from those computed with the multi-Poisson model.

With equations (11-14) the per cent and size of virus clumps have been computed for each of the three enteroviruses. These results are presented in Table 2-8, together with the values for n reported by the authors. The table demonstrates the much smaller (2-3) and consistent clump sizes computed by the multi-Poisson model than by the multi-hit equation. This indicates that the multi-Poisson model is suitable for the analysis of virus inactivation data.

Table 2-7. Rate Constant in Destruction *E. coli* by Free Chlorine (Analysis of data of Butterfield *et al.*[18])

Figure No.	Temperature (°C)	pH	Chlorine Concentration (ppm) Initial 0 min	Residual 60 min	k/min
9a	2-5	7	0.05		3.8
			0.04		3.0
			0.03		2.3
b		8.5	0.14	0.12	1.55
			0.07	0.06	0.8
			0.05	0.04	0.56
c		9.8	1.0	1.0	1.5
			0.72	0.72	1.05
			0.4	0.4	0.58
			0.15	0.15	0.22
d		10.7	1.0	1.0	0.46
			0.75	0.75	0.35
			0.5	0.5	0.24
			0.4	0.4	0.19
			0.3	0.3	0.14
10a	20-25	7	0.03	0.02	2.5
b		8.5	0.14	0.14	3.3
			0.07	0.07	1.6
			0.05	0.05	1.15
c		9.8	0.4	0.38	4.2
			0.3	0.26	3.1
			0.14	0.13	1.4
			0.08	0.08	0.8
			0.06	0.06	0.6
d		10.7	0.53	0.52	1.8
			0.4	0.4	1.4
			0.3	0.29	1.05
			0.16	0.16	0.56
			0.06	0.06	0.21
			0.05	0.05	0.18
			0.04	0.04	0.14

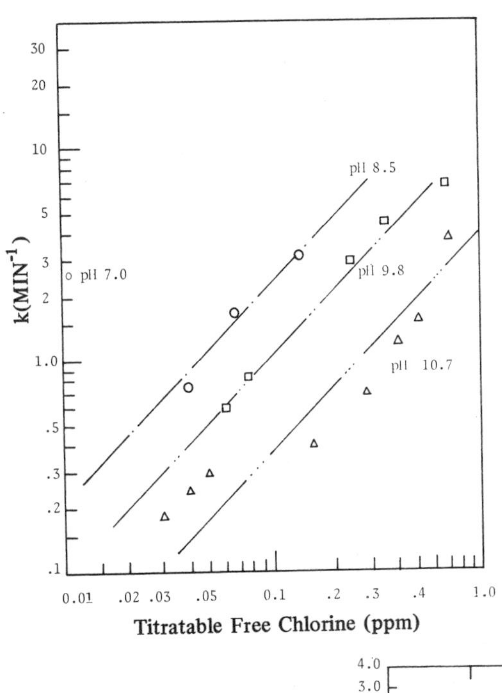

Figure 2-12. Rate constant vs. chlorine concentration in destruction of *E. coli* at 20-25°C

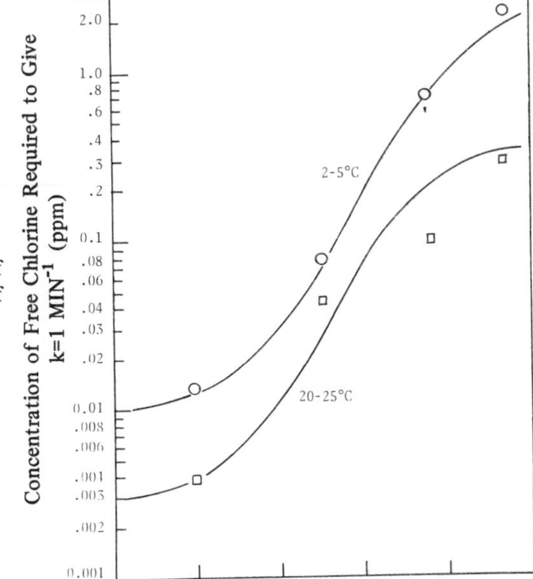

Figure 2-13. Effectiveness of free chlorine as a function of pH in destruction of *E. coli*

Table 2-8. A Comparison Between Values for Clump Size and Per Cent Ascertained by Multi-Poisson Distribution Model and Those Reported by Authors[10] in the Destruction of Enteroviruses by Iodine (I_2)

Experiment Number	Virus	Virus Clump Size (n) and Per Cent: Calculated by M-P Model n	%	Reported by Authors n	%
34	polio-1	2-3 singles	98 2	1.68-10.98	100[a]
35	echo-7	2 singles	92 8	2.04-3.09	100
36a		2 singles	97 3	0.63-2.21	100
36b	coxsackie A9	2 singles	97 3	1.03-4.22	100
36c		2 singles	98 2	1.06-4.32	100

[a]The multi-hit concept is based on clumps in one size only.

DISCUSSION

The multi-Poisson distribution model has been shown by this study to be valid and useful for treating disinfection data, especially data showing departure from exponential survival. The values for clump size and per cent distribution computed by the model were reasonably close to those observed or expected, and the values for k so computed produced good results in kinetics analysis.

The model is relatively complicated and requires a computer even if the number of clump size groups is limited to three. To increase the model's usefulness, a graphic analysis has been developed, which eliminates much of the mathematics. Theoretical survival curves have been generated with the model for values of n_i from 1 to 10 and values of 0.1 to 1.0 for k/min. These computed curves are available from S. L. Chang or J. H. Wei.

To analyze an observed survival curve graphically, the curve is plotted on tracing paper on the same scale as that used in the figures. An experimental survival curve with a shoulder is analyzed for the values of n and k by matching it with the computed curve that fits best. If the experimental curve has a linear tail from which the value of k can be computed, the match process is facilitated by using the computer-prepared figure with the same or nearly the same k value. For a sigmoid curve, the second shoulder can be matched with the computer-prepared curve in the same manner as the first shoulder for the value for n, the per cent of which can be read from the y intercept when the shoulder is extended to zero time. The value for k is approximated from the first shoulder matching, in which the length of the shoulder is more important than the "ramp" to obtain the best estimate.

For observed survival curves of the retardant die-away type, the values for k can be computed from the linear initial segment and from the concave segment as partitioned intuitively. By matching each partitioned segment with the predicted curves in the graph prepared with the same, or nearly the same k value (*e.g.*, Figures 2-9a, 2-9b and 2-9c), the values for n and their percentages can be obtained. However, these values are not nearly as accurate as those estimated for survival curves with a shoulder, mainly because the partitioned segments are usually too short for reliable matching. Such estimates would also lack accuracy by any other method. This drawback is not serious since, as stated before, the important thing in the analysis is to find the value for k, the rate constant in destruction of discrete organisms.

A smooth concave survival curve resembling that of a second order reaction cannot be satisfactorily analyzed for k and size and per cent distribution of clumps, because of the difficulty in partitioning without visible landmarks. It is also possible to have linear survival curves in destruction of organisms in clumps whose distribution follows a normal curve. Theoretically, the closer the clump size distribution approaches normality, the closer is the survival curve to linearity. But the clumping effect in such survival can be recognized only within unreasonably small value for k, compared with that found in destruction of discrete organisms.[4, 19] Most enteric viruses exhibit clumping size distribution skewed either toward the singles or clumps.[20]

Since most survival curves in virus and bacteria destruction have showed either a shoulder or a retardant die-away, or sometimes a sigmoid or stepladder, we believe that the multi-Poisson model should provide a useful and practical tool to analyze disinfection data.

SUMMARY AND CONCLUSIONS

A multi-Poisson distribution model has been developed to analyze disinfection data. The model's validity was established by using it to treat data on destruction of cysts of *N. gruberi* by I_2, in suspensions of known size and per cent distribution of cyst clumps; in these experiments rational values for k and good agreement between calculated and observed cyst clump size distribution were obtained. Application of the model to analyze data on destruction of *E. coli* by free chlorine reported by Butterfield *et al.*[18] obtained values of 2-5–mostly 4–for n. This agrees with electron microscopic findings obtained with young cultures of *E. coli* and with values for k that yielded good and reasonable values for various parameters in an analysis of the kinetics of destruction. Application of the model to data reported by Berg *et al.*[10] on inactivation of enteroviruses by I_2 yielded values of 2-3 for virus clump sizes, which are more reasonable and consistent than those reported by the authors.

Based on these analyses, it is concluded that the model is a useful tool to analyze disinfection data, especially data that show departure from exponential survival.

To use the model without the computer, single Poisson curves were prepared for graphic analysis of survival curves.

REFERENCES

1. Chick, H. "Investigations of the Laws of Disinfection," *J. Hyg.* **8**, 92-158 (1908).
2. Chick, H. "The Process of Disinfection by Chemical Agencies and Hot Water," *J. Hyg.* **10**, 237-286 (1910).
3. Chang, S. L. "Modern Concept of Disinfection," *J. Sanit. Eng. Div., Amer. Soc. Civil Eng.* **97**, 689-707 (1971).
4. Chang, S. L. *Statistics of the Infectious Units of Animal Viruses in Transmission of Viruses by the Water Route*, G. Berg, Ed. (New York: Interscience, 1967), pp. 219-234.
5. Kimball, A. W. "The Fitting of Multi-Hit Survival Curves," *Biometrics* **9**, 201-211 (1953).
6. Fair, G. M., J. C. Morris, and S. L. Chang. "The Dynamics of Water Chlorination," *J. New Engl. Water Works Assoc.* **61**, 285-301 (1947).
7. Chang, S. L. "Kinetics in the Thermodestruction of Cysts of *Endamoeba histolytica* in Water," *Amer. J. Hyg.* **52**, 82-90 (1950).
8. Chang, S. L. and M. Baxter. "Studies on Destruction of Cysts of *Endamoeba histolytica*. I. Establishment of the Order of Reaction in Destruction of Cysts of *E. histolytica* by Elemental Iodine and Silver Nitrate," *Amer. J. Hyg.* **61**, 121-132 (1955).
9. Chang, S. L., M. Baxter, and L. Eisner. "Studies on Destruction of Cysts of *Endamoeba histolytica.* II. Dynamics of Destruction of Cysts of *E. histolytica* in Water by Triiodide Ion," *Amer. J. Hyg.* **61**, 133-141 (1955).
10. Berg, G., S. L. Chang, and E. K. Harris. "Devitalization of Microorganisms by Iodine. I. Dynamics of the Devitalization of Enteroviruses by Elementary Iodine," *Virology* **22**, 469-481 (1964).
11. Taylor, D. and J. D. Johnson. "Kinetics of Bromine Disinfection of Viruses," in *Water Supply, Treatment and Distribution*, A. Rubin, Ed. (Ann Arbor, Mich.: Ann Arbor Science Publishers, Inc., 1973).

12. Clark, B. M. and J. F. Niehaus. *A Mathematical Model for Virus Devitalization. Transmission of Viruses by the Water Route*, G. Berg, Ed. (New York: Interscience, 1967), pp. 241-245.
13. Clark, R. M. "A Mathematical Model of the Kinetics of Virus Devitalization," *Math. Biosci.* 2, 413-423 (1968).
14. Feurzeig, W. and S. A. Taylor. "A Note on Exponential Fitting of Empirical Curves," *Argonne Natl. Lab. Quart. Report, Biol. Med. Div.*, Chicago, **ANL-4401**, 14-29 (1950).
15. Chang, S. L. "Cultural, Cytological and Ecological Observations on the Amoeba State of *Naegleria gruberi*," *Gen. Microbiol.* **18**, 565-578 (1958).
16. Chang, S. L. "Small Free-Living Amoebae: Classification, Cultivation, Identification, Pathogenesis and Resistance," *Current Topics in Comp. Pathobiol.* **1**, 201-253 (1971).
17. Chang, S. L. "The Use of Active Iodine as a Water Disinfectant," *J. Amer. Pharmac. Assoc. Sci. Ed.* **47**, 416-423 (1958).
18. Butterfield, C. T., E. Wattie, S. Megregian, and C. W. Chambers. "Influence of pH and Temperature on the Survival of Coliforms and Enteric Pathogens when Exposed to Free Chlorine," *Public Health Rep., U.S.* **61**, 157-193 (1943).
19. Sharp, D. G. "Electron Microscopy and Viral Particle Function," In *Transmission of Viruses by the Water Route*, G. Berg, Ed. (New York: Interscience, 1967), pp. 193-217.
20. Smith, K. O. and J. L. Melnick. "Electron Microscopic Counting of Virus Particles by Sedimentation on Alluminized Grids," *J. Immunol.* **89**, 279-284 (1962).

CHAPTER 3

RECYCLE–WHAT DISINFECTANT FOR SAFE WATER THEN?

Mark A. McClanahan

Department of Civil Engineering
Georgia Institute of Technology
Atlanta, Georgia 30332

INTRODUCTION

The general, widespread acceptance of public water quality speaks well of the successful operation of the many water treatment plants within the United States. The confidence of the American people in the quality of their drinking water has been based upon the demonstrated reduction of waterborne diseases, such as typhoid and cholera, by the application of proved treatment techniques for pathogen removal and destruction.

As man moves into the fourth quarter of the 20th century, the leading causes of death and serious disability among Americans are not diseases that are infestations of microorganisms, but are–for example, heart disease, cancer and emphysema–diseases directly or indirectly related to the environmental stress of the individual. The major exception to this is the virus diseases like flu and hepatitis. This change in the type of disease, from one with a causative microbiological agent that can be isolated, counted, measured, and that has a relatively short time between contact and clinical symptoms, to a type for which the causative agent is frequently only guessed at, and for which the time between initial contact and clinical symptoms may be decades, has vastly complicated the practice of preventive medicine. Many of these diseases may require not one contact with the causative agent but repeated or prolonged contact over a period of years.

For disinfection there are, in general, two considerations: What effects will the disinfectant have on the living, and on the nonliving, organic molecules present in our water supplies as we move toward greater recycle of wastewater, either intentionally in arid regions such as South Africa, Israel, and the southwestern United States, or, unintentionally, in large cities using well-used river water such as the Ohio, Mississippi and the Rhine as water supplies.

LIVING ORGANICS

It will be necessary to determine, as far as possible, the deliterious effects of consuming, for prolonged periods of time, water recovered from sewage. The most fruitful areas in which to make such evaluations are the cities that for many years have used heavily polluted surface waters as their raw water supply. Cities along the Ohio (reported to be about 15% sewage effluent at low flow[1]) and the Rhine (reported to be about 40% sewage effluent at low flow[1]) offer excellent populations for epidemiological studies to compare with similar populations from cities supplied by unpolluted surface or ground water. These two rivers have been heavily polluted for many years. The populations consuming their waters should show increased health impairment from this long-term exposure, compared with those using less polluted water, if consumption of water with heavy sewage pollution is detrimental to the individual's health. Recent undocumented reports of a study such as this have appeared.[2,3] Careful epidemiological studies on populations consuming water from such low quality raw water sources could possibly point out facts about the health hazard of drinking treated wastewater.[4]

Health hazards are also related to the bacteria and viruses present in wastewater. Laboratory experiments have demonstrated the recovery of viable nucleic acid from viruses inactivated by certain disinfectants.[5] This is caused by the reaction of the disinfectant with the protein material that surrounds the nucleic acid in such a way that the virion is no longer an infective particle but, at the same time, the nucleic acid remains infective. Under laboratory conditions the protein coat of these inactivated viruses can be removed and the intact nucleic acid made to infect a host cell, thereby producing new complete virus particles. Can the nucleic acid of viruses that are inactivated in disinfected wastewater be released from the protein coat once the inactivated virus enters the human body? And, once released, can the nucleic acid infect the host, causing clinical symptoms of disease?

A somewhat similar area of concern is whether the destruction of the usual indicator organism, *E. coli*, is sufficient to render the water free of disease-causing organisms. The enteroviruses are reported to be more resistant than *E. coli* to free chlorine.[6] Can there be persons infected by viruses from drinking water that has met the U.S. Drinking Water Standards for bacterial water quality?

At present there are no findings indicting present U.S. water treatment practice is presenting pathogenically unsafe water. However, the questions that are raised in this paper have not been satisfactorily answered. It is possible that there are many persons in the United States who have been infected by viruses in drinking water that received what was considered to be adequate treatment, or who have become sick because the viable nucleic acid from an inactivated virion released in their bodies. The relatively small number of persons infected in either of these ways would make the water vector of the disease difficult if not impossible for any physician to ascertain. Since the subclinical frequency of disease may be 10-1000 times higher than that of the clinical disease,[7] there could be many persons infected in either way even now.

Water that is safe for consumption must be treated to insure against possible deliterious effects to the consuming individual's health, and not only to prevent epidemics. That is why it is necessary to practice disinfection that will kill all the pathogenic particles present in the water. To do this, we must recognize the differences in efficacy of the various disinfectants towards the several classes of pathogenic organisms and apply this knowledge in the continuous operation of disinfection.

NONLIVING ORGANICS

Although attempting to eliminate all pathogenic organisms from recycled water is a worthy goal, it is not the only goal to consider. Treated wastewater will contain increased concentrations of organic molecules, some of which are presently known health hazards, such as toxicants, carcinogens, mutagens and teratogens.[1,8,9] Even disregarding the presence of these chemicals that are known health hazards, potable water may not be safe because of the chemical reactions that can occur between the disinfectant and the organic compounds, both natural and synthetic, that remain in the water after even the best of today's water and wastewater treatment processes.

The level of organic material remaining after tertiary treatment is about 10-20 mg/l COD at the South Lake Tahoe plant even with carbon adsorption.[10] With the Z-M process this plant effluent still had a COD of more than 6 mg/l.[10] Besik applying reverse osmosis to a waste found a dissolved organic carbon concentration in the product water of 2-3 mg/l, with other runs having values as high as 10-15 mg/l.[11] This demonstrates that even the best wastewater treatment does not remove all organic carbon compounds and, for that matter, no treatment short of distillation can reduce the dissolved organic carbon concentration below a few mg/l. Therefore, chemical reactions between residual organic compounds in the water and the disinfectant should be considered as a possible source of hazardous chemicals.

Such residual chemicals would be those that are difficult to degrade biologically, or compounds that passed through the treatment plant in a slug before the biomass became acclimated, or metabolic products from the degradation of chemicals in the wastewater, or cellular material from ruptured organisms in the biomass. All these organic compounds could possibly react with the disinfectant. If the disinfectant is chlorine, the formation of chlorinated organic compounds is possible.

A literature search for chemical definition of the organic compounds present in the fresh water, raw sewage and treated sewage yields only a few references, which were recently reviewed by Ongerth *et al.*[12] Less than 30% of the compounds isolated were identified, but several of these compounds were known to be carcinogenic.

To demonstrate that chlorination of organic chemicals is possible in aqueous solution under conditions similar to those found in disinfection, Barnhart and Campbell[13] treated 14 selected organic compounds with 10 mg/l chlorine in aqueous solution. Five of the compounds were found to be chlorinated after 30 minutes exposure. Breakpoint chlorination of Rhine river water has been shown to produce chloro- and bromo-methane derivatives.[14] Headspace gas chromatograms with mass spectral verification of identification showed from 6-54 μg/l of chloroform apparently produced from humic substances in the Rhine. A large variety of other volatile organics were found in diminished concentrations after breakpoint chlorination. With the untold thousands of organic chemicals deposited into wastewater by people and industry, there can be little doubt that many of these will react with chlorine. To get an estimate of the insult that the consumer's body might encounter from drinking this water, the following calculations are made:

Consider chlorine as a disinfectant of recycled water. Assume that the organic material is reduced by 99% from an original value of 300 mg/l which would leave only 3 mg/l. If only 0.1% of the residual organic matter were converted to chlorinated organic compounds, a human that consumed only 1 l/day for 20 years would ingest 22 mg of chlorinated hydrocarbons.

As a result of recycle, the concentration of these chlorinated organic compounds will increase. Assuming that only recycled water contains the chlorinated compounds and comprises 90% of the total flow, then according to the relation developed by Dick and Snoeyink[15] (Figure 3-1), the ultimate concentration in the recycled water under these conditions would be nine times the single pass value.

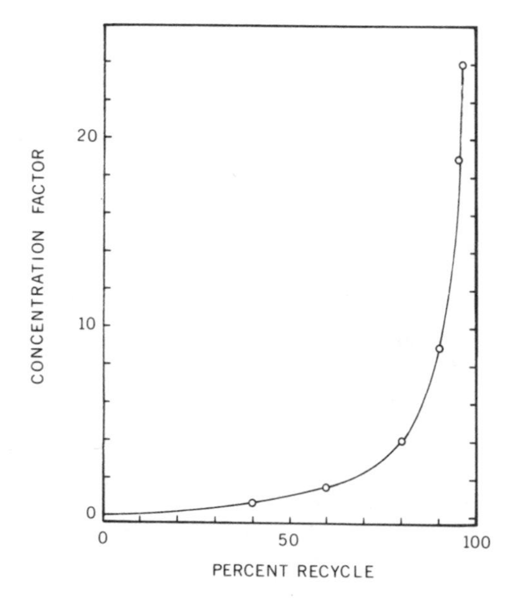

Figure 3-1. Maximum concentration factor for a substance as a function of per cent recycle, assuming makeup water contains none of the substance.[15]

Assuming this concentration factor, total 20-year ingestion would be 200 mg which, for a 70-kg person, would be about 3 mg/kg of body weight–a rather low but possibly not insignificant value. However, these compounds would tend to concentrate in fatty tissues where this value would increase by a factor of 3-5 times. As this would be a long-term exposure, the maximum concentration would not reach this level, because some of the compounds would be excreted or even decomposed.

From data reported for body fat storage of chlorinated pesticides,[16,17] it appears that a factor of about 20 is the limit between dosage and fat-storage concentration. Using this factor for the storage level of chlorinated compounds from drinking water, a recycle concentration factor of nine and a more realistic water consumption of 2 l/day, a wastewater that averages 10 mg/l organic carbon and a 0.1% yield would give a daily dosage of about 0.2 mg/l, or a fat-storage value of 3.6 mg/kg chlorinated organic compounds in the body fat of the consumer. (In making this calculation it is realized, of course, that the organic compounds that do not react with the chlorine may also be concentrated in the body in various organs and tissues, and that their concentration is 1000 times greater than that of the chlorinated compounds formed in the example cited. The presence of these organic compounds may be the most serious objection to the use of recycled water.) Again, a study to measure the total body burden of chlorinated organic compounds of persons drinking treated Ohio or Rhine River water would give some indication of the magnitude of body insult caused by long-term ingestion of chlorinated organic compounds produced in the disinfection of drinking water with a sewage origin.

Storage of chlorinated hydrocarbon pesticides in excess of this calculated level with no apparent health impairment has been reported.[17-20] After 30 years of exposing the general population to chlorinated hydrocarbon pesticides, there do not appear to be any significant health problems related to a general body burden of 10-100 μg/g of fat for specific pesticides, and a possible total body burden of chlorinated organic compounds that may be 10 times that high. Since the compounds formed in the chlorination of recycled wastewater are unknown, it is impossible to say whether the calculated level would be a serious health hazard.

The point to remember here is that because these compounds are chlorinated they will tend to resist degradation and therefore will remain in the body for a long time. It is this long-term insult that is of concern here, with little data available to make sound judgment. If all compounds were no more serious to the human physiology than the chlorinated hydrocarbon pesticides have proved to be during the last 30 years, there might be little reason to fear the consequences of consumption of chlorinated recycled wastewater. However, Epstein suggests that this is not the case.[7]

NEED FOR LONG-TERM STUDIES

Fundamental disinfection studies must define the type of chemicals that will devitalize the viral nucleic acid, not just inactivate the intact virus particle. They must also demonstrate that these disinfectants do not react with the organic molecules in the recycled water to form compounds that are harmful to man. It will take many years of experience before use of direct recycle water can be demonstrated to cause diseases with a long incubation time. There is little or no evidence even today that treated water is a vector of virus disease. However, it may well be that some of the diseases being diagnosed as resulting from environmental stress may have as a factor the water the person has consumed for the last 20, 40 or even 80 years. At present, there appears to be some correlation between the hardness of the drinking water and heart and blood vessel disease of the consumer.[21] A serious difficulty in developing correlations between disease and drinking water for persons in the United States is the transient nature of Americans. The initial innoculation of the causative agent for a disease with an incubation time of 20 years, may have been in a locale halfway around the world from where the person is presently living. This may make it nearly impossible to develop correlations between drinking water and disease for many years after complete recycle is an accomplished fact.

INTRODUCTION TO REACTIONS OF CHLORINE AND WATER

The following serves as an introduction to the chemistry of chlorine. This subject is treated in greater detail in subsequent chapters. Effective disinfection in recycled water must consider the chemical reactions that might occur in the water to alter the efficiency of the chemical as applied. One of the first areas to consider is the effect of pH on the disinfectant chemicals since some of the processes that will be used to treat wastewater for recycle will result in an elevated pH. If chlorine is chosen as the disinfectant, its simplest chemical reactions with water are described by the following equations:

$$Cl_2 + H_2O \rightleftharpoons H^+ + Cl^- + HOCl \tag{1}$$

$$HOCl \rightleftharpoons H^+ + OCl^- \tag{2}$$

Reaction (1) goes to completion. Reaction (2) is most important in disinfection of water for potable supplies. The equilibrium dissociation constant for the reaction, when the total salt or dissolved solids are low or when the activity coefficients are assumed to be equal to unity, is[22]

$$K = \frac{[H^+][OCl^-]}{[HOCl]} = 2.7 \times 10^{-8} \text{ (at } 20^\circ\text{C)} \tag{3}$$

Assuming no other reactions take place, an equation depicting the total concentration (C_T) or a mass balance can be written as

$$C_T = [HOCl] + [OCl^-] \tag{4}$$

For convenience the additional relationships can be defined as

$$\alpha_o = \frac{[HOCl]}{C_T} \tag{5}$$

$$\alpha_1 = \frac{[OCl^-]}{C_T} \tag{6}$$

where α_0 and α_1 are the fractions of free chlorine in the HOCl and OCl^- form. Since they are fractions:

$$\alpha_0 + \alpha_1 = 1 \tag{7}$$

Combining equations (3) and (4), and substituting them into equations (5) and (6) yields

$$\alpha_0 = \frac{[H^+]}{[H^+] + K} \tag{5a}$$

$$\alpha_1 = \frac{K}{[H^+] + K} \tag{6a}$$

Equations (5a) and (6a) show the α coefficients are functions of the hydrogen ion concentration and the dissociation constant of the acid. Since the dissociation constant does not change for constant temperature and ionic strength, the α coefficients will be a function only of the hydrogen ion concentration, or the pH. Thus, setting the two definitions of each of the α terms equal to each other and solving for the respective chemical species yields:

$$[HOCl] = \frac{[H^+] C_T}{[H^+] + K} \tag{8}$$

$$[OCl^-] = \frac{K\, C_T}{[H^+] + K} \tag{9}$$

Once the free chlorine concentration, C_T, is determined in the system, variation of the pH will dictate the concentration of the respective species at a given dissolved solids and temperature. A log-log plot of equations (8) and (9) demonstrates the dependence of concentration of HOCl and OCl^- on pH for a given total concentration. Figure 3-2 shows these relationships for C_T of 5×10^{-3} molar, M. Other log concentration scales could equally be used.

Figure 3-3 is a log diagram of $HOCl\text{-}H_2O$ system. In Figure 3-2, C_T was maintained constant while in Figure 3-3 the concentration of HOCl was maintained constant. The rationale for holding the concentration of HOCl constant was based on the experimental data obtained for disinfection of various microorganisms. It has been observed that as the pH increased beyond the pK value of HOCl, the disinfection efficiency of the system decreased. To maintain the same disinfection efficiency, the total concentration in the system had to be increased, leading to the assumption that OCl^- is not a disinfectant. If this assumption were valid, the concentration values for the disinfection data would be expected to fall on the line representing C_T in Figure 3-3. In other words, to maintain a constant rate of kill for

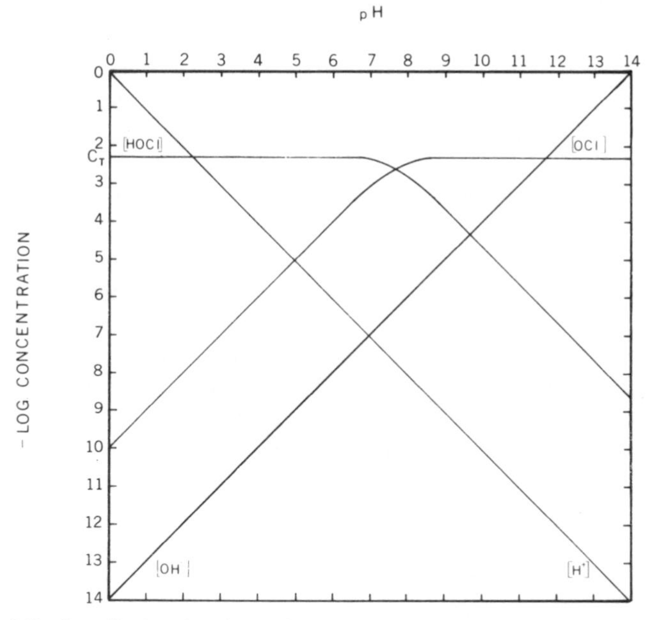

Figure 3-2. Log diagram for the $HOCl\text{-}H_2O$ system with total concentration (C_T) constant at 5×10^{-3} M/l.

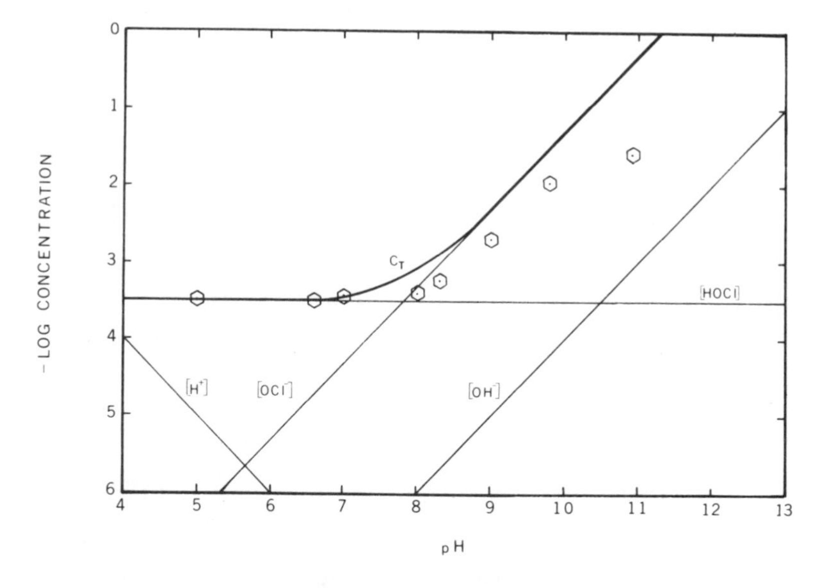

Figure 3-3. Log diagram for the system $HOCl\text{-}H_2O$ with the concentration of HOCl constant. Data points show effect of pH on the concentration required to inactivate cysts of *E. histolytica* (99.995% in 10 minutes).[23]

the organisms, C_T, would have had to increase to compensate for the effect of pH on the dissociation of the disinfectant. The data for the destruction of *E. histolytica* cysts by chlorine reported by Chang[23] agrees well with the assumption up to about pH 10; beyond this point the destruction of the cysts exceeds that which would be predicted from the total free chlorine concentration of disinfectant. At this point the experimental value of the concentration was only about 0.1 of the value predicted from the log diagram plot. The graph shows that the concentration of OCl^- was about 100 times more than that of the HOCl required for the same end point, indicating that HOCl is about 100 times as effective a disinfectant as OCl^- in destruction of cysts.

Figure 3-4 is a similar plot with data for virus (at 0°C[24]), bacteria (at 2-5°C[25]), and cysts (at 22-25°C[25]). These data also closely follow the C_T plot over the pH range for which data are available. These data indicate

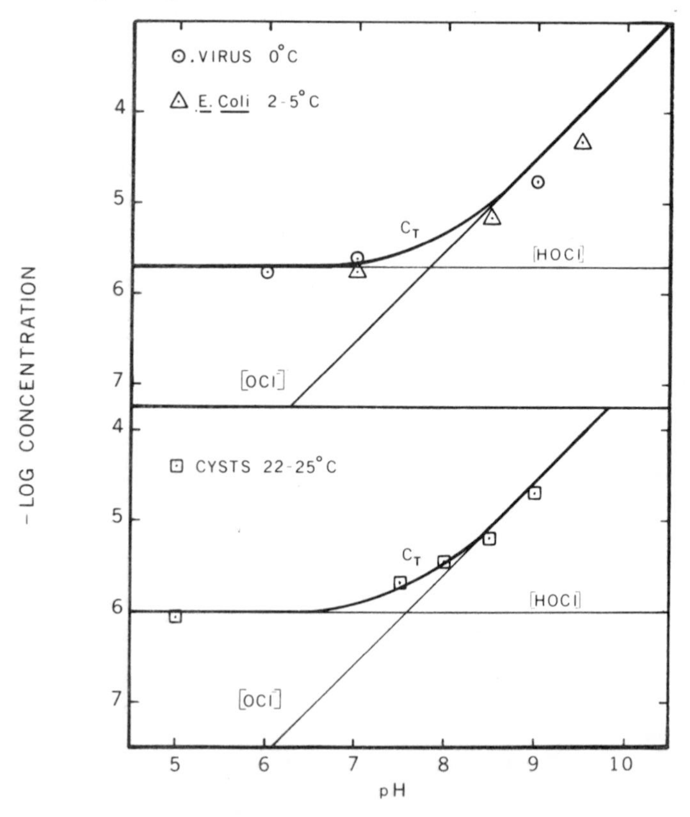

Figure 3-4. Log diagram for the system $HOCl\text{-}H_2O$ with the concentration of HOCl constant. Data points show effect of pH on the concentration required to inactivate virus (99.6%),[24] *E. coli* (99.999%) and *E. histolytica* cysts (99.999%).[25]

that the organisms of importance in waterborne disease have a similar response to the free chlorine required as the effect of pH on the dissociation of HOCl below pH 9. The conclusion is that HOCl should be held constant, not total free chlorine. Since elevated pH may well be one of the characteristics of treated wastewater available for recycle, this fact must play an important role in considering chlorine as the disinfectant to be used.

INTRODUCTION TO REACTIONS OF CHLORINE AND AMMONIA

Chlorine also undergoes other reactions with ammonia that could present a serious problem in disinfection of recycle water:

$$HOCl + NH_3 \rightarrow H_2NCl + H_2O \quad (10)$$

$$HOCl + H_2NCl \rightarrow HNCl_2 + H_2O \quad (11)$$

$$HOCl + HNCl_2 \rightarrow NCl_3 + H_2O \quad (12)$$

These combined chlorine species have been demonstrated to be much less effective than free HOCl in disinfection of bacteria, viruses and cysts.[26-28] It was shown that when free HOCl disappeared from the reaction system, viral inactivation nearly stopped even though there was nearly 30 mg/l combined chlorine in the system.[26] The fact that viruses are not inactivated by the chloramines explains the infectious hepatitis epidemic in New Delhi, where the combined residual chlorine was sufficient to destroy the indicator bacterial pollution but not the viral population.[27,29] Viruses are more resistant than bacteria to combined chlorine forms These chemical reactions between chlorine and ammonia are most conveniently overcome by the technique of breakpoint chlorination (Figure 3-5), by which the ammonia is destroyed and sufficient chlorine is added to maintain a residual of free available chlorine.

COMPARISON OF DISINFECTANTS

To compare disinfectants and their effect on various types of organisms, survival curves are needed for the various organisms with the different disinfectants. If the survival curves can be linearized by methods such as described in Chapters 1 and 2, a graphical comparison can be made. Hom[30] presented a simple mathematical model that can be used to linearize most disinfection data. This model was of the general form:

$$\frac{dN}{dt} = -kNt^mC^n \quad (13)$$

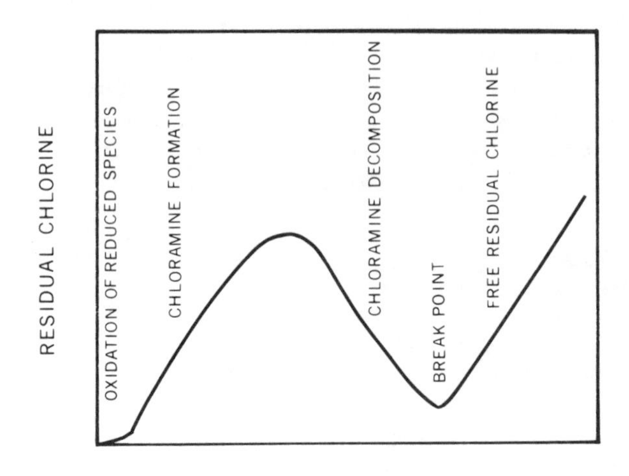

Figure 3-5. Relationship between chlorine dosage and residual chlorine for breakpoint chlorination.

Chick's Law,[31] the specialized condition for first-order kinetics, can be obtained from equation (13) by setting m and n equal to zero. The integrated form of the specific equations are:

for Chick's Law $$\ell n\ N/N_o = -kt \quad (14)$$

which gives a linear plot on semi-log paper;

for m = 0 and n ≠ 0 $$\ln N/N = -k' \ln t/t_o \quad (15)$$

which gives a linear plot on log-log paper;

and for m ≠ 0 and n ≠ 0

$$\ln N/N_0 = \frac{-k''\ t^m}{m} \quad (16)$$

which gives a linear plot of log (log N/N_o) versus log t, where k, k′ and k″ are the specific rate constants for the respective equations, N is the number of organisms at any time t, N_o is the initial population, and C is the concentration of disinfectant.

Other simple linearizing transformations[24] have been used such as:

$$\ln N/N_0 = k''\ t^2 \quad (17)$$

This gives a linear plot on semi-log paper although its form is simply a specialized form of equation (13) with m = 2 and n = 0. Linearization of the data allows a log-log plot of concentration versus time to be made so the disinfectants can be compared in terms of efficiencies (Figure 3-6), from data reported by Chang.[25] For these comparisons to be valid, the same end point must be used, *i.e.*, the same percentage reduction in organisms, and the same environmental conditions. As seen in Figure 3-6, the change of pH causes a major change in the location of the concentration-contact time relationship. Figure 3-6 also shows another comparison–that of the same disinfectant under the same environmental conditions, but versus a different test organism. This type of information makes it possible to determine the relative resistances of the microorganisms to various disinfectants and to aid in the choice of the correct disinfectant for the problem at hand.

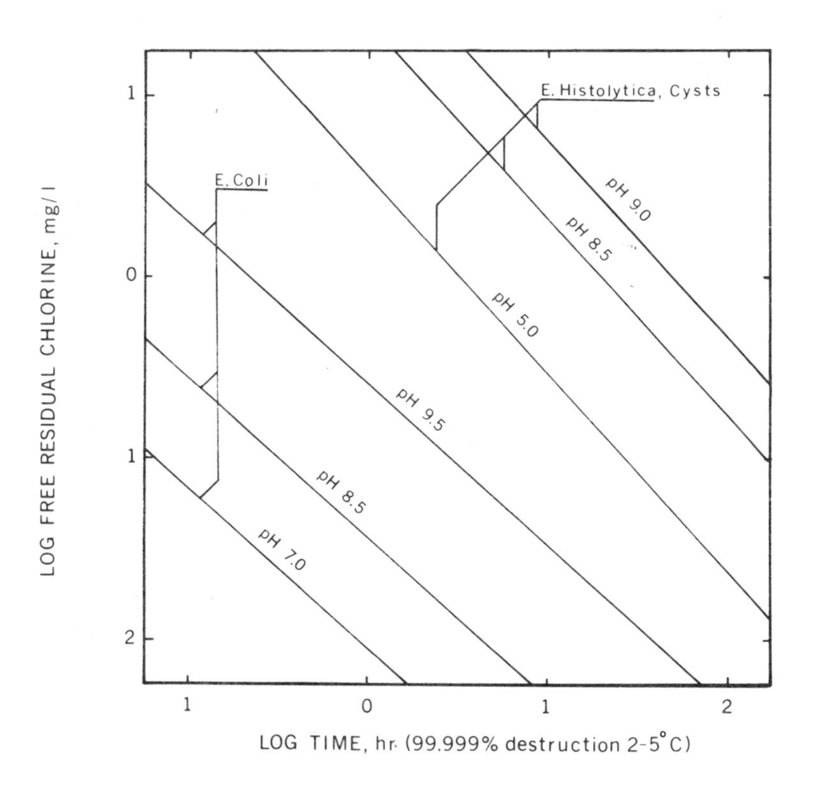

Figure 3-6. Relationship between contact time and disinfectant concentration for *E. coli* and *E. histolytica* cysts at various pH levels.[25]

Better and more complete methods of comparing disinfectant data are presented in the previous two chapters. Controlled laboratory experiments to obtain information on disinfection efficiency with recycled water are needed to ensure that the actual environmental conditions will be approximated.

INTRODUCTION TO OZONE–A PARTIAL ANSWER?

One of the most promising disinfectants for recycled water is ozone. In a review of the literature, Venosa reported ozone to be effective against bacteria, virus and cysts under various environmental conditions.[32] The recovery of viable nucleic acid from ozonated viral suspensions has not yet been reported. Ozone does not undergo dissociation reactions with pH in water as does chlorine, with the resulting loss in disinfection efficiency; nor does it react with ammonia to form a less efficient disinfectant like a chloramine. It is a very strong oxidizing agent; therefore, a high ozone demand is exerted by reduced material, both organic and inorganic, present in the water. This demand can be expected to be greater than for chlorine because ozone is the more powerful oxidizing agent. A major benefit of ozone disinfection is that ozonated organic compounds are more biodegradable than either the chlorinated compounds or the parent compound. This has been demonstrated by the increase in Biochemical Oxygen Demand (BOD) from pre- to post-ozonation solutions with various systems, from detergent solutions[33] to secondary waste effluent.[34,35] The major deficiency of ozone as a disinfectant is that, because of its high oxidizing power, it is very unstable in water; it has a very short life-time. Because of this, disinfection with ozone does not provide a residual to serve as an indicator of continued microbiological safety, in case the water becomes contaminated during distribution.

OPERATIONAL CONSIDERATIONS

Possibly the major danger encountered in the use of recycled water is the failure of the waste treatment process. If it fails catastrophically, the waste would logically be diverted, and another source of raw water used. It is the minor changes in wastewater plant efficiencies that will cause the greatest problems. For instance, what if the ammonia removal process released 3 mg/l NH_3-N instead of 1 mg/l? Unless there were a continuous monitoring of the NH_3-N and chlorine were being used as the disinfectant, the water would not be adequately disinfected for virus, because the amount of chlorine being added to breakpoint water with 1 mg/l NH_3-N would not be enough for 3 mg/l. The result would be disinfection with chloramines rather than free chlorine. Depending upon the actual contact times, this might not be sufficient disinfection even for the bacteria in the water, and certainly not satisfactory for disinfection of the virus.

Direct connection to a major source of pathogenic organisms, as will occur in recycle, necessitates the use of the best disinfection practice available. It appears that chlorination alone will not adequately protect the consumer under the varying quality of effluent from the waste treatment processes that will be used. Several possibilities that could be used to solve this problem, not the least of which would be improved reliability of the water and waste treatment processes; another would be the use of a dual disinfectant system. The most promising are ozone-chlorine and chlorine-bromine. With the combination of ozone and chlorine, since the ozone does not have the interferences of dissociation and reaction with ammonia that chlorine does, these would destroy the organisms in the raw recycled water under varying pH and ammonia levels. Because of the short life of ozone in water, chlorine, the second disinfectant, would have to be added to maintain water quality throughout the treatment plant and distribution system. Use of ozone first also would reduce the amount of organic compounds that could be chlorinated by the subsequent chlorination of the water. With the combination of bromine and chlorine, the chlorine would be added first to economically remove the majority of the reducing demand. Bromine would then be added to oxidize the chloramines formed or produce bromamines that are good disinfectants. Also, bromine is not subject to the interference of pH as OBr^- isn't formed in the normal pH range. Bromine will not have the advantage of forming nontoxic easily oxidized organics since these are likely to be similar to chlororganics. Bromine would be less expensive and more reliably dosed than ozone due to the low solubility of ozone.

The multiplied health hazards presented to the consumer by a malfunctioning recycle water plant require that the plant have much larger capacity for treatment and storage than a similar plant treating water from a higher quality source. Increased storage of treated water will be necessary because the water source could possibly be an unacceptable quality for several hours, or possibly even days at a time. A city that has no alternative source of supply of raw water equal to one week's demand would be at the mercy of the wastewater treatment plant operation.

It is obvious that as dilution water is eliminated from the waste effluent and the time of passage between the waste effluent outfall and the potable water intake is reduced by direct connection of the two, the need for expert operation of both plants will be required. The cost of water treatment under the constraints of recycle may dictate that water used for drinking purposes be supplied in a separate distribution system to minimize the volume required, and to assure that it is safe to drink. Needless to say, drinking water standards for recycle water must be much more stringent than are those of the present USPHS Drinking Water Standards which assume a high quality water source.

In the end, direct recycle, where the waste treatment plant effluent line is directly connected to the intake line of the potable water plant with zero retention time, will prove to be less than desirable no matter what disinfectant system is chosen. The most practicable recycle system will be one in which a series of holding ponds provide some days or weeks of storage prior to reuse of the water. With this constraint the economic considerations may eliminate direct recycle, except in emergencies, to those fortuitous circumstances in which the required land is available.

SUMMARY

From the data available it appears that chlorine may not be the best disinfectant to use for drinking water where poor quality raw water or complete recycle water is used. The possibility that disinfection by chloramines will allow virus to remain viable, or that the inactivated virus particle has viable nucleic acid within it that may be released within the human host, the reduction of disinfection efficiency with elevated pH, and the formation of persistent chlorinated organic compounds are all reasons to consider alternative disinfectants. Disinfection by ozonation is expensive but may not have these problems associated with chlorine. Even bromine has many of these desirable properties and is only slightly more expensive. The short life of ozone in water requires another disinfectant be added to provide residual protection throughout the distribution system.

REFERENCES

1. Shuval, H. I. and N. Gruener. "Health Considerations in Renovating Wastewater for Domestic Use," *Environ. Sci. Tech.* 7, 600-604 (1973).
2. Harris, R. H. and E. M. Brecher. "Is Water Safe to Drink," In *Consumer Reports* (Mt. Vernon, N.Y.: Consumers Union, 1974)
3. Harris, R. H. *The Implications of Cancer-Causing Substances in Mississippi River Water* (Washington, D.C.: Environmental Defense Fund, 1974).
4. McCabe, L. J. and J. Symons. "Chlorinated Organic Compounds in Water," Workshop Session, In *Natl. Meeting Amer. Water Works Assoc.*, Dallas, Texas, December 3-4, 1974.
5. Hsu, Y. C., S. Nomura and C. W. Krusé. "Some Bactericidal and Virucidal Properties of Iodine Not Infecting Infectious RNA and DNA," *Amer. J. Epidemiol.* 82, 317-328 (1965).
6. Chang, S. L. "Modern Concept of Disinfection," *J. San. Eng. Div. Amer. Soc. Civil. Eng.* 97 (SA5) 689-707 (1971).

7. Melnick, J. L. "Detection of Virus Spread by the Water Route," *Proc. 13th Water Qual. Conf.*, Urbana, Illinois (February 1971).
8. Long, W. N. and F. A. Bell, Jr. "Health Factors and Reused Water," *J. Amer. Water Works Assoc.* **64**, 220-225 (1972).
9. Epstein, S. S. "Environmental Determinants of Human Cancer," *Cancer Res.* **34**, 2425-2435 (1974).
10. Zuckerman, M. M., I. E. Joya and T. M. Zaferatos. "Plant Scale Demonstration of the Z-M Process for Wastewater Treatment," *Water Sewage Works* **119** (10), 87-91 (1972).
11. Besik, F. "Some Aspects of Reverse Osmosis in Treatment of Domestic Sewage," *Water Sewage Works* **119**, 72-79, 82-85 (1972).
12. Ongerth, J. J., D. P. Spath, J. Crook and A. E. Greenberg. "Public Health Aspects of Organics in Water," *J. Amer. Water Works Assoc.* **65**, 495-498 (1973).
13. Barnhart, E. L. and G. R. Campbell. "Effect of Chlorination on Selected Organic Chemicals," *U. S. Natl. Tech. Inform. Serv. P. B. Rep. No. 211160* (1972).
14. Rook, J. J. "Formation of Haloforms During Chlorination of Natural Waters," *Water Treat. Exam.* **23**, 234-243 (1974).
15. Dick, R. I. and V. L. Snoeyink. "Equilibrium Concentrations in Reuse Systems," *J. Amer. Water Works Assoc.* **65**, 504-505 (1973).
16. Hayes, W. J., Jr. *Scientific Aspects of Pest Control,* Pub. No. 1402 (Washington, D. C.: National Academy of Sciences, National Research Council, 1966).
17. Durham, W. F. "Body Burden of Pesticides in Man," *Ann. N. Y. Acad. Sci.* **160**, 183-195 (1969).
18. Hoffman, W. S., H. Adler, W. I. Fishbein and F. C. Baner. "Relation of Pesticide Concentrations in Fat to Pathological Changes in Tissues," *Arch. Environ. Health* **15**, 758-765 (1967).
19. Morgan, D. P. and C. C. Roan. "Chlorinated Hydrocarbon Pesticide Residue in Human Tissues," *Arch. Environ. Health* **20**, 452-457 (1970).
20. Ecobichon, D. J. "Chlorinated Hydrocarbon Insecticides: Recent Annual Data of Potential Significance to Man," *Can. Med. Assoc. J.* **103**, 711-716 (1970).
21. Winton, E. F. and L. J. McCabe. "Studies Relating to Water Mineralization and Health," *J. Amer. Water Works Assoc.* **62**, 26-30 (1970).
22. Morris, J. C. "The Acid Ionization Constant of HOCl from 5 to 35," *J. Phys. Chem.* **70**, 3798 (1966).
23. Chang, S. L. "Studies on *Endamoeba histolytica*," *War Med.* **5**, 46 (1944).
24. Weidenkopf, S. J. "Inactivation of Type 1 Poliomyelitis Virus with Chlorine," *Virology* **5**, 56-57 (1958).
25. Chang, S. L. *The Present Status of Knowledge on Destruction and Removal of Enteric Pathogens in the Water Environment.* (Cincinnati: Environmental Health Service, Public Health Service, Department of Health, Education and Welfare, 1969).
26. Olivieri, V. P., T. K. Donovan and K. Kawata. "Inactivation of Virus in Sewage," *Proc. Natl. Specialty Conf. on Disinfection.* (New York: American Society of Civil Engineers, 1970), pp. 365-383.

27. Stringer, R. and C. W. Krusé. "Amoebic Cysticidal Properties of Halogens in Water," *J. San. Eng. Div. Amer. Soc. Civil Eng.* **97** (SA6) 801-811 (1971).
28. Krusé, C. W., Y. Hsu, A. C. Griffiths and R. Stringer. "Halogen Action on Bacteria, Viruses and Protozoa," *Proc. Natl. Specialty Conf. on Disinfection* (New York: American Society of Civil Engineers, 1970), pp. 113-136.
29. Viswanathan, R. "Infectious Hepatitis in Delhi (1955-56), Epidemiology," *Ind. J. Med. Res.* **45**, 1-29 (1957).
30. Hom, L. W. "Kinetics of Chlorine Disinfection in an Ecosystem," *Proc. Natl. Specialty Conf. on Disinfection* (New York: American Society of Civil Engineers, 1970), pp. 515-537.
31. Chick, H. "An Investigation of the Laws of Disinfection," *J. Hyg.* **8**, 92-158 (1908).
32. Venosa, A. D. In *Ozone in Water and Wastewater Treatment,* F. L. Evans, III, Ed. (Ann Arbor, Mich.: Ann Arbor Science Publishers, Inc., 1972), pp. 83-100.
33. Evans, F. L., III, and D. W. Ryckman. "Ozonation Treatment of Wastes Containing ABS," *Proc. 18th Ind. Waste Conf.,* Purdue University, 141-157 (1963).
34. Stuber, L. M. *Tertiary Treatment ofCarpet Dye Wastewater Using Ozone Gas and Its Comparison to Activated Carbon.* Special Research Problem (Atlanta, Georgia: Georgia Institute of Technology, 1973).
35. Nebel, C., R. D. Gottsching and J. H. O'Neill. *Ozone Decolorization of Pulp and Paper Mill Secondary Effluents* (Philadelphia: Welsbach Ozone Systems Corp., 1973).

CHAPTER 4

WATER DISINFECTION–CHEMICAL ASPECTS AND ANALYTICAL CONTROL

A. T. Palin

Newcastle and Gateshead Water Company
Newcastle-upon-Tyne, England

INTRODUCTION

Water disinfection generally involves the use of chlorine. Besides disinfection, other benefits can result such as color removal, correction of tastes and odors, and the suppression of unwanted biological growths.

This paper presents a guide to the chemistry of modern chlorination processes and discusses chlorine test procedures. Regular testing for chlorine level, that is the residual chlorine, maintains the desired antimicrobial concentration without producing such undesirable results as chlorinous tastes and odors in drinking water or unpleasant bathing conditions in swimming pools. These unwanted side effects could arise not only from excessive concentrations of chlorine itself, but also from the presence of obnoxious chlorocompounds resulting from interactions between chlorine and ammoniacal compounds or nitrogenous organic matters. To utilize fully the economy and efficiency of modern chlorination techniques and eliminate all problems of incorrect control and application, a reliable and simple testing system is essential.

Great advances in our knowledge of water chlorination chemistry over the past two or three decades resulted from the development of suitable analytical techniques. The apparent complexities of chlorine chemistry, as exemplified by the breakpoint phenomenon, could not have been elucidated without some means of determining the nature and amounts of the different types of residual chlorine compounds involved in the various reactions that occur at relatively minute concentrations in water. The behavior of such compounds, and especially their ability to react among themselves and with free chlorine, provided a rewarding field for continuing exploration made possible by the analytical methods.

In addition to this fundamental research, the methods developed have been successfully adapted for use in the usual test-kit form (see Chapter 5). Other chemicals that could be used in water disinfection include chlorine dioxide, bromine, iodine and ozone. Here again, control by residual determination is normally required.

THE CHEMISTRY OF CHLORINATION

History

Toward the end of the nineteenth century there were several instances of chlorine disinfection of water. The introduction of water chlorination as a continuous process occurred soon after the turn of the century, both in England and in America. Developments since that time are as follows:[1]

1905-1915	A period mainly of hypochlorite disinfection, with some skepticism and prejudice against chlorination.
1915-1925	The evolution of gaseous chlorination and continued education of the public.
1925-1935	A time of very great interest in chlorination. Bacterial and taste problems were of primary concern, and the use of chloramine was much advocated.
1935-1945	A period of greater flexibility in chlorination methods, with increased use of semi-automatic apparatus.
1945-1955	Increased understanding of basic principles coupled with much fundamental research. Greatly improved methods of chlorination control and also development of chlorine dioxide treatment of water.

Since 1955 attention has focused on the refinement of test procedures for more accurate determination of residual chlorine compounds, to meet every treatment control requirement and ensure optimum results under all conditions.

Chlorine Dissolved in Water

In approaching chlorination chemistry it is necessary to first consider chlorine reactions in pure water. First, chlorine rapidly hydrolyzes to form hydrochloric acid and hypochlorous acid thus

$$Cl_2 + H_2O \rightleftharpoons HCl + HOCl \quad (1)$$

The hypochlorous acid then partly dissociates to give hydrogen ions and hypochlorite ions

$$HOCl \rightleftharpoons H^+ + OCl^- \qquad (2)$$

The three forms of free available chlorine involved in these reactions, molecular chlorine (Cl_2), unionized hypochlorous acid (HOCl) and the hypochlorite ion (OCl^-), exist together in equilibrium. Their relative proportions are determined by pH value, temperature and dissolved solids. Furthermore, the proportions are the same for any set of conditions whether the chlorine is introduced as chlorine gas or a hypochlorite.

In water chlorination pH is the all important factor governing the relative proportions, as shown in Figure 4-1. It is evident that, as the pH falls below 2, the predominant form is Cl_2. Between pH 2 and pH 7 the equilibrium overwhelmingly favors HOCl. At 20°C, pH 7.4, and low dissolved solids HOCl and OCl^- are about equal, while above this pH increasing proportions of OCl^- are present. Above pH 8.5 nearly all available chlorine is present in this poor disinfectant form. The preceding chapter presents a similar discussion based on logarithmic diagrams. At colder temperature these pH values shift to values nearly 0.5 pH units higher.

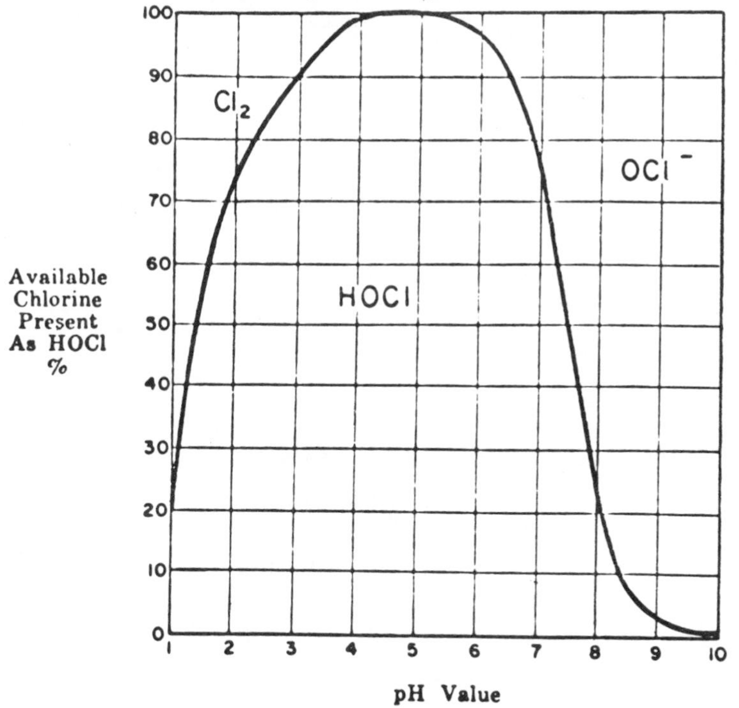

Figure 4-1. Effect of pH value on form of free available chlorine in water at 20°C and less than 100 mg/l dissolved solids.

Chlorine, which exists in water as Cl_2, HOCl or OCl^-, or in any mixture of these, is defined as "free available chlorine." Clearly, at the pH ranges encountered in water chlorination, free chlorine consists of a mixture of HOCl and OCl^-.

It will be apparent from the reactions considered above that adding gas chlorine to water produces hydrochloric acid (HCl) that will reduce the alkalinity. In practice, however, the amounts involved are so small that except in very soft waters with little buffering capacity or in cases involving very high chlorine doses, the resulting effect on the pH of the water is inappreciable. In fact the application to water of chlorine in a dose of 1 mg/l will reduce the alkalinity by approximately 1 mg/l which would be insignificant in a natural water of 100-200 mg/l alkalinity.

Breakpoint Reactions

Having considered the behavior of chlorine dissolved in pure water, it is now necessary to examine how the reactions are affected by those impurities that may be encountered in the chlorination of natural and polluted waters. It is now established that the most profound influence on water chlorination chemistry is exerted by the ammonia generally associated with pollutants.[2]

By simple substitution reactions it is possible to proceed as follows:

NH_3	→	NH_2Cl	→	$NHCl_2$	→	NCl_3
ammonia		monochloramine		dichloramine		nitrogen trichloride

Generally, the lower the pH and the higher the chlorine:ammonia ratio, the greater the tendency to produce more highly chlorinated derivatives. The rates of these reactions become progressively slower with greater substitution of chlorine or ammonia (see also Chapter 14). But with increasing chlorine:ammonia ratio secondary reactions occur, the study of which has produced results of the greatest possible significance. The rate of these secondary oxidation reactions depends upon pH and is greatest in the pH range of 7.0 to 8.0.

At the normal pH range of drinking and swimming pool water, the product of the reaction between chlorine and excess ammonia is almost entirely monochloramine, which forms very quickly—probably in less than one minute. The reaction may be shown as follows:

$$NH_3 + Cl_2 = NH_2Cl + HCl \quad (3)$$

From this reaction it may be calculated that one part by weight of ammonia-nitrogen requires five parts by weight of chlorine. Thus ignoring any loss of chlorine from other causes, we can say that provided the chlorine dose does not exceed 5 times the ammonia-nitrogen in the water, all chlorine will go toward producing stable monochloramine in the neutral and alkaline pH range.

But if more chlorine has been added than is required for this rapid initial reaction, a continuing oxidation reaction occurs. At pH values above 7.5 this oxidation goes faster than substitution to form dichloramine. Eventually mainly nitrogen is produced as follows:

$$2NH_2Cl + Cl_2 = N_2 + 4HCl \tag{4}$$

Whenever the chlorine dose exceeds the ammonia-nitrogen in the water by more than 5:1 (again ignoring the effect of other chlorine-consuming substances) this type of reaction will occur resulting in a loss of available chlorine. There is usually some appearance of dichloramine and nitrogen trichloride in the zone where the chlorine dose exceeded the 5:1 ratio, but the final result of increased chlorine eventually corresponds to the following overall equation:

$$2NH_3 + 3Cl_2 = N_2 + 6HCl \tag{5}$$

From this we calculate that 7.6 parts by weight of chlorine are required to oxidize 1 part by weight of ammonia-nitrogen, at which point loss of available chlorine would be maximal. Because of certain side reactions leading for instance to the formation of trace amounts of nitrate, the observed ratio is approximately 8.3:1 at pH 7-8. In practice one must allow for other substances in natural waters capable of absorbing chlorine, so that the usual ratio for the point of maximal chlorine loss, now called the *breakpoint*, is about 10:1, but of course for grossly contaminated waters it could be 25:1 or even higher.

If a series of water aliquots containing the same ammonia concentration is treated with increasing amounts of chlorine, and total residual chlorine values after a period of contact are plotted against the corresponding chlorine doses, a *hump and dip* type of curve is produced, known as the breakpoint curve. This curve has three distinctive stages related to the chemistry already considered:

1. An initial rise of residual chlorine, in which the compound present is monochloramine.
2. A secondary fall in residual, corresponding to the unstable zone containing chlorine in excess of that required to produce complete formation of monochloramine by Reaction 3. Here this excess has entered into mutual decomposition reactions with the initially formed chloramine (Reaction 4).
3. A zone characterized by a final rise in residual chlorine, corresponding to nearly complete NH_3 oxidation, mainly to nitrogen (Reaction 5). Here the continuing addition of chlorine to the water gives a *pro rata* increase in the free residual $HOCl + OCl^-$. Significant quantities of NCl_3 generally less than 15% of chlorine are found in this region below pH 7.

The minimum point between stages 2 and 3 is the breakpoint, and adding sufficient chlorine to exceed this point provides many advantages. Even after several hours the minimum does not reach zero because of stable organic chloramines. The relation of breakpoint in chlorine-ammonia reactions to pH is shown in Figures 4-2a, 4-2b and 4-2c.

Other nitrogenous compounds produce breakpoint curves in which the hump and dip are clearly distinguishable, although generally not so marked as in the case of chlorine and ammonia. All can be explained on the same basis of initial formation of chloramine residual, followed by loss of residual by oxidation by excess chlorine and eventual remainder of free chlorine.

Apart from the early stages of the breakpoint reactions, when free chlorine may exist in the water for only a short time in conditions corresponding to those at the dip portion of the curve, the vital distinction before breakpoint is that the residual available chlorine is present in the form of chloramines and related compounds. This is termed *combined chlorine.* Beyond the breakpoint chlorine is present as free chlorine, a mixture at normal pH values of hypochlorous acid, HOCl, and hypochlorite ion, OCl^-.

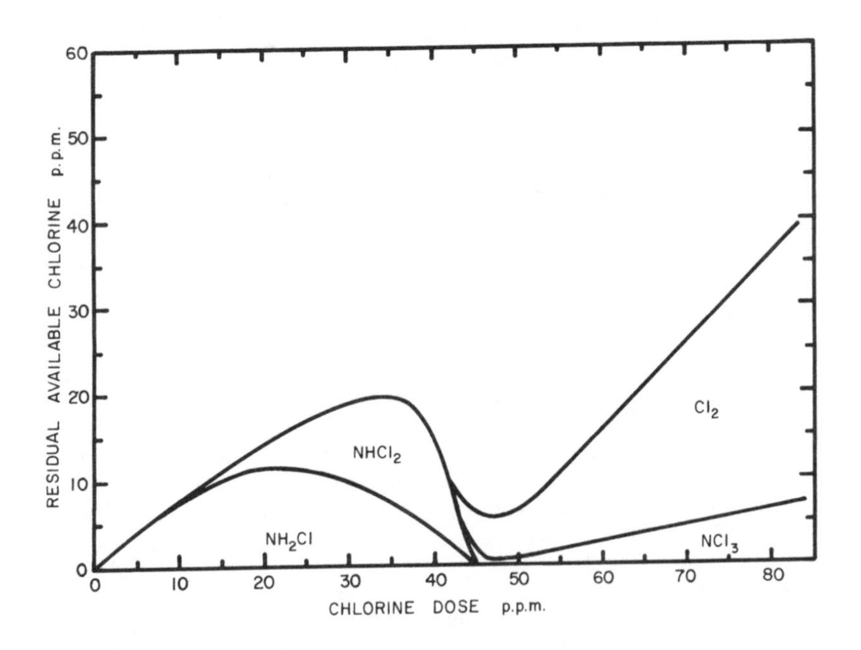

Figure 4-2a. Chlorine dose–residual curve at pH 6.0 after 1 day. Initial ammonia 0.5 ppm (N).

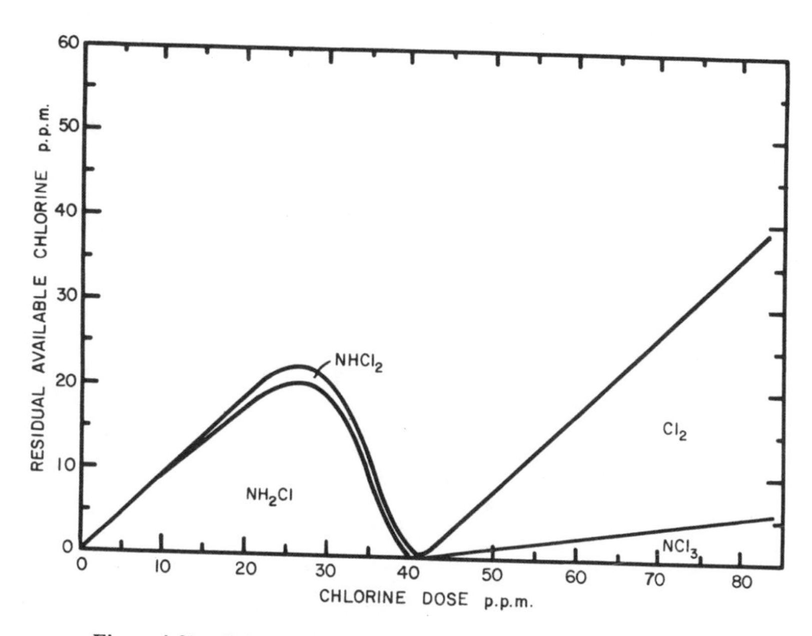

Figure 4-2b. Chlorine dose–residual curve at pH 7.0 after 1 day. Initial ammonia 0.5 ppm (N).

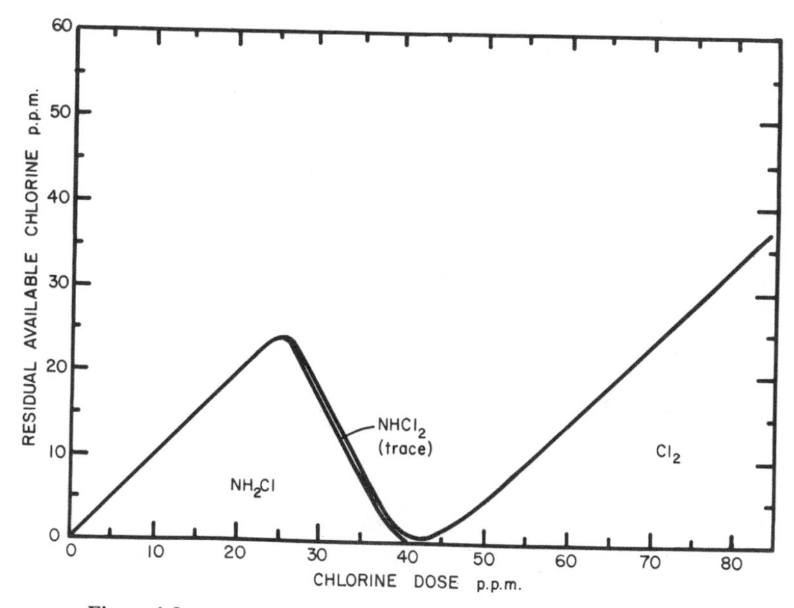

Figure 4-2c. Chlorine dose–residual curve at pH 8.0 after 1 day. Initial ammonia 0.5 ppm (N).

Monitoring these two forms of residual chlorine is of the greatest importance since the chemical, bactericidal and virucidal properties of free chlorine as HOCl are vastly superior to those of combined chlorine. In the presence of ammonia-type impurities, chlorination should be adequate to ensure that continuing reaction with combined residual chlorine compounds is minimized; *i.e.*, the water must be chlorinated beyond breakpoint. That the treatment has been properly carried out can only be established using test procedures capable of distinguishing free chlorine from combined chlorine.

A disadvantage of breakpoint chlorination is that the free chlorine residuals produced in the water might, unless the pH is fairly high, be accompanied by traces of nitrogen trichloride. In some situations this has proved a problem in view of the objectionable chlorinous-type odors thereby imparted to the water.[3] It might then be necessary to dechlorinate completely followed by rechlorination or ammonia-chlorine as a final treatment.

Traces of nitrogen trichloride may also be encountered in a swimming pool,[4] especially if the pH value is allowed to fall. In indoor pools this is the cause of eye irritation and since the volatile NCl_3 escapes into the atmosphere it causes the chlorine odor associated with pools. Consequently, satisfactory application of modern chlorination techniques to swimming pools requires not only differential testing for free and combined chlorine, but a measure of pH and alkalinity control. Providing adequate chlorine to pass breakpoint and maintaining the pH above 7 eliminates the chloramine problem.

Modern Chlorination—Definitions and Classification

Against this background of chemistry, we may now set down the fundamental definitions of modern chlorination.

Free available residual chlorine is defined as that residual chlorine existing in water as the sum of hypochlorous acid and hypochlorite ion.

Combined available residual chlorine is defined as that residual chlorine existing in water in chemical combination with ammonia or organic nitrogen compounds.

Chlorine demand of a water is defined as that portion of the applied chlorine dose which has been converted to non-available chlorine that is, by definition, not able to liberate iodine from an acidified iodide solution, and similarly giving no response in the usual residual chlorine tests. For a precise definition of chlorine demand, however, it is necessary to specify the chlorine dose, time of contact, temperature and nature of the residual aimed at, whether free or combined-available chlorine. Available chlorine residual includes all free and combined or chloramine chemical forms able to react with iodide but not necessarily available as an effective disinfectant.

General acceptance of the microbiological superiority of free residual chlorine as HOCl compared with combined residual chlorine indicates that for most cases the preferred definition of *chlorine demand* is: the difference between the amount of chlorine added to water and the amount of free available chlorine remaining at the end of a specified contact period. On the breakpoint curve this corresponds to the chlorine required to reach breakpoint. Further chlorine addition beyond the amount required to satisfy this demand will produce the desired level of free residual chlorine beyond breakpoint.

However, many waters might have a chlorine demand due to non-nitrogenous organic matter or oxidizable inorganic substances with relatively insignificant amounts of ammoniacal or nitrogenous impurity. In such cases the dose-residual curve will not exhibit the typical hump and dip of the breakpoint curve, but show a low level during the stage of satisfying the demand; afterward, free residual chlorine will appear and show the *pro rata* increase with increasing dose. The common feature is that sufficient chlorine is applied to produce free, as opposed to combined, residual chlorine.

The modern approach to the classification of chlorination processes is based upon this important distinction between free and combined chlorine, thus giving two main types of processes. These are termed free residual chlorination and combined residual chlorination. In applying free residual chlorination to some water supplies, it could be desirable, either because of very short contact time or variable pollution loads, to operate at levels of free chlorine so high that dechlorination is subsequently required before delivery to the consumer. This process, called superchlorination and dechlorination, may then be applied without particular reference to a breakpoint, to ensure that ample free chlorine is produced.

In other circumstances where time is adequate to achieve disinfection, there may be some advantage in working with combined chlorine residuals produced by reaction between chlorine and nitrogenous constituents naturally present in the water (simple or marginal chlorination). Alternatively, ammonia may be deliberately applied, usually before the chlorine, as part of the process (chloramine or ammonia-chlorine treatment).

The classification of chlorine processes may be set down as in Table 4-1.

Table 4-1. Classification of Chlorination Processes

Chlorination			
Free Residual		Combined Residual	
Breakpoint Chlorination	Superchlorination and Dechlorination	Simple (or Marginal) Chlorination	Ammonia-Chlorine Treatment

It is possible, of course, to introduce variations to suit particular requirements; for instance an operator may apply free residual chlorination through his plant with final conversion to combined residual (by applying ammonia) to maintain chloramine residuals in the distribution network. Again, a novel application of breakpoint reactions is to use ammonia as the dechlorinating agent. Dechlorination is further discussed in Chapter 16.

Whatever modification of chlorine treatment is used, however, the results obtained under any given conditions always depend on the nature and amount of residual chlorine produced in the water. Production of residuals of known composition by suitable control of chlorination has been made possible by the chlorine residual differentiation methods now available.

DISINFECTION BY CHLORINE

The Effect of pH

While this chapter is primarily concerned with the chemistry rather than the bacteriology of water chlorination, it is necessary to understand those physical and chemical factors that influence the bactericidal and virucidal power of chlorine.

Figure 4-1 and Reaction 2 show that within the range of pH 6.0-9.0 free available chlorine comprises a mixture in varying proportions of hypochlorous acid and hypochlorite ion. With pH rising above 6.0 the proportion as HOCl declines from virtually 100% down to almost zero at pH 9.0. As a bactericide HOCl is approximately 80 times more powerful than the OCl^- ion. Consequently, in free residual chlorination the higher the pH value the less active the residual because of its lower proportion of HOCl. Higher temperature and dissolved solids also lower the proportion of HOCl. Apart from the desirability of distinguishing free from combined chlorine in the residual chlorine test, there would also seem to be a need for an analytical method to differentiate between these two forms of free chlorine, HOCl and OCl^-. Until recently it has not been possible to measure HOCl alone in the presence of OCl^-. Johnson has developed an amperometric electrode similar to the oxygen probe to measure the HOCl portion of free chlorine.[5] The author of this chapter is currently developing a test specifically for HOCl, based on his DPD method for free chlorine (see page 84). In water disinfection to date, it has not been considered practical to determine the HOCl concentration, and usually the recommendation simply has been to work to a higher free chlorine residual at the higher pH values. Some even specify the level of free chlorine residual for different pH and temperature values, to ensure that the same amount, say 1 mg/l, of HOCl is always present. Table 4-2 shows the levels of free

Table 4-2.

pH Value	Total Free Residual Chlorine to Give 1 mg/l HOCl	
	Temperature, 0°C	Temperature, 20°C
6.0	1.0	1.0
6.5	1.1	1.1
7.0	1.2	1.4
7.5	1.7	2.2
8.0	3.2	4.3
8.5	8.0	12.0
9.0	22.0	40.0

chlorine required to achieve this result at less than 100 mg/l total dissolved solids. The effect of dissolved solids is to require slightly larger concentrations of total free residual. For example at pH 7.5 2.4 mg/l would be required at 20°C and 500 mg/l total dissolved solids.

Extensive work on the bactericidal efficiency of free available chlorine substantially confirms the retarding influence of high pH. Possibly the most comprehensive studies were those conducted with *E. coli* by the Public Health Service,[6,7] from 1943 to 1948. They clearly showed only the HOCl fraction of free chlorine as a good bactericide. They also demonstrated that the killing power of chloramine is much poorer than HOCl and that the chloramines' colicidal efficiency decreases with increasing pH.

Effect of Temperature and Time of Contact

The rate of bactericidal action of both free chlorine and chloramines decreases with temperature despite the increase in the HOCl fraction (Table 4-2). In any situation in which the effects of lowered temperature and high pH value are combined, reduced efficiency of free chlorine and chloramines is very marked.

These factors have an important bearing, therefore, on the exposure time necessary to achieve satisfactory disinfection. Under favorable conditions, the contact time required with free available chlorine may be only a few minutes; combined available chlorine under the same conditions might require one or two hours. Whatever the conditions, however, the final test of disinfection resides in bacteriological examination to ensure that the treated water complies with recognized standards such as the USPHS Standards of Drinking Water Quality.[8]

Nature of Residual Chlorine and Minimum Safe Levels

In the Public Health Service studies, a comparison was made between the bactericidal activities of free chlorine and chloramine. The experiments were conducted so that the chloramine would be preformed in the water before adding bacterial suspensions, which is not truly representative of actual water treatment practice. Nevertheless, the conclusions were highly significant in showing that to completely destroy bacteria (a) with the same exposure period, about 25 times as much chloramine as free chlorine was required, and (b) with the same amounts of residual, about 100 times the exposure period was required with chloramine as with free residual.

Data obtained in these studies, using 10 minutes' contact with free available chlorine and 60 minutes' contact with combined available chlorine, provided the basis for recommended minimum safe residuals. In 1956 the National Research Council[9] reanalyzed the Public Health Service data and submitted revised recommendations based on a 30-minute exposure period. At the same time additional recommendations were made in connection with minimum cysticidal chlorine residuals. These later recommendations form the basis of Table 4-3.

Table 4-3. USPHS Minimum Cysticidal and Bactericidal Residuals (After 30 minutes contact)

	Free Chlorine, $HOCl + OCl^-$			Combined Chlorine, NH_2Cl
pH Value	Bactericidal 0-25°C	Cysticidal 22-25°C	Cysticidal 2-5°C	Bactericidal 0-25°C
6.0	0.2	2.0	7.5	2.0
7.0	0.2	2.5	10.0	2.5
8.0	0.2	5.0	20.0	3.0
9.0	0.6	20.0	70.00	3.5

It may be assumed that, in practice, the performance of ammonia-chlorine treatment, where chloramine formation occurs in the water itself rather than being preformed, will be considerably better than is suggested by the above figures. This is explained by the fact that the reaction between chlorine and ammonia requires up to 1 minute at pH 7 for completion, actual time being strongly dependent on pH, temperature and chlorine and ammonia concentrations. During this period the proportion of unreacted chlorine, although decreasing quite rapidly, especially at higher pH and temperature, nevertheless retains the bactericidal power of free available chlorine. As a result there is a short initial period in ammonia-chlorine treatment during which an enhanced rate of bacterial kill may be expected, especially on the acid side. After chlorine and ammonia have reacted, the rate falls to the very much lower level characteristic of combined chlorine. Thus, even if full allowance is made for the initial period of rapid action in ammonia-chlorine treatment, the differences in the bactericidal and other properties of free chlorine and combined chlorine remain so marked that modern trends are more and more toward free residual chlorination.

DETERMINATION OF FREE CHLORINE AND OTHER RESIDUALS

Introduction

The following general discussion is mainly concerned with those methods that appear in the 1971 edition of Standard Methods[10] published jointly by American Public Health Service, American Works Association and Water Pollution Control Federation. The major emphasis is on the Diethylene Paraphenylene Diamine (DPD) method developed by this author. Field-test systems and some other newer approaches are discussed in Chapter 5. The DPD method remains unique, however, in providing a range of differential procedures, not only for the various forms of residual chlorine but for their mixtures with such other types of residual as chlorine dioxide, chlorite, bromine and ozone.

Iodometric Titration

This is essentially a laboratory procedure for higher concentrations, in which a standard solution of sodium thiosulfate is used to titrate the iodine liberated from potassium iodide by the chlorine, usually in acid solution, unless interfering substances are present—then neutral solution is preferred. Starch, which gives a blue color with iodine, is used as the indicator. For high chlorine residuals the method is more precise than colorimetric procedures, but for lower residuals it becomes inaccurate. The method gives total residual-available chlorine.

There are indications that chloramines tend not to respond if the titration is performed in neutral solution, so that by carrying out two titrations, one in acid and the other in neutral solution, some approximate free-combined residual separation is obtained. For this purpose, however, superior methods are available.

Amperometric Titration

This is not practical for field use as it requires a source of 110-V current. However, it does provide an accurate titration method for laboratory use and permits separate determination of free and combined chlorine. The principle is based on the fact that a suitable bimetallic cell immersed in a solution containing an oxidizing agent such as chlorine produces a current flow, indicated on a microammeter. When a reducing agent is added to decrease the concentration of the oxidizing agent, the current falls. Eventually an end point is reached where the current is reduced to a minimum, beyond which further addition of reducing agent causes no change. For accurate results considerable operator skill is required, and certain essential precautions, as discussed later, must be observed. The presence of oxidized manganese and copper may give false readings but otherwise, for potable water, the method is substantially free from interference.

Orthotolidine Methods

There are four versions of the orthotolidine method. The first (1913), with various modifications introduced from time to time, has been widely used for the routine measurement of residual chlorine.

A) The Simple Acid Test utilizes the reaction between chlorine and orthotolidine in acid solution to give a yellow color. For correct color development it is necessary that the solution be acidified to give a pH of 1.3 or lower, the ratio by weight of o-tolidine to chlorine be at least 3:1, and the chlorine concentration not exceed 10 mg/l. For precise results addition of the reagent to the water sample must be carried out in the prescribed manner, that is, sample to o-tolidine in the tube followed by rapid mixing.

B) To remedy the deficiencies of the original test, a Modified Acid Reagent was introduced in the 9th edition of APHA Standard Methods (1946), and has remained the standard acid o-tolidine reagent. Unfortunately, in correcting some of the deficiencies of the original reagent, the value of the test for the differential determination of free and combined chlorine was affected adversely. Increased o-tolidine in the color mixture coupled with the lower pH, speeded up the rate of color development with combined chlorine to the point where the test was inadequate for this purpose.

The use of the modified acid o-tolidine reagent for total available chlorine specified 20°C for maximum response of combined chlorine and minimum interference. Unfortunately the 3-5 minutes required at this temperature for maximum color also produces significant fading. In the standard modified acid orthotolidine procedure it was established that the color should be read at the time of maximum color development.

C) Attempts have been made to improve the o-tolidine test to provide reasonable accuracy in the determination of free chlorine separate from combined chlorine. The introduction of arsenite in the form of a solution of sodium arsenite formed the basis of the Ortho-Tolidine Arsenite (OTA) test. The purpose of the arsenite is to destroy the combined chlorine immediately (*i.e.*, within 5 seconds) after the free chlorine has reacted with the o-tolidine. In a second test arsenite is not used, so that the resulting color indicates total available chlorine. Comparing the two readings gives an estimate of both free and combined chlorine. Further, it is possible to allow for the effect of any interfering substances that might give false colors, by adding arsenite before the o-tolidine. This destroys all forms of available chlorine and leaves the interference so that previous readings may be suitably corrected. These modifications correct for manganese but do not eliminate the combined chlorine interference and, in general, reduce the precision of the method.

D) Because the range of these o-tolidine methods is limited to available chlorine concentrations not greater than 10 mg/l, a Drop Dilution Method is provided for field use at high concentration. This consists simply in bringing the chlorine concentration within the prescribed range by diluting the sample with distilled water. Sufficient water is dropped from a calibrated pipette to a prepared tube holding the o-tolidine reagent and distilled water. Serial addition of drops is not permitted. If an increased number of drops is required to produce an easily readable color, the procedure must be repeated using a fresh tube of o-tolidine plus distilled water. The method is not recommended when accuracy is desired.

E) The Neutral O-tolidine Titration Method was developed by the author during the course of his studies of the formation and decomposition of the chloroderivatives of ammonia and related compounds.[11] This new method subsequently appeared in the 12th edition of Standard Methods (1965), but was replaced in the 13th edition by the authors of the DPD Method. The normal yellow color produced in the acid o-tolidine test is due to holoquinone, a compound formed by complete oxidation of the o-tolidine. In neutral solution there is only partial oxidation, giving products known as meriquinones. The resulting colors are generally greenish, but in the presence of stabilizers such as hexametaphosphate, only the pure blue meriquinone color is formed.

In the titration version of this method, the author introduced ferrous ammonium sulfate as a standard solution. The strength was adjusted so that the volume required to destroy the blue color indicated mg/l chlorine directly. To differentiate free and combined chlorine, potassium iodide was used as an activating agent to develop color from the combined chlorine. This method of inducing a response from combined available chlorine in colorimetric testing was discovered by the author in previous work using the indicator p-aminodimethylaniline (dimethyl-p-phenylene diamine).[12]

In further studies using the iodometric method for standard comparison, it was evident that with some waters, usually rather acid in character, a full response from combined chlorine was not being obtained. This elusive fraction was due to dichloramine. A modified procedure with a third stage involving acidification and subsequent neutralization was therefore introduced, permitting differentiation of monochloramine from dichloramine. A supplementary procedure was devised for determinating nitrogen trichloride, as it had then become apparent, contrary to what had always been assumed, that there could well be conditions in waters treated by breakpoint chlorination in which this compound could form and remain co-existent with free chlorine.

Stabilization of the colors in the neutral o-tolidine test by hexametaphosphate, while adequate for a titration procedure, was hardly adequate for a colorimetric method since judgment of the color match against standards often takes time. Therefore, for this purpose, an improved stabilizer was introduced in 1957 consisting of an anionic surface active agent.[13]

A colorimetric procedure for free and combined chlorine was developed in 1969 by J. D. Johnson, *et al.*[14] using Aerosol OT as the stabilizer, known as the SNORT method (Stabilized Neutral Ortho-Tolidine Test). They showed the general stabilizing effect of anionic and especially highly branched chain surfactants. Their improved stabilizer permits operation at pH 7.0 where the interference of monochloramine in the free chlorine measurement is slowed five times that at pH 6.3, used previously.[13] Still this interference can be 0.5% of the monochloramine per minute, read as free chlorine. To eliminate this interference the SNORTA or SNORT plus arsenite method has recently been introduced as described in Chapter 5. Provision is made for the separate determination of monochloramine by reaction with iodide in neutral solution, and of dichloramine by reaction with iodide in acid solution as in the original neutral o-tolidine method. The author's findings that the action of iodide is catalytic in nature was confirmed by the workers at the University of North Carolina. It is stated in Standard Methods (1971) that because of this catalytic effect, lesser amounts of iodide are required compared to amperometric titration, which improves the separation of the residual mono and dichloramine.

Leuco Crystal Violet

Another colorimetric method mentioned in Standard Methods (1971) utilizes leuco crystal violet to measure separately free and total available chlorine, thus giving combined available chlorine by difference. Two different sets of standards are required for the two stages of the differential test, and correct color development depends on a number of factors and conditions that must be observed carefully. As described in 1971, dilution is required for samples containing more than 2 mg/l available chlorine. A new improved procedure has recently been introduced which extends the range 10 to 15 mg/l. Preparation of the buffer and indicator solutions is a complicated procedure, and extreme care is essential when handling ingredients such as mercuric chloride. Reagent stability is questionable; for instance, it is not good practice to incorporate potassium iodide in a buffer solution of pH 4.0, as iodides in acid medium become more susceptible to oxidation through exposure to oxygen giving traces of free iodine and consequently high readings in the test for total available chlorine. But the main problem in using the Leuco Crystal Violet Method has been the near impossibility of maintaining the glass or plastic cells in test kits free from staining by the crystal violet dye, doing without more frequent cleaning than is practically acceptable. A modified reagent has been proposed to reduce this problem.[15]

Methyl Orange

Methyl orange has been adapted to colorimetric testing for free chlorine and chloramines, and several different procedures have emerged since 1928. The principle of the test is that free chlorine bleaches an acid solution of methyl orange quantitatively; chloramines can then be made to react by adding excess bromide. The optimum pH of the color mixture is 2.0; a higher pH results in incomplete color development, and a lower pH increases interference by chloramines in the free chlorine test. The latest modification of the methyl orange method appeared in 1965 and forms the basis for the procedure in Standard Methods (1971). There are inherent disadvantages in any method based upon a bleaching action. Care is required in mixing sample and reagent and in measuring the amount of indicator to be added. An amount of indicator that gives reasonable sensitivity at low chlorine levels will be bleached completely at high levels. Conversely, an amount of indicator that provides the necessary excess for testing high chlorine levels becomes too insensitive for low levels. In addition to the methyl orange reagent, the test requires an acid reagent to give the desired pH, and sodium bromide solution for chloramine activation. For the reasons stated it has received little or no general acceptance by water analysts.

Syringaldazine Procedures

A test for estimating free chlorine in water has been reported[16] but is not yet in Standard Methods. This method uses a mixed indicator consisting of syringaldazine and vanillinazine in conjunction with a pH 6.0 buffer. (The role of the vanillinazine is not fully understood and has been dropped from the new procedure described in Chapter 5, p. 104. The solutions are used to impregnate strips of paper which, after drying, can be used during field testing as a simple means of detecting free chlorine. Test conditions preclude the accuracy of normal colorimetric techniques. Further attention has been given to the development of a comparator utilizing two reagents in liquid form—a buffer and an indicator solution. The range is limited to 2 mg/l and higher concentrations require dilution of the sample with distilled water, which must have zero chlorine demand since the reagents are added after dilution. For field testing a dilution procedure is, in any event, impracticable and further work is described in Chapter 5, p. 105, which attempts to widen the range. Only limited experience is available in the test application under all conditions likely to occur; the stability of the liquid reagents is also relatively unknown. It would appear that staining presents the same problem as in the leuco crystal violet method since it has been recommended that cells should be rinsed with acetone after each use. Further developments in this method are discussed in Chapter 5.

In this work, however, the effect of NCl_3 on syringaldazine does not appear to have been studied. In this author's preliminary experiments it was found that waters containing NCl_3, prepared by Nicolson's[20] method, gave a strong response to syringaldazine, using the FACTS version of the method, with little or no response after removal of NCl_3 from the water sample by CCl_4 extraction. Any claim for specificity for free chlorine would thus be invalidated unless the possibility of interferences by NCl_3 can be eliminated.

DPD Methods

In the search for the ideal residual chlorine method, attention was first focused on the use of *p*-aminodimethylaniline in 1940. Further work convinced the author that neutral o-tolidine had more to offer. At this stage ferrous ammonium sulfate was introduced as a standard solution to provide a titration method, in addition to a colorimetric procedure, for the separate determination of free chlorine, monochloramine, dichloramine and nitrogen trichloride. Independent comparative trials in England, in 1949, against the amperometric titrator of that time revealed that the latter did not respond to dichloramine, contrary to original claims in 1942. To correct this deficiency the amperometric titration was modified by the introduction of an acidifying stage to induce a response from dichloramine and, at the same time, a change was made from sodium arsenite to phenylarsine oxide as the titrant.[17]

Testing methods for chlorine in water that require acidification at one stage or another cannot be viewed as completely acceptable however, because the pH shift can change the conditions of chlorine equilibria existing in the water as sampled; either between the different components of the chlorine-chloramine system, or between the chlorine and the chlorine-consuming constituents present. Ideally the reagents used in the residual chlorine test should not appreciably alter the pH value of the water being tested.

In the test for free available chlorine the following methods do not meet this requirement: orthotolidine, OTA, leuco crystal violet and methyl orange. In testing for total available chlorine, neutral orthotolidine, SNORT, and the amperometric titration require acidification to bring out dichloramine.

Recognizing the possible disadvantage of his own neutral o-tolidine titration method and the need for reducing the number of reagents for differential chlorine-chloramine testing, the author re-examined the behavior of the phenylenediamine group of indicators. The dimethyl derivative (dimethyl-*p*-phenylene diamine, or *p*-aminodimethylaniline) had been the first indicator used, and a return to this earlier work revealed that by switching to the corresponding diethyl compound (diethyl-*p*-phenylene diamine) an improved residual chlorine test could be achieved.[18]

Following extensive trials by many independent workers and particularly by the Analytical Reference Service of the Environmental Protection Agency,[19] the new method (DPD) subsequently appeared in Standard Methods (1971), providing both titrimetric and colorimetric procedures to meet laboratory and field requirements for chlorination control and investigation.

Selection of Procedure

In pure laboratory solutions the iodometric method is the standard against which other methods are judged, but is rarely used as a practical method for water chlorination control. Interference by nitrites, ferric iron and oxidized manganese can be quite considerable, and other disadvantages, such as uncertain end points at low concentration and the necessity for rather large sample volumes, have contributed to its unpopularity. For special purposes, including standardization of chlorine water used in the preparation of temporary colorimetric standards and in laboratory studies of chlorine demand, it retains a place, and also provides the means of determining available chlorine strengths of chemicals used for chlorination.

As a titrimetric method for general use, amperometric titration has been widely preferred in America to the iodometric method, although acceptance in Britain and other European countries is minimal. The Water Research Association in England reported adversely on it after investigating methods for determining free chlorine in water.[20] The agitation caused by the stirrer resulted in chlorine losses, and thus low results. Doubt was also expressed whether the reaction with phenylarsine oxide proceeded to completion. In these comparative studies the performance of the DPD titrimetric method was judged superior to that of the amperometric titration.

In a recent report by the Medical Environmental Engineering Research Unit of the U.S. Army Medical Research and Development Command,[21] precautions taken in using the amperometric titration included:

(a) Exposing the samples to least possible light.

(b) Adding about 90% of the titrating solution before turning on the cell stirrer.

(c) Completing the titration as quickly as possible.

When an amperometric analysis for combined chlorine, which entails addition of potassium iodide, was followed by amperometric analysis for free available chlorine, iodide residues caused erratic results. As a further precaution, therefore, this report recommended that the cell be rinsed with four changes of chlorine demand free water and the stirrer be operated for 30 seconds with each change. Attention was drawn to the desirability of circumventing this source of error by using two titrators, one for combined and one for free chlorine.

The Water Research Association work was subsequently extended to cover combined as well as free chlorine. It was eventually concluded that both titrimetric and colorimetric DPD methods were the best of those examined.[20] Research supported by the U.S. Army Medical Research and Development Command at Syracuse University, N.Y., confirmed that the DPD method was applicable to determine free residual chlorine, and accurate results could be obtained by a relatively simple procedure.[22] More recent U.S. Army studies of field test kits led to the conclusion that DPD was more accurate and precise over temperature and pH variation than the other test kit procedures studied.[21] Chapter 5 discusses these test kit results in greater detail.

Furthermore, the DPD method has been adopted by the Ministry of the Environment of Ontario, Canada, for its Basic Gas Chlorination Workshop Manual.[23]

Of the remaining colorimetric methods, only those using orthotolidine can claim to be sufficiently well tested to warrant further discussion at this time: in general the various limitations of the others have precluded any serious consideration for acceptance as practical residual chlorine tests. Even the well established acid orthotolidine procedures are now recognized as having serious deficiencies and have been adversely criticized, particularly where free-combined differentiation is concerned. To achieve any degree of accuracy, the sample must be chilled to about 1°C. In the OTA or arsenite version, the reaction between arsenite and chloramine is relatively slow, resulting in higher readings for free chlorine than are actually present, additionally, there are reports that other interfering substances continue to give erroneous results.

Another factor causing the swing to DPD was the discovery that o-tolidine could cause tumors in the urinary tract. Consequently, this chemical is on the English, Factories Act list of carcinogenic agents and its manufacture is discouraged.[24] In some countries it has been virtually banned as a reagent for chlorine testing and the DPD method has been recommended as replacement by the governmental authorities. For instance, in the recommended methods for water analysis published by the British Department of the Environment,[25] only the DPD method is now given for the colorimetric determination of residual chlorine. In America, the method has received substantial official support by its inclusion in the current edition of Standard Methods.[10]

Possibly the most searching investigations of methods for determining chlorine in water were those carried out by the Analytical Reference Service of the U.S. Environmental Protection Agency.[19] In their first report, published in 1969, it was concluded that "The best overall accuracy and precision was shown by the DPD titrimetric method. The lack of accuracy of the old orthotolidine methods makes them the least acceptable." In continuing investigations all acid o-tolidine methods except OTA were deleted. SNORT and DPD colorimetric procedures were added to those selected for further study. Conclusions of the second report, published in 1971, included the following statements:

> The apparently excellent performance of leuco crystal violet is somewhat inconclusive because of the very small amount of data for this method. Beyond all doubt, the SNORT procedure performed well and produced acceptable results for the three most meaningful determinations; namely, free and total chlorine. The two DPD procedures, colorimetric and titrimetric, were nearly equal in overall performance. The overall performance of the amperometric titration ranked below the DPD methods and an acceptable performance was obtained only for the total chlorine measurement.
>
> The poorest results were obtained with the orthotolidine-arsenite (OTA) procedure. The data shows the method to be the least in precision and next to last in accuracy and is unacceptable for all determinations. The total error for OTA is one third more than the next poorest method; methyl orange.

In the application of colorimetric methods to the determination of available chlorine in wastewaters it has been stated that only the DPD method is worthy of consideration.[26]

In a recent study by Wei and Morris[27] on the dynamics of breakpoint chlorination the DPD-FAS method was finally adopted as the analytical means for differentiation of free chlorine and components of chloramines.

Selection of this method was based on a number of considerations from which the following are quoted:

> DPD-FAS can determine free chlorine without interference from NCl_3. Unambiguous determinations of free chlorine were considered highly important because it is the most potent species available for disinfection.
>
> Determinations by the DPD-FAS method are all done at a relatively neutral pH of 6.4.
>
> DPD-FAS method was extensively tested and found to give adequately accurate and discriminatory results.

This author concludes, therefore, that the general authoritative opinion overwhelmingly favors DPD for both titrimetric and colorimetric procedures.

REFERENCES

1. Waddington, A. H. *Proc. Soc. Water Treat. Exam.* **4**, 71 (1955).
2. Palin, A. T. *Water Works Eng.* **54**, 151, 189, 248 (1950).
3. Williams, D. B. *J. Amer. Water Works Assoc.* **41**, 441 (1949).
4. Palin, A. T. *Proc. Nat. Assoc. Baths Superintendents* **20**, 73 (1950).
5. Johnson, J. D. and J. W. Edwards. *Proc. Environ. Chem. Div., Amer. Chem. Soc.* **14**(1), 169 (1974).
6. Butterfield, C. T. and E. Wattie. *Public Health Rep. (U.S.)* **58**, 1837 (1943), **61**, 157 (1946).
7. Butterfield, C. T. *Public Health Rep. (U.S.)* **63**, 934 (1948).
8. *U.S. Public Health Service Drinking Water Standards.* (Washington, D.C., 1962).
9. Snow, W. B. *J. Amer. Water Works Assoc.* **48**, 1510 (1956).
10. *APHS, AWWA and WPCF Standard Methods for the Examination of Water and Wastewater,* 13th edition (New York, 1971).
11. Palin, A. T. *J. Inst. Water Engr.* **3**, 100 (1949).
12. Palin, A. T. *Analyst* **70**, 203 (1945).
13. Aitken, R. W. and D. Mercer. *J. Inst. Water Engr.* **5**, 321 (1957).
14. Johnson, J. D. and R. Overby. *Anal. Chem.* **41**, 1744-1750 (1969).
15. Whittle, G. P. and A. Lapteff, Jr. "New Analytical Techniques for the Study of Water Disinfection," In *Chemistry of Water Supply, Treatment, and Distribution,* A. J. Rubin, ed. (Ann Arbor, Mich.: Ann Arbor Science Publishers, Inc., 1974), p. 63.
16. Bauer, R., B. F. Phillips, and C. O. Rupe. *J. Amer. Water Works Assoc.* **64**, 787 (1972).
17. Marks, H. C., D. B. Williams, and T. U. Glasgow. *J. Amer. Water Works Assoc.* **43**, 201 (1951).
18. Palin, A. T. *J. Amer. Water Works Assoc.* **49**, 873 (1957).
19. "Analytical Reference Service, Environmental Protection Agency," Study No. 35 (1969), Study No. 40, Cincinnati, Ohio (1971).
20. Nicolson, N. J. *Water Res. Assoc. England, Tech. Rep.* **29** (1963), **47** (1965), **53** (1966).
21. Guter, K. J. and W. J. Cooper. U.S. Army Medical Environmental Eng. Res. Unit, Report No. 73-03, AD No. 752440, Edgewood Arsenal, MD (October 1972).

22. Bjorkland, J. G. and M. C. Rand. *J. Amer. Water Works Assoc.* **60**, 6038 (1968).
23. Ministry of the Environment, Ontario, Canada. *Basic Gas Chlorination Workshop Manual,* 3rd ed. (June 1972).
24. Chester Beatty Research Institute, Royal Cancer Hospital, London (April 1966).
25. Department of the Environment (U.K.). *Analysis of Raw, Potable and Waste Waters.* (London, 1972).
26. White, G. W. *Handbook of Chlorination.* (New York, 1972), p. 436.
27. Wei, I. W. and J. C. Morris. In *Chemistry of Water Supply, Treatment and Distribution,* A. J. Rubin, ed. (Ann Arbor, Mich.: Ann Arbor Science Publishers, 1973), p. 297.

CHAPTER 5

SELECTION OF A FIELD METHOD FOR FREE AVAILABLE CHLORINE

Charles Sorber, William Cooper and Eugene Meier

U.S. Army Medical Bioengineering Research
and Development Laboratories
Fort Detrick, Frederick, Maryland

INTRODUCTION

Water treatment facility operators and engineers who field test water supplies require a simple, specific rapid technique for determining chlorine residuals and, in particular, free chlorine, $Cl_2/HOCl/OCl^-$. Several field test kits using simplified laboratory procedures for determining available chlorine are available commercially. Each kit, depending on its design and the chemistry it employs, has particular advantages or disadvantages regarding the tests being conducted and the water being tested. To elucidate the differences between commercially available test kits, several studies were conducted by the Environmental Quality Division of the U.S. Army Medical Bioengineering Research and Development Laboratories.[1-3] Results of these studies should help chlorine residual test kit users decide which kit best suits their particular application.

To properly evaluate a test kit, users must first be aware of the properties of a good field test kit procedure and also of the stresses imposed on the procedure by the test conditions and sample.

TEST KIT CHARACTERISTICS

Qualities

Specificity for free available chlorine (FAC) is the most essential characteristic. In water treatment it is often important that FAC concentration be maintained at a certain level for a specified time to ensure that the treated water has been properly disinfected. In these cases, the operator must be confident that he is measuring free available chlorine and not some other species. Combined chlorine is a much less active disinfectant and usually is not desirable for drinking water treatment. Thus, the test used must respond only to the most active free chlorine species, $Cl_2/HOCl/OCl^-$. Historically, false positive readings (*i.e.,* positive indications of FAC when none is present) have been the major problem with test kit procedures.

The operator must also be confident that the FAC level he measures is accurate to ensure that he is not adding too much or too little chlorine to achieve the desired results. In addition, he must be able to reproduce his results several times during the day with minimum time allocated to the analysis. Thus, he requires an accurate, precise, relatively simple procedure, thereby permitting him to do several analyses in a short time. Accuracy and precision criteria for test kits were established to compare the results of this evaluation with previous studies. Analytical quality data in the thirteenth edition of *Standard Methods for the Examination of Water and Wastewater*[4] and data from a previous study[5] were reviewed. The accuracy and precision values listed in Table 5-1 represent desired performance limits for a field test kit specific for FAS as established for this study.

Table 5-1. FAC Test Kit Accuracy and Precision Limits

	Accuracy	
Relative Error	**FAC Range**	**Example**
50%	0.0- 0.4 mg/l	@ 0.2 mg/l ± 0.1
30%	0.4- 2.0 mg/l	@ 1.0 mg/l ± 0.3
20%	2.0-10.0 mg/l	@ 5.0 mg/l ± 1.0
	Precision	
Relative Standard Deviation	**FAC Range**	**Example**
40%	0.0- 0.4 mg/l	@ 0.2 mg/l ± 0.08
25%	0.4- 2.0 mg/l	@ 1.0 mg/l ± 0.25
15%	2.0-10.0 mg/l	@ 5.0 mg/l ± 0.75

Configuration

The test kit configuration is also an important consideration. A dip stick method, by which reagents are attached to a stick or paper strip, is convenient and rapid for individual samples; however, it may be more susceptable to environmental stresses than the other configurations. Methods using tablet reagents or powder pillows eliminate the requirement for measuring reagents or counting drops; however, with tablets, the operator must be careful to crush the table and take the time to ensure complete dissolution. In some instances undissolved portions may interfere with the test. When several samples are measured solid reagent methods are thus slower and less convenient. Although liquid procedures generally produce the best results, these methods require volumetric measurement or drop counting of the reagent added. Thus, in selecting a test kit configuration, the operator must decide whether he desires speed and convenience or improved accuracy and precision.

Consideration must also be given to the individual color comparator. The number of color standards, accuracy of the color and engineering design of the comparator are important factors in determining overall test kit precision and accuracy.

Conditions

The waters being tested can have a wide range of temperature and pH. A good test kit should show satisfactory performance for temperatures from 5 to 35°C over the pH range 5-10. The buffer capacity requirement of a test kit depends on the chemistry it employs. A test kit must be capable of producing acceptable results with waters having pH 5-10 with acidity or alkalinity to 500 mg/l as $CaCO_3$.

The environment where the tests are performed is also important. In the laboratory power is available and conditions such as temperature and humidity can be controlled so more sophisticated procedures can be employed. In the field, however, test conditions can vary considerably from day to day. These fluctuations must not interefere with test procedures.

WATER CHARACTERISTICS

Most published analytical procedures will perform well in relatively clean systems free from interferences. A good test kit analytical procedure must perform under a wide range of possible field situations. The nature of the water being tested is an important consideration. Waters with color, turbidity, high alkalinity or acidity, and organic contamination can present problems.

In these studies both synthetic and natural waters were employed. Synthetic water samples were prepared in the laboratory by adding known amounts of interfering substances. In addition, five natural waters were obtained in an attempt to represent a wide range of natural conditions, including the presence of trace quantities of elements or compounds that can alter a test kit performance in ways other than those observed with the synthetic water samples.

The natural water samples were collected in Maryland and Pennsylvania from various sources including streams, a bog and a sewage treatment plant. The physical, chemical and biological description of the natural waters is presented in Table 5-2. A general description of the waters and the specific sample collection locations were as follows:

Colored Water

This water contained 50-100 standard units of color. Total dissolved solids (TDS) were less than 500 mg/l and interfering ions, such as manganese and iron, were not present in significant concentrations. The colored water sample was obtained from the stream which drains Bear Meadows, an acid bog located approximately ten miles east of State College, Pennsylvania.

Turbid Water

Turbidity was considered primarily a physical optical interference for colorimetric analyses. The turbidity of this water was between 50-100 JTU and the TDS concentration was less than 500 mg/l. This water also contained high iron concentration. The turbid water sample was collected from a tributary to Winters Run in Edgewood, Maryland.

High Total Dissolved Solids Water

Interference due to high ionic strength and an attempt to stress the buffering capacity of the various analyses were two reasons for using this type of sample. The TDS limit recommended for military field water supplies is 1500 mg/l. A review of the chemical composition of worldwide waters indicated that bicarbonate is the dominant anion present in surface water supplies.[6] A surface water with approximately 1500 mg/l TDS was used with bicarbonate as the dominant anion in the samples tested. The high TDS water sample was collected from the Susquehanna River above tidal influence. Because this water contained only 100 mg/l in solids, additional salts were added to increase the TDS to approximately 1500 mg/l. The supplemental salts and their proportions were as follows: 20% NaCl, 40% Na_2SO_4 and 40% $NaHCO_3$, by weight.

Table 5-2. Characteristics of Natural Water Samples

Characteristic		Characteristic Name Given to Each Sample				
		Colored	Turbid	High TDS	Inter-ference	Organically Contaminated
Acidity (mg/l as $CaCO_3$)		9	3	8	9	11
Alkalinity, total (mg/l as $CaCO_3$)		13	23	405	12	54
BOD_5 (mg/l)		<1	3	3	1	11
Chlorides (mg/l)		1	4	192	7	74
Color (Pt units)	Apparent	90	50	15	<5	30
	True	70	30	<5	<5	20
Hardness (mg/l)		16	36	65	70	55
Iron (mg/l)	Total	0.22	3.68	0.43	0.20	0.15
	Filtered	0.20	0.92	0.30	0.18	0.15
Manganese (mg/l)	Total	0.05	0.06	0.10	0.80	0.06
	Filtered	0.04	<0.01	0.09	0.72	0.05
Nitrogen, Ammonia (mg/l)		<0.1	0.2	0.2	0.4	4.5
Nitrogen, Nitrite (mg/l)		<0.1	<0.1	<0.1	<0.1	0.6
Nitrogen, Nitrate (mg/l)		<0.1	1.5	0.9	1.2	2.7
pH		4.6	6.5	7.6	4.8	7.1
Total Dissolved Solids (mg/l)		120	213	1444	204	179
Total Organic Carbon (mg/l)		8.5	3.2	3.9	0.5	–
Turbidity (JTU)		<1	62	1	<1	1

Interference Water

Two elements that frequently interfere in residual chlorine analyses are manganese (especially oxidized forms) and iron. A natural water system containing both these ions was sought. The interference water sample was collected from the Still Creek Reservoir Branch of the Schuylkill River near Drenersville, Pennsylvania.

Organically Contaminated Water

Organically contaminated water was defined as that water discharged before chlorination from secondary treatment at an appropriate sewage treatment plant, then mixed in varying proportions with aerated tap water. The organically contaminated water was taken from the wastewater treatment plant effluent in Joppa, Maryland. The effluent had undergone primary settling and secondary activated sludge treatment but had not been disinfected by the addition of chlorine.

Chlorination breakpoint curves were determined on these natural water samples, and test kit evaluations were conducted at various points along the breakpoint curve. Ideally, these points would be selected as shown in Figure 5-1. This procedure produced sufficient data to demonstrate a test kit's specificity for FAC under a variety of natural water conditions.

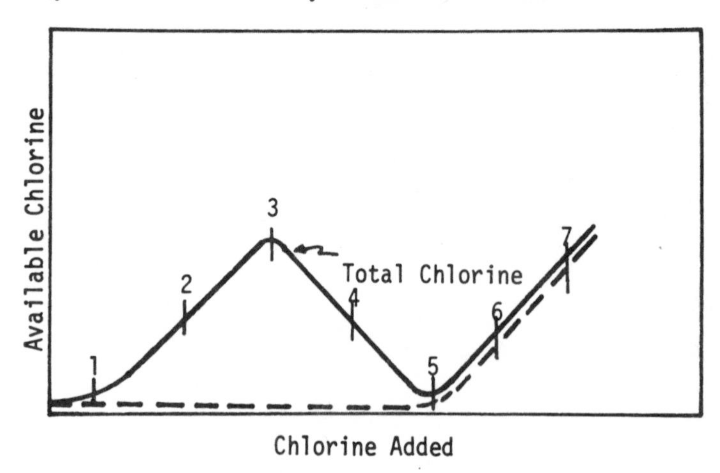

Figure 5-1. Expected breakpoint curve and analysis points.

SELECTION OF TEST KITS

Several studies have been reported in which different methods for determining FAC in water were examined. Nicolson[7] compared 12 laboratory methods for determining FAC to evaluate reproducibility, sensitivity, and the effect of interfering substances upon the methods. The accuracy and precision of FAC test methods were evaluated in studies by Lishka *et al.*[8,9] In these three studies the methods tested were those used in the laboratory, and the data was not directly applicable to the evaluation of field test kits. However, from the information in these reports and a review of current commercially available field test kits, it was determined that the following techniques should be evaluated:

Syringaldazine

The syringaldazine procedure was first developed and manufactured by the Ames Company, Elkhart, Indiana, as a simple FAC dip stick test called AQUA-CHECK®. This product has a range of 0-2 mg/l and was initially intended for swimming pool use. The test is based on a white-to-purple color change of the compounds syringaldazine and vanillinazine which are impregnated on a pad attached to a plastic stick. (The use of vanillinazine has been eliminated in current kits.) The reagents and reactions have been described by Bauer and Rupe.[10] The First U.S. Army Medical Laboratory performed a preliminary investigation of AQUA-CHECK[11] and Black and Thompson,[12] acting as independent researchers, evaluated the test strip in more detail. The Ames Company has more recently developed a wide range test strip (0-10 mg/l) utilizing syringaldazine. In addition, a color comparator utilizing reagents in liquid form has also been developed by the Ames Company. The liquid form of the test as manufactured has a range of 0-1 mg/l FAC, with an expanded range of 0-10 mg/l obtained by dilution of the test sample. The latter two techniques were selected for evaluation and the methods referred to as syringaldazine (high range) and syringaldazine (liquid), respectively.

Leuco Crystal Violet (LCV)

The LCV procedure was presented by Black and Whittle in 1967[13] and, when buffered to a pH of approximately 4, will develop a bluish color in the presence of FAC. This procedure is commercially available in kit form from Taylor Chemicals, Inc., Baltimore, MD and from Hack Chemical Co., Ames, Iowa. The maximum FAC concentration that could be measured with the kit evaluated was approximately 1.2 mg/l. The range of this procedure has been extended to 10 mg/l FAC in a recent study.[14]

DPD

The N,N-diethyl-p-phenylenediamine (DPD) procedure was first presented by Palin.[15] The initial procedure has been modified for simplicity and reagent stability to the point where all reagents required for the FAC procedure are contained in one tablet. This tablet is added to a sample containing FAC and produces a red color that can be matched to a standard color comparator. Kits using this technique with a range of 0-10 mg/l are available from LaMotte Chemical Products Company, Chestertown, Maryland.

SNORT (Liquid)

The reagents for the stabilized neutral orthotolidine test (SNORT) were developed by Johnson and Overby.[16] Two color comparators were used with this procedure. The standard Army color comparator has no zero window. This can result in an increased number of false positives since the user would be inclined to report the lowest reading on the comparator for a solution containing no FAC. Therefore, a modified color disc produced by the Delta Scientific Corporation, Lindenhurst, N.Y., having FAC window values of 0.0, 0.2, 0.5, 1.0, 2.0, 3.0, 5.0, 7.0 and 10.0 mg/l was also employed in this evaluation. This procedure had an FAC range of 0-10 mg/l. A kit for SNORT is also available from LaMotte Chemical Products Company, Chestertown, Maryland.

SNORT (Dry)

As a result of a contract by the U.S. Army Mobility Equipment Research & Development Center,[17] the reagents developed by Johnson and Overby[16] were slightly modified and attached to sticks in powder form, by Delta Scientific Corporation. One stick carries the buffer-stabilizer and a second carries the orthotolidine reagent. The original stabilizer, Aerosol OT (American Cyanamid Co.),[16] was changed to Aerosol OT-B in the stick test. In addition, oxalic acid was added to the orthotolidine, thereby increasing reagent shelf life. The range of this kit and the color comparator used were the same as for the liquid procedure.

MOTA

The Army standard modified orthotolidine-arsenite (MOTA) procedure was evaluated. It had a range from 0-10 mg/l FAC.

Methyl Orange

This procedure was considered for inclusion in this study. However, a review of available information[18] led to its exclusion because (a) it was insensitive at low range if adequate methyl orange was present for a test to 10 mg/l, (b) reagents must be added in very exact quantity because it is a bleaching method, and (c) the sample must be added to the reagent with rapid mixing.

TEST KIT SPECIFICITY FOR FREE AVAILABLE CHLORINE

Interference studies with synthetic waters and the results obtained with natural waters containing no FAC defined the specificity of each test kit for free available chlorine. The major concern in these studies was the interference of combined chlorine, primarily chloramines. A summary of the data from this study is found in Table 5-3. Note that the organically contaminated water produced the largest stress on test kit specificity. This

Table 5-3. Summary of False Positive Data for Test Kit Procedures in Both Synthetic and Natural Waters, Expressed as Ratio of False Positives to Total Tests

	Synthetic Water		Natural Water
	pH and Temperature Variations	Interfering Substances Present	Organically Contaminated
Syringaldazine (Liquid)	0.04	0.00	0.04
Syringaldazine (High Range)	0.11	0.00	0.44
Leuco crystal violet	0.22	0.75	0.81
DPD	0.07	0.67	0.87
SNORT (Dry)	0.02	0.67	0.65
SNORT (Liquid) (no 0.0 mg/l FAC window)	a	a	0.91
SNORT (Liquid) (with 0.0 mg/l FAC window)	0.00	0.15	0.29
MOTA	a	a	0.90

[a]No tests performed on synthetic water.

was not unexpected, since this water contained a number of interferences common to FAC analysis. Combined chlorine (*e.g.*, mono-, di-, and trichloramine), organic chloramines (*e.g.*, N-chloroamino acids) and oxidized forms of manganese are the predominate interfering species. The effect of temperature and pH on the test kits was minimal; however, the syringaldazine (high range), LCV and DPD procedures yielded a significant number of false positives. In the initial phase of the study the DPD and LCV procedures showed comparable specificity. Syringaldazine (liquid) was the most specific procedure tested. Data from MOTA procedure on natural waters show that it is not as specific as either the SNORT (liquid-with zero window) or the syringaldazine (liquid) procedures. When SNORT was used with modified color disc, false positive readings were reduced significantly. Under these conditions SNORT (liquid) is more specific than the SNORT (dry) kit; however, it remains subject to interference above 10 mg/l monochloramine as Cl_2 and is not as specific as the syringaldazine (liquid procedure). It was more specific than the DPD, LCV or MOTA methods tested. It was shown in spectrophotometric studies, that the SNORT (liquid) procedure's specificity is dependent on the concentration of monochloramine in the test sample and the actual reading of FAC concentration with the comparator. The number of false positives observed increases with higher monochloramine concentration and longer time between reagent addition and color comparison. This agrees with Johnson's[16] originally reported monochloramine interference of 1.1% per minute at 25°C. Thus, to be reasonably specific, the operator must take his reading on the comparator immediately after reagent addition. Spectrophotometric studies have shown that concentrations of monochloramine as high as 33 mg/l have no effect on the syringaldazine procedure up to 10 minutes after reagent addition.

The Leuco Crystal Violet procedure underwent limited testing. The effect of temperature and pH on the specificity of this procedure was similar to that observed with the DPD test kit; however, other drawbacks such as residual color in the sample cells, non-specificity in organically contaminated waters, and a range limited to 1.2 mg/l resulted in its elimination from further study. A modified version of this procedure has been reported by Whittle[14] to have eliminated many of the shortcomings observed in this study.

TEST KIT ACCURACY AND PRECISION

No procedure evaluated in this study was within the predetermined accuracy and precision limits at all concentrations of FAC tested. However, the deviations from the predetermined limits were not great enough to eliminate the tests as acceptable field test kits.

The LCV and syringaldazine (liquid) procedures performed well within the 0-1 mg/l range for which they were designed; however, both procedures experience some problems at higher FAC levels. Both procedures yielded results consistently lower than the true FAC concentration. Syringaldazine (high range), LCV, DPD and SNORT (dry) were all within the accuracy limits specified in Table 5-1 when tested on natural waters. SNORT (liquid) and MOTA were not tested for accuracy with natural waters.

No kit was superior because of its accuracy or precision.

MODIFIED PROCEDURES

After evaluating the commercial kits, the study was continued in an attempt to develop a method with all the desired qualities of a good field test kit. Thus, the three most promising test kit procedures–DPD, SNORT, and syringaldazine (liquid)–were selected for modification.

DPD

This procedure was found to be accurate and precise, but not specific, for the determination of FAC. Several procedural changes were proposed by Palin[19] to eliminate the interference from chloramines. These changes are summarized below:

Color Compensation Technique

This change adds a "DPD Ammonia" tablet and requires an additional tube, as compared with the original procedure. Addition of ammonia to tube 1 converts all FAC to combined chlorine. Thus this tube has a sample which is identical to the original sample without FAC. The only color produced is that which results from the DPD interference reaction with combined chlorine. Tube 1 is used as a blank and its reading (the result of combined chlorine interference) is subtracted from the reading observed in the original procedure with tube 2. Tube 2 color is produced by both FAC and the combined chlorine interference. Thus, theoretically, the difference in the tube colors is that produced by FAC. A potential error with this method is accentuated when the FAC concentration is significant compared with the original combined chlorine concentration. Palin cites the case where FAC is 6 mg/l and combined chlorine is 4 mg/l.[19] Assuming the chloramine false positive result is equivalent to 3% of the combined chlorine present (Johnson and Overby found 3.6% per minute DPD interference at 25°C[16]), this would result in an overcorrection of 0.18 mg/l (*i.e.*, 6 mg/l FAC $\xrightarrow{NH_3}$ 6 mg/l combined chlorine x 3% = 0.18 mg/l additional color produced).

Timing Method

Under test conditions the reaction between monochloramine and DPD was found to be linear with time. By taking two readings at equal time intervals and subtracting the second from twice the first, a modification of the DPD test for FAC in heavily chlorinated polluted waters is possible. This procedure would work well under laboratory conditions where a spectrophotometer is available to measure adsorbance. However, the precision and accuracy tolerances of the color comparator of the DPD test kit are large enough to cause considerable error when used as a field procedure.

Method Using Less DPD

The rate of the reaction of monochloramine with DPD is proportional to the DPD concentration. Palin observed that reducing the amount of DPD in a solution containing monochloramine decreased the magnitude of the color interference.[15] Using less DPD reagent with low concentrations does reduce the interference from chloramines. At the same time, however, it increases the complexity of the procedure.

Glycine Addition

Palin found "that the addition of glycine after the initial reaction between FAC and DPD markedly retards any continuing development of color by reaction between combined chlorine and DPD."[19] This procedure is more complicated than the original and does not completely eliminate the interference due to chloramines.

It is apparent that each of the four modifications proposed by Palin decreases, but does not necessarily eliminate the interference caused by chloramines. Unfortunately, each modification also increases the complexity of the DPD procedure and thereby reduces its desirability as a field test method. Although not studied, manganese in the Mn(IV) state is known to also interfere with DPD. An arsenite modification is available to overcome this problem.[4]

Another possible modification would involve changing the pH of the buffer to prevent reaction of DPD with chloramines, an approach similar to that employed by Johnson and Overby in the development of SNORT.[16] Preliminary results obtained by Palin show that no pH change sufficiently retards the reaction between monochloramine and DPD to permit a satisfactory test at all anticipated combinations of FAC and chloramines.

Leuco Crystal Violet, Modified

In the test procedure evaluated in this study a range of 0-1.2 mg/l FAC was obtainable. However, in a recent publication[14] several improvements in the procedure were presented. It has been reported that by allowing reaction between the FAC and leuco crystal violet at pH 7.0 for one minute, followed by acidification to pH 3.7, the entire range 0-10 mg/l FAC is measurable. The specificity for FAC in the presence of up to 10 mg/l monochloramine has been improved by the addition of hydroxylamine hydrochloride, acting much the same as arsenite in the MOTA method.[14]

Lastly, it has been reported that the problem with color adsorption on the comparator cell walls has been overcome by use of n-propanol in the pH 7 buffer.[14] Manganese in the Mn(IV) state interferes with FAC determinations by this modified LCV method. An arsenite blank correction does correct this interference.[4] It should be noted that this procedure is not available commercially nor has it undergone extensive comparative testing.

SNORT

It was shown that the SNORT (liquid) procedure was more specific for FAC than the SNORT (dry) procedure. It was also demonstrated that chloramines and oxidized forms of manganese interfere with the SNORT (liquid) procedure to produce false positive results. Since the reaction of orthotolidine with FAC is much faster than its reaction with chloramines, an arsenite addition step, which was successful in decreasing the magnitude and number of false positives in the MOTA procedure, was attempted.

SNORTA

An arsenite modification of the SNORT procedure was tested on solutions containing three different levels of monochloramine: 6.7, 13.4 and 33.5 mg/l as Cl_2. The SNORT liquid procedure was compared spectrophotometrically with the modified procedure. The results are summarized in Figure 5-2, which shows that the arsenite modifications (SNORTA) significantly decreased the interference caused by monochloramine. However, significant interference remained at the higher monochloramine concentrations tested (Curves B and C). At 13.4 mg/l of monochloramine, the arsenite procedure was able to overcome the interference for approximately 85 seconds after addition of SNORT reagent. It is apparent that addition of an increased amount of arsenite in this procedure may eliminate the interference at even higher concentrations of monochloramine. The limiting factor for the success of this procedure would be the rate of the reaction between monochloramine and orthotolidine, which is proportional to the concentration of monochloramine.[16] At high monochloramine concentrations the reaction may be so fast that the arsenite cannot be added in time to prevent the interference. Using Johnson's rate,[16] the arsenite must be added before 20 seconds at 25°C to keep the interference less than 0.1 mg/l from 30 mg/l monochloramine as Cl_2. This agrees with our data (Figure 5-2, C).

No interference with the SNORT procedure was observed for dichloramine concentrations as high as 9.2 mg/l.

The effect of oxidized forms of manganese on SNORTA was also tested. Suspensions of MnO_2 were prepared to simulate conditions in natural waters.

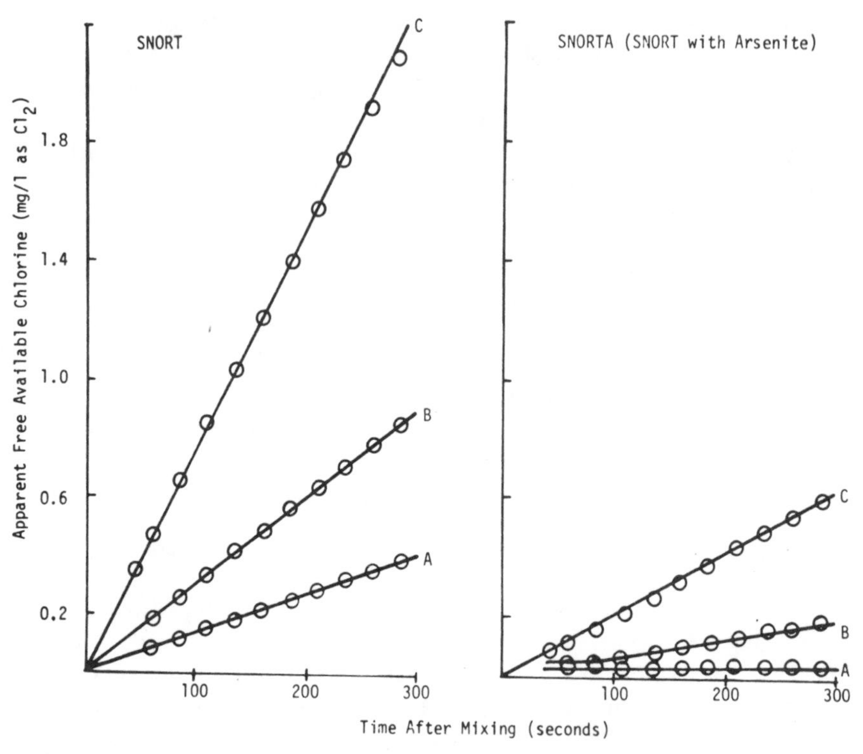

Figure 5-2. Response of the SNORT and SNORTA procedures to monochloramine (6.7, 13.4 and 33.5 mg/l, curves A, B and C respectively).

The reaction of manganese (1 mg/l) with the SNORT ortholidine was so fast that interference color was produced before arsenite could be added. Thus, the addition of arsenite after the sample in the SNORTA procedure could not prevent the interference caused by the oxidized form of manganese. One possible method to correct for oxidized manganese requires a blank with arsenite.[4] Addition of arsenite to the sample before it is added to the reagents, reduces free chlorine but does not reduce manganese. Arsenite added before the reagents rather than after, as in SNORTA, produces the color from manganese(IV) interference alone. This blank can then be subtracted from the chlorine plus interference to correct for manganese in the direct SNORTA procedure. This approach was not attempted in our study.

Syringaldazine

The syringaldazine (liquid) method of Bauer and Rupe[10] was the only procedure evaluated found to be specific for FAC.[1,2] This procedure requires a dilution step to extend its range to 10 mg/l FAC, a step unacceptable in an Army field test. A study with this test procedure was initiated to eliminate the dilution step while retaining its specificity for FAC.

Two modified procedures were developed for a Free Available Chlorine Test with Syringaldazine (FACTS I and II).

FACTS I

This procedure was designed for increased sensitivity at 0.1 mg/l FAC and obeys Beer's Law through the range of 0 to 5 mg/l. It involves the following steps:

1) Add 3 ml of sample to be tested for FAC to a test tube.
2) Add 0.1 ml of buffer (0.5 M phosphate, pH = 6.5).
3) Add 1 ml of syringaldazine indicator (115 mg/l in 2-propanol), cap the tube and invert twice to mix.
4) Read the absorbance at 530 nm with a spectrophotometer. (In a test kit this step would involve color comparison using a standardized color comparator.)

Calibration curves (Beer's Law plots) were obtained to show the response of this procedure to FAC concentrations ranging from 0-10 mg/l. A least squares linear regression was used to obtain equations that best represent the linear portion of each curve obtained. In all cases the standard error of the estimate was less than 0.07. The resulting equations are plotted in Figure 5-3. Although the response of this procedure is linear to 5 mg/l, FAC concentrations between 5 and 10 mg/l can be easily distinguished. The data for Figure 5-3 was obtained over a period of several weeks with three different indicator solutions (114, 115 and 116 mg/l syringaldazine). All data for any one curve were obtained on the same day. These data show that results obtained with this technique are very reproducible, even with small fluctuations in the concentration of the indicator solution.

FACTS II

This procedure was developed for an effective range of 0 to 10 mg/l FAC and was designed for increased sensitivity and linearity in the 5-10 mg/l FAC region. The test procedure involves the following steps:

1) Add 5 ml of sample to be tested for FAC to a test tube.
2) Add 0.2 ml of buffer (0.5 M phosphate, pH 6.5).
3) Add 2 ml of syringaldazine indicator (115 mg/l in 2-propanol), cap the tube and invert twice to mix.
4) Read the absorbance at 530 nm with a spectrophotometer. (In a test kit this step would involve color comparison using a standardized color comparator).

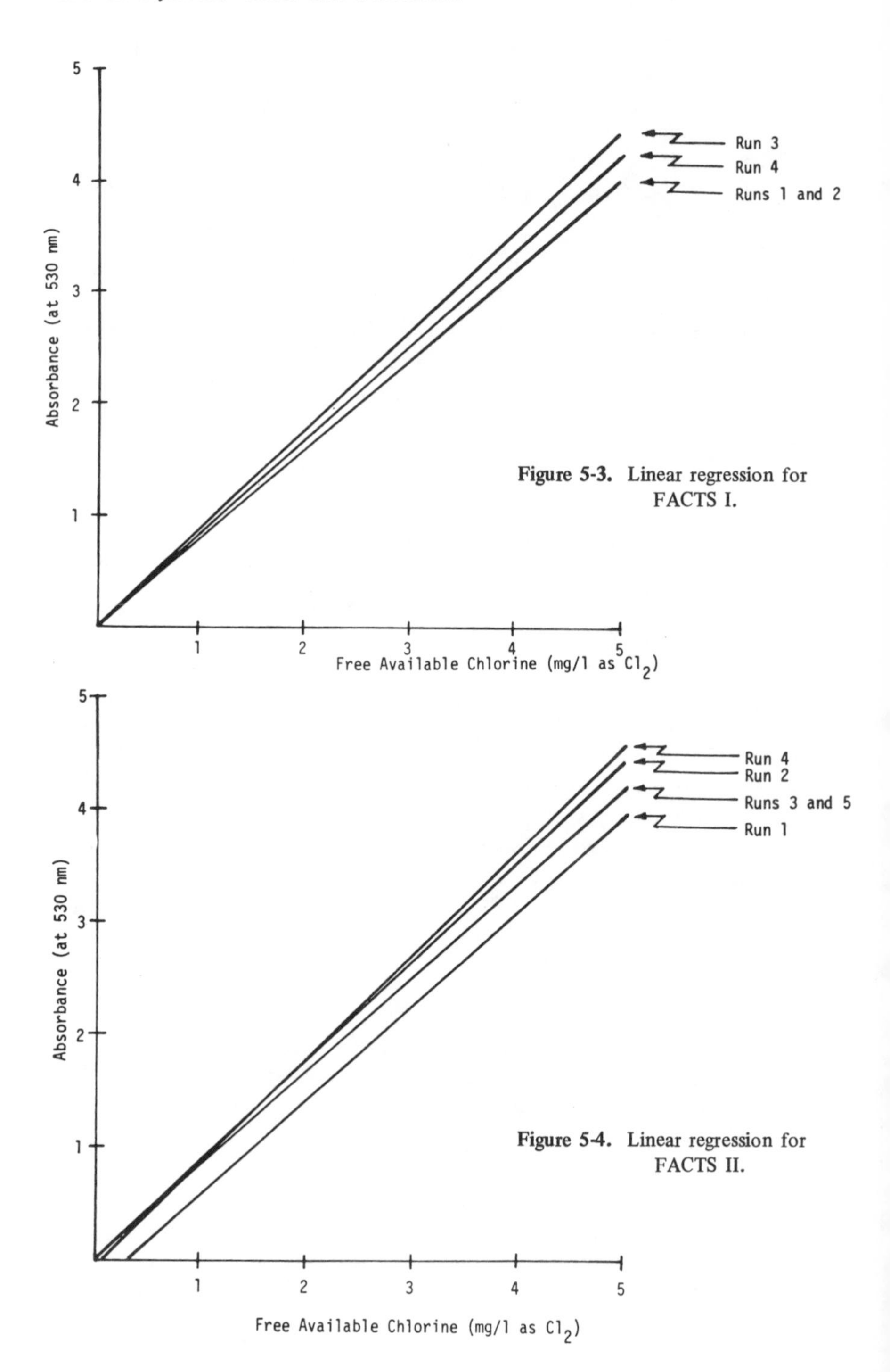

Figure 5-3. Linear regression for FACTS I.

Figure 5-4. Linear regression for FACTS II.

This procedure underwent testing similar to that of FACTS I. Its response is linear for FAC concentrations up to 7 mg/l; however, concentrations up to 10 mg/l can easily be detected visually using this technique. Duplicate determinations were made at each FAC concentration. The data were analyzed by least squares linear regression to obtain equations best representing the linear portion of each curve. The standard error of the estimates was less than 0.07. This analysis yielded the equations plotted in Figure 5-4.

FACTS II permitted increased sensitivity for the syringaldazine test at higher FAC concentrations. However, it also slightly reduced the sensitivity at the lowest FAC levels. FACTS I had a lower detection limit of 0.1 mg/l, whereas the lower detection limit of FACTS II was approximately 0.15 mg/l. This lower detection limit is the result of the increased yellow background color due to the higher syringaldazine concentration in FACTS II. This background tends to mask the light pink color produced by the FAC-syringaldazine reaction at the 0.1 mg/l FAC level.

Based on the above results, the FACTS II procedure would be acceptable for determining FAC in the concentration range of 0.15 to 10 mg/l, and could be adapted easily to a field test kit.

FACTS Evaluation

Since the FACTS method was most specific for FAC, several additional experiments were performed to help evaluate FACTS II for the field determination of FAC. These included investigation of the effect of temperature on the procedure, reagent aging studies, interference studies and the development and qualitative testing of a prototype test kit.

Temperature studies were performed at 5°, 25° and 35°C. The results show that a decrease in temperature slightly increases the color intensity but, overall, has a negligible effect on the color reaction. At 35°C the results show a slight decrease in color intensity with increasing temperature; however, this change would have a negligible effect on results obtained with a field test procedure.

Aging studies have been initiated to test the shelf life of the FACTS reagents. Results to date show no difference in the data obtained with freshly prepared indicator and that obtained with indicator that has aged under ambient laboratory conditions for three months. In addition, the results obtained in the previous studies with the original syringaldazine reagent indicate that the reagents are stable for at least 1.5 years under ambient laboratory conditions.

FACTS II was employed to test solutions containing species known to interfere with FAC determinations. The results are summarized in Table 5-4. Although iron in high concentrations is known to react with syringaldazine, no interference was observed for concentrations as high as 10 mg/l

Table 5-4. Interference Test with Facts II

Interference	Concentration (mg/l)	Result
NO_2^-	250	No false positive
$SO_4^=$	1000	No false positive
NO_3^-	100	No false positive
Cl^-	1000	No false positive
Fe(III)	10	No false positive
Cu(II)	30	No false positive
Mn(IV)	2.6	False positive within 5 minutes
Mn(IV)	1.0	No false positive
Hardness	1000[a]	Heavy precipitate, no false positive
Hardness	500[a]	Light precipitate, no false positive
Hardness	250[a]	No false positive or precipitate
Alkalinity	1000[a]	Dark yellow solution, no false positive
Alkalinity	300[a]	No false positive
Color	120 cu	Colorless, no false positive
Monochloramine	70[b]	False positive within 5 minutes
Monochloramine	35[b]	False positive after 5 minutes, no false positive within 2 minutes
Monochloramine	18[b]	No false positive
Dichloramine	9.2[b]	No false positive

[a]As $CaCO_3$

[b]As Cl_2

of Fe(III). Manganese, which interferes with the SNORT, DPD and LCV procedures, produced no interference with FACTS II for Mn(IV) concentrations as high as 1.0 mg/l. Although hardness at a concentration of 500 mg/l did produce a precipitate with this procedure, it did not interfere with the color produced nor did these concentrations result in false positive readings. Although alkalinity at 1000 mg/l interferes with color production, this can be overcome by an increase in the buffer capacity of the buffering solution. Monochloramine, which interferes with the DPD test procedure, did not interfere with the FACTS II procedure at monochloramine concentrations as high as 18 mg/l (as Cl_2). Finally, no interference was produced by dichloramine concentrations as high as 9.2 mg/l (as Cl_2).

A crude prototype color comparator and test kit were prepared to evaluate the applicability of FACTS II to a field test kit procedure. Standard color solutions for the color comparator were prepared by dilution of an Alizarin Red S solution containing phosphate buffer. The hue and color intensity of each standard was adjusted by visual comparison of that standard with the appropriate FAC standard. The color standards were sealed into glass tubes and placed in a stand with a white background. FAC concentrations of 0, 0.2, 0.5, 1.0, 2.0, 3.0, 5.0, 7.0 and 10.0 mg/l were represented.

Prior to assembly of the comparator, the nine tubes were used for the following test: The tubes were randomized and all nine were given to each of 16 non-technical people for rearrangement in order of increasing color intensity. Each individual could distinguish the various color intensities and arrange the nine standards in proper order. Thus, an operator using the FACTS II procedure and a standardized color comparator should easily be able to distinguish relative FAC concentrations from 0 to 10 mg/l.

The prototype kit was tested against solutions containing from 0-12 mg/l FAC. Results showed that the FACTS procedure could be adaopted successfully as field test for FAC.

Procedures using syringaldazine can be applied to the determination of other compounds having available chlorine. This has been demonstrated in the laboratory. Chloramines can be determined if an iodide addition step, similar to that used with the amperometric titrator, is included. This procedure would have the following steps:

1) Adjust the pH to the proper value of the chloramine of interest.
2) Add an excess of KI (iodide reacts with available chlorine to produce iodine and chloride).
3) Adjust the pH with FACTS buffer (pH 6.5).
4) Follow the FACTS I or II procedure to obtain a colored solution.
5) Compare this color with color standards.

This procedure works with mono- and dichloramine. Thus, syringaldazine and the FACTS procedure should be considered when it is necessary to develop methods to analyze for compounds containing total available chlorine.

CONCLUSIONS

None of the test kits evaluated met all criteria for a good field test kit. However, modifications to the three most promising procedures did result in improved techniques.

Table 5-5 summarizes and compares the different test procedures. Syringaldazine procedures offer the most specific test for FAC. However, an acceptable syringaldazine test kit is not yet commercially available. Several strip and stick methods are available for testing swimming pool waters but, as shown here, are not completely acceptable as field test kits for FAC. FACTS procedures offer the most promising approach in the development of a specific field test kit for FAC. Although preliminary data indicate that FACTS is an acceptable procedure for FAC, this approach has not been as extensively used and tested as the SNORT, DPD and LCV test kit procedures.

Table 5-5. Summary and Comparison of Test Procedures Free Available Chlorine

Test Procedure	Reagent Form	Range (mg/l) Without Dilution	Identified Interferences	Commercially Available	Principal Reagents
Syringaldazine					
Syringaldazine (liquid)	Liquid	0.1	None	Being developed	Syringaldazine Solution Buffer Solution
Syringaldazine (high range)	Strip	0-10	Monochloramine	Yes	Syringaldazine
FACTS	Liquid	0-10	None	Being developed	Syringaldazine Solution Buffer Solution
Orthotolidine					
MOTA	Liquid	0-10	Manganese(IV), Combined Chlorine	Yes	Orthotolidine Arsenite Solution
SNORT (liquid) no zero window w/ zero window	Liquid	0-10	Manganese(IV), Combined Chlorine	Yes	Orthotolidine Stabilizer/buffer Solution
SNORT (dry)	Stick w/tablet	0-10	Manganese(IV), Combined Chlorine	No	Buffer Tablet Orthotolidine Stick
SNORTA	Liquid	0-10	Manganese(IV)	Use SNORT kit with arsenite reagent	Orthotolidine Solution, Buffer Solution, Arsenite Solution
Leuco Crystal Violet					
LCV	Liquid	0-3	Manganese(IV), Combined Chlorine	Yes	LCV Buffer Solution
LCV (modified)	Liquid	0-10	Manganese(IV), None Reported	No	LCV Buffer Solutions (pH 7 and 3.7) Hydroxylamine Hydrochloride
N,N-diethyl-p-phenylenediamine					
DPD	Tablet	0-10	Manganese(IV), Monochloramine	Yes	DPD Tablet
DPD (modified)	Tablet	0-10	Manganese(IV), Monochloramine	Use DPD kit in modified manner	DPD Tablet

The SNORT (liquid) procedure suffers from interferences from high concentrations of monochloramine and oxidized forms of manganese. It is more specific for FAC than for LCV or DPD, however. The arsenite modification, SNORTA, eliminates monochloramine interference. Manganese interference can be corrected by a blank with arsenite, added before reagents. One consideration in using this or any other procedure, is to ensure that the comparator has a zero sample window to prevent the operator from reporting false positives, simply because the lowest reading available is a positive value.

DPD test kits have all reagents in one table, and this procedure is probably the simplest available, if the time required to dissolve tablets is not a problem. Even with the modifications suggested by Palin,[19] this procedure is subject to interference from monochloramine. Oxidized manganese also interferes but can be corrected by an arsenite blank.[4] This method has been tested and used extensively. However, when used in the field, the operator should be aware of possible false positives due to combined chlorine.

The LCV kit tested was comparable to DPD in specificity, accuracy and precision. Whittle's recent modifications to this procedure indicate that several of the procedure's shortcomings have been eliminated.[14] Unfortunately, these modifications have neither appeared in commercial form nor undergone extensive testing.

Finally, it must be pointed out that the modified orthotolidine-arsenite procedure is the least acceptable for FAC analysis.[5,9,20] The most acceptable procedures, Syringaldazine (liquid) and FACTS, are not yet commercially available. Thus, operators presently have the choice of preparing a syringaldazine kit or selecting a commercial procedure (DPD, LCV or SNORT) which may have some disadvantages. Of the three, the SNORT procedure is most specific for FAC. DPD is a simple, versatile procedure and has received the most testing. However, if one of these three procedures is used, the operator must know its disadvantages and the effect of interferences on the test results.

REFERENCES

1. Guter, K. J. and W. J. Cooper. "U.S. Army Medical Environmental Engineering Research Unit, Report No. 7303," AD No. 752440, Edgewood Arsenal, MD (October 1972).
2. Cooper, W. J., E. P. Meier, J. W. Highfill and C. A. Sorber. "U.S. Army Medical Bioengineering Research & Development Laboratory Technical Report 7402," AD No. 780054, Aberdeen Proving Ground, MD (April, 1974).
3. Meier, E. P., W. J. Cooper and C. A. Sorber. "U.S. Army Medical Bioengineering Research & Development Laboratory Technical Report 7405," AD No. 780857, Aberdeen Proving Ground, MD (May, 1974).

4. *Standard Methods for the Examination of Water and Wastewater*, 13th ed. (New York: American Public Health Association, 1971).
5. "Evaluation of the Stabilized Neutral Orthotolidine Test (SNORT) for Determining Free Chlorine Residuals in Aqueous Solutions," *Sanitary Engineering Special Study No. 24-3-68/70*, U.S. Army Environmental Hygiene Agency, Edgewood Arsenal, MD (1970).
6. Livingstone, D. A. "Data of Geochemistry," In *Chemical Composition of Rivers and Lakes.* U.S. Department of the Interior, Geological Survey Professional Paper, 440-G, Chapter G, (1963).
7. Nicolson, N. J. "Technical Paper 29," Water Research Association, Medmenham, Marlow, Bucks, England (1963).
8. Lishka, R. J., E. F. McFarren and J. H. Parker. "Water Chlorine (Residual) No. 1," Analytical Reference Service, U.S. Dept. Health, Education and Welfare, Study No. 35 (1969).
9. Lishka, R. J. and E. F. McFarren. "Water Chlorine (Residual) No. 2," Analytical Reference Service, Environmental Protection Agency, Report No. 40, Cincinnati, Ohio (1971).
10. Bauer, R. and C. Rupe. *Anal. Chem.* **43**, 421 (1971).
11. "Report of Preliminary Evaluation of 'AQUA-CHECK' Chlorine Test Kit," Environmental Health Engineering Services, First U.S. Army Medical Laboratory, Fort Meade, MD (1971).
12. Black, A. P. and C. G. Thompson. "An Evaluation of 'AQUA-CHECK' Reagent Strips for the Determination of Free Chlorine Residuals in Swimming Pool Water," Unpublished data.
13. Black, A. P. and G. P. Whittle. *J. Amer. Water Works Assoc.* **59**, 607 (1967).
14. Whittle, G. P. and A. Lapteff, Jr. "New Analytical Techniques for the Study of Water Disinfection," In *Chemistry of Water Supply, Treatment, and Distribution.* (Ann Arbor, Mich.: Ann Arbor Science Publishers, Inc., 1974), Chapter 4, p. 63.
15. Palin, A. T. *J. Amer. Water Works Assoc.* **49**, 873 (1957).
16. Johnson, J. D. and R. Overby. *Anal. Chem.* **41**, 1744 (1969).
17. "Development of Dry Reagent Stick," Delta Scientific Corporation, USAMERDC Contract No. DAAK 02-07-C-0289 (1971).
18. Sollo, F. W. and T. E. Larson. *J. Amer. Water Works Assoc.* **57**, 1575 (1965).
19. Palin, A. T., Newcastle and Gateshead Water Company, Newcastle Upon Tyne, England, personal communication, 1973.
20. "Evaluation of the Orthotolidine Tests for Determining Free Chlorine Residuals in Aqueous Solutions," Project No. 3879E6-61/63, U.S. Army Environmental Hygiene Agency, Edgewood Arsenal, MD (1963).

CHAPTER 6

INTERHALOGENS AND HALOGEN MIXTURES AS DISINFECTANTS

Jack F. Mills

Halogens Research Laboratory
The Dow Chemical Company
Midland, Michigan

INTRODUCTION

The properties and disinfectant use of the interhalogens bromine chloride (BrCl), iodine chloride (ICl), and iodine bromide (IBr), as well as various halogen mixtures, have been explored in a number of investigations.[1-3] Some of the advantages of these disinfectants have been reported in areas such as biocidal activity, physical properties, environmental properties, and economics.

These interhalogen compounds may be considered either bromine- or iodine-based materials since the larger, more positive halogen is the reactive portion of the interhalogen molecule during the disinfection process. Therefore, this chapter will discuss the unique properties of bromine and iodine in these compounds and mixtures. Due to the demand for wastewater disinfectants and other applications, the production and availability of bromine and iodine will also be discussed.

In particular, much of this chapter will discuss the properties and disinfectant use of bromine chloride which shows considerable promise for future use as an active, environmentally safe disinfectant.[4,5]

AVAILABILITY OF BROMINE AND IODINE FOR DISINFECTION

Bromine

Bromine does not occur in nature as a free element, but is found mostly in bromide form, widely distributed and in relatively small proportions. Bromides available for extraction occur in the ocean and salt lakes, brines or saline deposits left after these waters evaporated during earlier geological periods. The bromine content of average ocean water is 65 ppm by weight, or 308,000 tons of bromine per cubic mile of sea water. Total bromine content in the earth's crust is estimated at 10^{15}-10^{16} tons.[6]

The Dead Sea is one of the richest commercial sources of bromine in the world, containing nearly 0.4% at the surface and up to 0.6% at deeper levels. Principal sources of bromine in the United States are the brine wells in Arkansas, Ohio and Michigan with bromide content ranging from 0.2-0.4%. Those in Michigan underlie a large area of the Great Lakes region and occur in various sandstone strata at depths of 700-8000 ft. Here bromine contents vary from 0.05-0.3% and are generally higher in deeper brines.

Iodine

Iodine is widely distributed in nature, occurring in rocks, soils, and underground brines in much smaller quantities (20-50 ppm), while seawater contains only about 0.05 ppm. The mineral lautarite (anhydrous calcium iodate) found in the large Chilean nitrate deposits, is probably the most important iodine source.

Despite the low iodine concentration in seawater, some brown seaweeds such as Laminaria and Fucus can extract and accumulate up to 0.45% iodine on a dry basis. The world's principal commercial sources of iodine are the Chilean nitrate deposits, Japanese oil well brines, and the Michigan and Arkansas brine wells.

INDUSTRIAL PROCESSES

Bromine Production

To extract bromine from bromide containing solutions, four steps are necessary: (1) oxidation of bromide to free bromine, (2) separation of bromine from solution, (3) bromine vapor condensation or isolation, and (4) purification. The major problem is caused by the relatively small concentrations of bromide in the most common sources–ocean water and natural brines.

Oxidation of bromide to bromine can be accomplished by chemical or electrochemical oxidation; however, the electrochemical methods are no longer commercially important. Of the chemical oxidizing agents, which include manganese dioxide, hypochlorite and chlorine, only chlorine is now used commercially.

Current bromine production methods are based on the modified Kubierschky or steaming-out processes and H. H. Dow's blowing-out process. In the steaming-out process, raw brine is preheated to about 90°C, treated with chlorine in a packed tower, then enters the steaming-out tower into which steam and additional chlorine are injected. The outgoing brine is neutralized with caustic and used to preheat raw brine.

From the top of the steaming-out tower, the halogen and steam vapor passes into a condenser, then into a gravity separator. From the separator vent gases return to the chlorinator, the upper water layer containing Br_2 and Cl_2 is returned to the steaming-out tower, and the lower crude bromine layer passes on to a stripping column. Bromine from the stripping column is purified in a fractionating column from which 99.8% pure liquid bromine is withdrawn as product.

Air is used in the blowing-out process since it is a more economical extracting agent than steam, especially when the bromine source is as dilute as ocean water. In this process, the halogens are absorbed from the air in a sodium carbonate solution or by sulfur dioxide reduction.

$$Br_2\ (Cl_2) + SO_2 + 2H_2O \rightarrow 2HBr\ (2HCl) + H_2SO_4$$

Bromine can be separated by chlorinating the mixed acids in a steaming-out tower.

In 1973 estimated total bromine production in the United States was 441 million pounds per year.[7] About 70% was used to make ethylene dibromide, a component of antiknock fluid for gasoline. However, if future pollution controls require the lead alkyls and subsequently ethylene dibromide (the lead scavenger) to be removed from gasoline, a large portion of the bromine produced could become available for other uses such as disinfection.

Iodine Production

The major world source of iodine is the Chilean nitrate deposits.[8] The nitrate ore, called "Caliche," contains 0.05%-0.3% iodine. Iodine in the form of calcium iodate is dissolved with sodium nitrate when Caliche ore is leached with hot water for the recovery of the latter substance. The leach solutions are filtered and cooled to precipitate most of the nitrate salts. The iodates are allowed to build up in the leach solutions, which are recycled until a reasonable concentration is obtained, before withdrawing to treat with sodium bisulfite to precipitate the free iodine.

$$2NaIO_3 + 5NaHSO_3 \rightarrow I_2 + 3NaHSO_4 + 2Na_2SO_4 + H_2O$$

The amount of bisulfite added is critical since an excess will reduce the iodine to iodide. The free iodine is separated and purified by sublimation.

The major domestic source of crude iodine is from natural brines underlying much of lower Michigan. Brines are pumped from a subterranean sandstone basin at a depth of about 5000 ft. Michigan brines contain 30-50 ppm iodine as iodide salts, in addition to other salts that are a source of several commercial products. In the Dow process, hot brine is acidified with HCl and oxidized with chlorine to liberate the elemental iodine. The chlorinated brine is then passed over a packed desorption tower where a stream of air desorbs the iodine from the brine and carries it to the HI absorber tower. Hot brine is then forwarded to the bromine steaming-out process tower (see bromine production, p, 115). Water and sulfur dioxide are fed into the HI tower to reduce and absorb the iodine resulting in a HI/H_2SO_4 solution. The more concentrated HI solution is drawn off continuously and forwarded to a precipitator where the addition of gaseous chlorine precipitates elemental iodine from the solution. The precipitated iodine is melted under concentrated sulfuric acid and then removed in the liquid state at purity as high as 99.8%.

The major problem that limits iodine production is the low concentrations in which it is distributed in natural brines and ores and its relatively low abundance in the earth's crust. Most of the world's production has been as a by-product of other more abundant products. For this reason, it is impractical to use iodine as a disinfectant for sewage.

In 1970 Japan emerged as the world's largest producer of iodine, supplying about 60% of the 16 million pounds produced.[9]

ECONOMICS OF BROMINE AND IODINE

Historically bromine and iodine have been more expensive chemicals by weight than chlorine. As a general rule, bromine is about 3 times and iodine about 40 times the market price for chlorine. During 1970 to 1973 the market prices in cents per pound were: chlorine (3.75-7.6), bromine (16.75-18.0) and iodine (180-227). However, considering their activity and other beneficial properties, iodine and bromine compounds may be cost-competitive with chlorine in some disinfectant applications.

Relative energy requirements for producing the halogens is another important consideration affecting both their cost and end use. Both bromine and iodine are somewhat more energy-sensitive than chlorine, requiring between 2 and 3 times as much energy to produce as chlorine. By contrast, ozone requires between 4 and 16 times as much energy to produce as chlorine, depending on the generator size and oxygen source.

INTERHALOGEN COMPOUNDS

The interhalogens are compounds formed from two different halogens. They resemble the halogens themselves in their physical and chemical properties, except where differences in electronegativity are marked. This is shown clearly by the polar ICl compound having a boiling point nearly 40°C above that of bromine although it has the same molecular weight. Also, interhalogens have lower bond energies than do the halogens. Thus in most cases, interhalogens are more reactive. These and other unique properties of the interhalogens influence their function as germicides.

Bromine Chloride

Bromine chloride at equilibrium is a fuming dark-red liquid below 5°C; it exists in the solid state only at relatively low temperatures.[10] Precise, uniform quantities of liquid BrCl can be routinely withdrawn from storage vessels containing dip pipes for removing the liquid under its own pressure (30 psig, 25°C). Liquid BrCl can be vaporized and metered as a vapor in equipment similar to that used for chlorine.

Bromine chloride is much less corrosive than bromine and may be stored in steel containers; it is shipped under the classification of a corrosive liquid. Like bromine, liquid BrCl rapidly attacks the skin and other tissues producing irritation and burns, and even low concentrations of vapor are lacrimatory and irritate the respiratory tract. Ventilation should be sufficient to keep the concentration of BrCl in air within the generally accepted limit for bromine of 1 ppm by volume for an 8-hr working period.

Preparation of Bromine Chloride

Bromine chloride is prepared by adding an equivalent amount of chlorine (as gas or liquid) to bromine until the solution has increased in weight by 44.3%.

$$Br_2 + Cl_2 \rightleftharpoons 2\,BrCl$$

Temperature change during the reaction is quite small, within 1°C of the temperature obtained upon mixing and before any reaction occurs.

Bromine chloride may also be prepared by the reaction of bromine and chlorine in the gas phase[11] or in aqueous hydrochloric acid solution.[12] It is conveniently prepared in the laboratory by oxidizing a bromide salt in a solution containing hydrochloric acid.[13]

$$KBrO_3 + 2KBr + 6\,HCl \rightarrow 3BrCl + 3KCl + 3H_2O$$

Physical Properties of Bromine Chloride

Bromine chloride exists in equilibrium with bromine and chlorine in both gas and liquid phases.[14,15]

$$2BrCl \rightleftharpoons Br_2 + Cl_2$$

The equilibrium in the gas phase, and in solvents such as carbon tetrachloride, has been well established by spectrophotometric measurements.[16] There is little information on the equilibrium in the liquid state, but a recent study using chemical methods indicates significantly less dissociation (<20%) in the liquid phase.[17] The equilibrium constant for the vapor-phase dissociation of BrCl is close to 0.34, corresponding to a degree of dissociation of 40.3% at 25°C.[18] The dissociation at 500°C is 46.4%, showing that the equilibrium constant varies little with temperature. In carbon tetrachloride solution, the equilibrium constant was determined spectrophotometrically to be 0.381 ± 0.008, corresponding to a degree of dissociation of 43.2% at 25°C.[16]

Table 6-1 compares a number of the reported physical properties of bromine chloride with those of iodine bromide and iodine monochloride.

Table 6-1. Physical Constants of BrCl, ICl and IBr[1,2]

	BrCl	ICl	IBr
Molecular weight	115.37	162.38	206.83
Melting point, °C	-66	27.2	40-41
Boiling point, °C	5	100-102	119
Density, g/cc	2.34(20°C)	3.13(45°C)	4.42(0°C)
Heat of fusion, cal/g	17.6	11.4	12.4 (Subl)
Heat of vaporization, cal/g, b p	53.2	61.3	
Heat of formation, Kcal/mol, H_o° (g)	0.233	3.125	1.41
Heat capacity, cal/deg. mole, 298°K	8.38	8.49	8.72
Entropy, cal/deg. mole, 298°K	57.34	59.12	61.75
Dipole moment, D	0.56	0.65	–
Electrical conductivity, $ohm^{-1}\ cm^{-1}$	–	4.6×10^{-3}(45°)	3×10^{-4}(55°)
Degree of dissociation, vapor 25°C	21%	0.38%	8.8%

Precise determinations of the physical properties of bromine chloride are prevented by its dissociation. The values given for the melting point and boiling point of bromine chloride are in reality equilibrium temperatures for the phase changes. An important advantage of its freezing point (-65°C), lower than that of bromine (-7°C), is that heating and insulating is not required during the winter. Figure 6-1 compares the vapor pressure-temperature curves for bromine chloride, bromine, and chlorine. The vessel gauge pressure for liquid bromine chloride is 30 psig at 25°C.

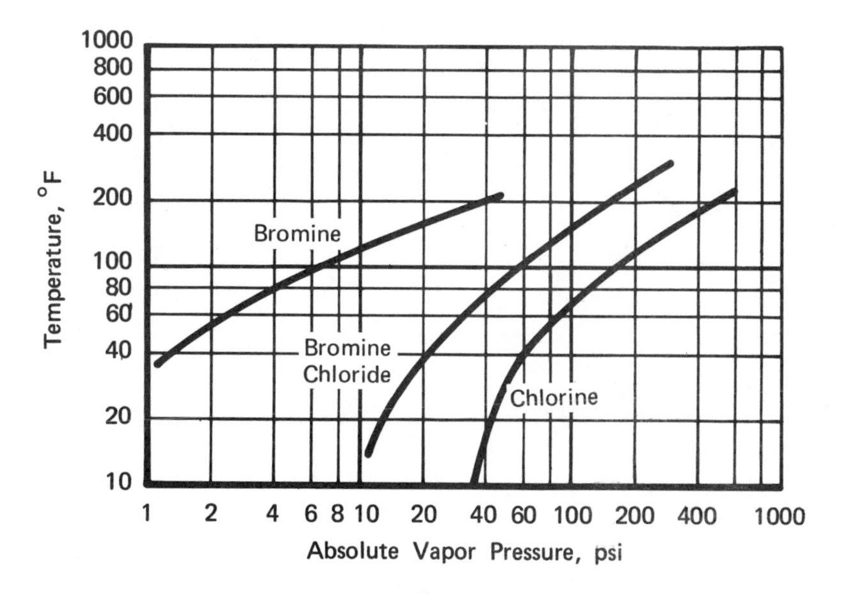

Figure 6-1. Vapor pressure-temperature curves for bromine, bromine chloride and chlorine.

Due to its polarity, BrCl shows greater solubility than bromine in polar solvents. In water, BrCl has a solubility of 8.5 g/100 g at 20°C, which is 2.5 times the solubility of bromine, 11 times that of chlorine and 3600 times that of ozone. Its solubility in water is increased markedly by adding chloride ion to form the complex chlorobromate ion, $BrCl_2^-$. At lower temperatures bromine chloride forms a yellow crystalline hydrate, $BrCl \cdot 7.34\ H_2O$, mp 18°C (1 atm).[19]

Chemical Reactions of Bromine Chloride

There is increasing concern about the chemical products formed during disinfection of wastewater effluents.[20,21] The halogen demand of wastewater must be satisfied before reliable disinfection residuals can be obtained. In other words various organic and inorganic reducing agents react with and destroy free halogen residuals before and during reaction with the microorganisms. Some of these competitive reactions in the wastewater disinfection are illustrated in Figure 6-2.

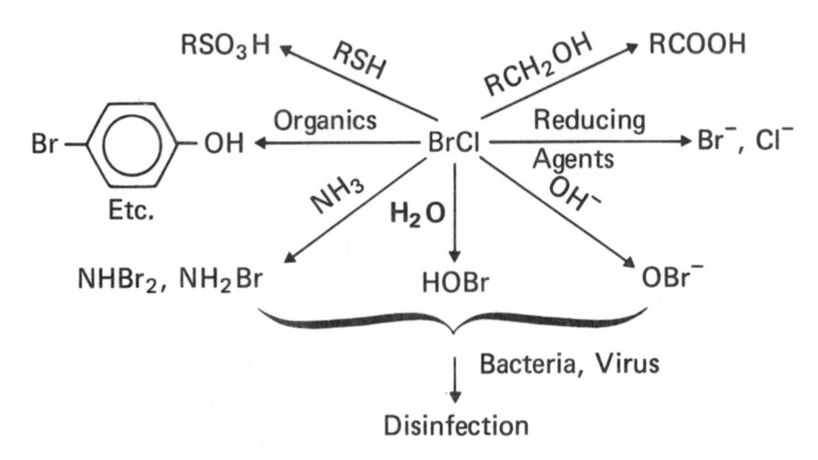

Figure 6-2. Competitive reactions of bromine chloride in wastewater disinfection.

These competitive reactions depend on the reactivity of the chemical species, concentrations, pH, temperature, and contact time. Under practical conditions, the quality of the effluent and the method of introducing the disinfecting agent are very important in determining reaction pathways, *i.e.*, the quality of recycled water at the chlorinator and "flash mixing"[22] can lead to improved reaction with microorganisms.

Although bromine chloride is about 40% dissociated into bromine and chlorine in most solvents, its high reactivity and fast equilibrium often produces bromine products resulting almost exclusively from BrCl. Figure 6-2 shows the major bromine chloride reactions. In all cases the disinfectant product formed contains bromine. With the exception of addition and substitution reactions with organic compounds, the major portion of the BrCl is ultimately reduced to inorganic bromides and chlorides. Table 6-2 shows relative amounts of bromide and non-bromide residuals in sewage treated with BrCl.

Table 6-2. Formation of Bromide Residues From Chlorobromination of Midland, Michigan Secondary Sewage Effluent

BrCl Dosage (ppm)	Bromide Residual (ppm)	Nonbromide Residual (ppm)	Conversion to Bromide (%)
5	2.8	0.7	80
10	6.4	0.6	91
18	10.0	0.5	95

An Orion bromide-specific electrode registered over 90% of the residual bromine in the form of bromide ion after disinfection of this effluent.

Reaction of BrCl with either ammonia or water in which hydrogen chloride is the major by-product may be compared with similar reactions of bromine resulting in by-product hydrogen bromide. The latter represents considerable waste of bromine. Although BrCl is primarily a brominating agent competitive with bromine, its chemical reactivity suggests uses similar to those of chlorine, such as disinfection, oxidation and bleaching.

Reaction with Water

Bromine chloride appears to hydrolyze exclusively to hypobromous acid.

$$BrCl + H_2O \rightarrow HOBr + HCl$$

If any hydrobromic acid (HBr) were formed by hydrolysis of the dissociated bromine, it would be quickly oxidized by hypochlorous acid to hypobromous acid.

$$HBr + HOCl \rightarrow HOBr + HCl$$

The hydrolysis constant for BrCl in water is 2.94×10^{-5} at 0°C,[23] which is 40,000 times greater than the hydrolysis of bromine (0.70×10^{-9} at 0°C).[24] The hydrolysis constant for chlorine in water is 1.45×10^{-4} at 0°C.[25]

Since the hypohalous acids are much more active disinfectants than the hypohalite ions, the effect of pH on their ionization is very important. The lower ionization value for hypobromous acid (pK = 8.70) compared with that of hypochlorous acid (pK = 7.45)[26] contributes to the higher disinfectant activity of BrCl compared with chlorine.[27] For example, at pH 8.0, a typical aqueous solution disinfected with BrCl would yield 90% of its bromine as the active hypobromous acid (HOBr) species whereas, under similar conditions, chlorine would produce only 19% hypochlorous acid (see Table 6-3).

Table 6-3. Distribution of Halogen Species at pH 8.0 in Water at 0°C

Halogen Species	pK_A	% HOX	% OX^-	% NH_2X
HOCl	7.45	19	81	–
HOBr	8.7	90	10	–
NH_2Cl	~1.0	–	–	100
NH_2Br	~6.5	82	8	<10

Reaction with Ammonia

Because of the difference in activities, the equilibria among the various species of halogens shown in Figure 6-3 play an important role in the disinfection processes.[28] Both monobromamine and dibromamine, the major products formed in the reaction between bromine chloride and ammonia, are unstable in conventional waste treatment plant effluents. The half-life of bromamine residuals was less than 10 minutes duration in secondary sewage effluent and about 60 minutes in demand-free water. Comparing bromamine and chloramine activities, the effects of ammonia and high pH improve the bromamine performance whereas chloramine activity is greatly reduced. Table 6-3 shows the more active bromine species predominate at higher pH[29] in water without excess ammonia.

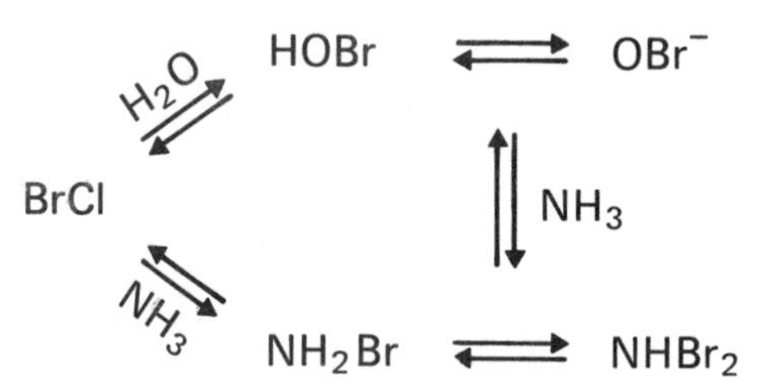

Figure 6-3. Active bromine species in aqueous ammonia solutions.

Ammonia's reaction with either chlorine or BrCl to form the corresponding halamine is rapid and essentially complete in most sewage effluents. Thus, the presence of ammonia is a major factor in the disinfectant properties and fish toxicity of halogen residuals.[4] In most sewage effluents where the ammonia level is relatively high (5-20 ppm) the predominant bromine species at pH 7-8 are monobromamine and dibromamine, about equally distributed. Both Morris[29] and Johnson[30] and their co-workers have contributed significantly to understanding the stability and properties of bromamines and chloramines.

Reaction with Organics

A number of organic chemicals can undergo halogenation during disinfection of wastewater.[20] An undesirable reaction is the combination of chlorine and phenol to produce chlorophenols. These compounds are a source of obnoxious tastes and fish toxicities even at very low concentrations. The brominated phenolic adducts produced in the chlorobromination of wastewater are more easily degraded and less obnoxious than their chlorinated analogs. There is evidence that brominated derivatives are more readily degraded photochemically and biologically.[31]

Some of the most important reactions of BrCl are electrophilic brominations of aromatic compounds. BrCl reactivity and its formation of by-product HCl are particularly useful. However, most aromatic compounds do not react in aqueous solutions unless the reaction involves an activated aromatic compound such as phenol.[32,33]

In free-radical reactions bromine is very selective but also very unreactive compared to chlorine. Bromine chloride undergoes free-radical reactions much more easily than bromine. A surprising aspect is that alkyl bromides are formed almost exclusively, with very little chloride formation. The reaction probably proceeds through a free-radical chain reaction mechanism as described for methane.[34]

$$BrCl \rightarrow Br^{\cdot} + Cl^{\cdot}$$

$$\left.\begin{array}{ll} Cl^{\cdot} + CH_4 \rightarrow CH_3^{\cdot} + HCl & -4 \\ Br^{\cdot} + CH_4 \overset{nil}{\rightarrow} CH_3^{\cdot} + HBr & +12 \\ CH_3^{\cdot} + BrCl \rightarrow CH_4Br + Cl^{\cdot} & -13 \end{array}\right\} \quad \Delta H, \text{kcal}^{34}$$

The positive heat of reaction for hydrogen abstraction by the bromine radical helps explain the comparatively low bromine activity as well as the tendency favoring HCl by-product for the BrCl reaction. Free-radical reactions in disinfection could be important if initiated by ultraviolet light or sunlight.

Chemical Reactivity of Bromine Chloride

Several studies have shown that BrCl reacts considerably faster than bromine or chlorine, probably due to the polarity of the compound which can be shown as $^{+}Br\text{-}Cl^{-}$. This polarity helps explain why very little chlorination is observed in electrophilic substitution reactions. Schulek and Burger[35] studied the reaction of BrCl with several phenolic compounds and compared the rates to those obtained with bromine. In each case BrCl was a more rapid brominating agent than molecular bromine. In another study comparing the reaction rates of BrCl to bromine with a variety of aromatic hydrocarbons,[36] the rapidity ratio, defined as K_{BrCl}/K_{Br_2}, ranged from a low of 4.31 for phenetole at 25°C, to 286 for p-cresol at 35°C. On the average, BrCl reacted more than 100 times faster than molecular bromine.

White and Robertson[37] established that BrCl added faster to olefinic compounds than any other halogen or interhalogen; *e.g.*, the addition of BrCl to *cis*-cinnamic acid was reported to be 400 times faster than the rate of bromine addition (see Table 6-7). BrCl also shows high reactivity in oxidation reactions. A comparison of oxidation rates of primary and secondary alcohols by aqueous BrCl and bromine was reported by Konishi and co-workers.[38]

Reaction with Inorganics and Metals

Metal ions such as iron and manganese in their reduced form can react with bromine chloride.

$$Fe^{++} + BrCl \rightarrow Fe^{+++} + Br^{-} + Cl^{-}$$

Hydrogen sulfide and nitrites occasionally found in wastewater contribute to higher halogen demand. Most of these reactions reduce halogens to their halide salts.

The reaction with metals, or corrosivity, of BrCl is not as great as that of bromine, but is closer to that of chlorine.[39] Dry BrCl is 1 to 2 orders of magnitude less corrosive than dry bromine to metals such as low-carbon steel, stainless steel, nickel, and monel (Table 6-4). Moist BrCl is not so strikingly superior to moist bromine, but it is still significantly less corrosive to these metals. Unlike chlorine, BrCl shows no evidence of stress-corrosion cracking of stainless steel with or without added moisture. For practical comparisons, moist BrCl is less corrosive to low-carbon steel than bromine is to stainless steel or Monel 400 under similar conditions of temperature and moisture. Like chlorine, BrCl can be stored and shipped in steel containers. Teflon,® Kynar and Viton plastics are preferred over PVC where liquid or concentrated vapors are contacted.

Table 6-4. Corrosion Rates (mil/yr)
(30 Days at 20°C)[a]

Metals	Dry		0.01% H_2O	
	Br_2	BrCl	Br_2	BrCl
Low-carbon steel	13	0.93	17	1.75
"SS" 316	1.7	0.14	8.8	2.2
Monel 400	0.46	0.017	1.8	0.41
Nickel 200	0.38	0.044	0.96	0.37

[a]Duplicate coupons were 50% immersed in the liquid in sealed bottles.

Disinfectant Properties of Bromine Chloride

Compared to chlorine, the results of competitive reactions of bromine chloride with water, ammonia, and various reducing agents (including organics) are more complementary to disinfection. Chlorine's reaction with ammonia to form chloramines greatly reduces its bactericidal and virucidal effectiveness. The biocidal activity of monochloramine has been found to be only 1/50th to 1/100th as great as that of free chlorine.[22,40] The ammonia found in typical secondary sewage (5-20 ppm) is one order of magnitude greater than the amount required to form monochloramine from normal chlorination dosages (5-10 ppm). Consequently, monochloramine is the major active chlorine species in chlorinated sewage plant effluents.

By contrast to chlorination, the reaction between BrCl and ammonia gives as the major product bromamines which have significantly different disinfectant properties than chloramines. Both the bactericidal[41] and virucidal[42] properties of bromamines have been shown to be superior to chloramines. The bactericidal activity of monobromamine is reported approximately the same as that of free bromine or hypobromous acid (HOBr).[43] Polio II virus with 10 ppm ammonia present was sterilized in less than five minutes using BrCl, whereas equivalent chlorine concentrations failed to kill all the virus within 60 minutes as shown in Figure 6-4.[42] However, at unusually high ammonia concentrations bromamines were reported by Krusé to be less effective virucides.[44]

In a test of sewage disinfection effectiveness total bacteria counts for bromamines were consistently lower than those for chloramines.[4] Kamlet reported that BrCl was more effective than either bromine or chlorine in killing *B. coli* (see Table 6-8).[27]

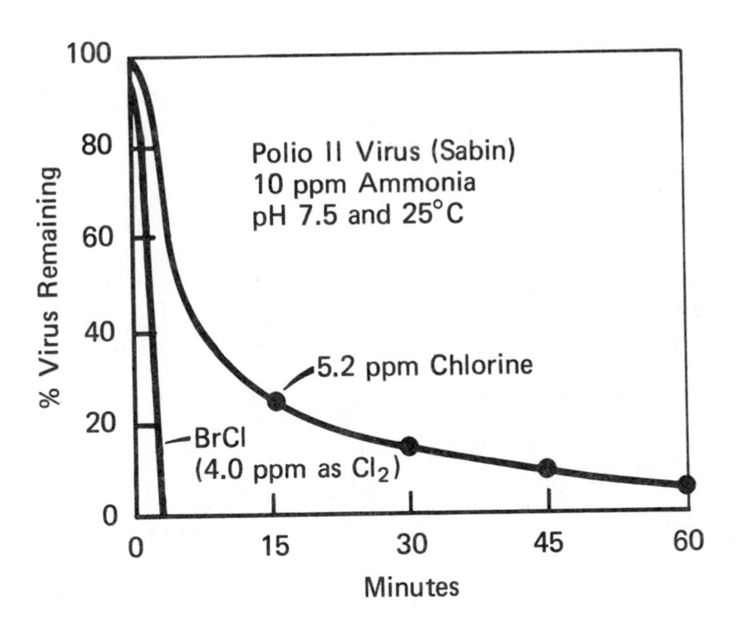

Figure 6-4. Comparative virucidal activity of BrCl vs Cl_2 in NH_3 solutions.

Cooling Water Treatment

Many of bromine chloride's properties are similar to bromine's—others are unique. This combination makes bromine chloride suitable for many water-treatment applications, including cooling towers and condensers.[45,46] It is an economical method of using bromine and produces the same results, since both yield highly reactive hypobromous acid as their major product.

The use of bromine to treat cooling water dates from the 1940s when various types of installations sought a more effective and economical alternative to chlorine. Among the early users were refineries and petrochemical plants whose experience demonstrated the efficacy of bromine in removing algal deposits as well as preventing new growth. Excellent control of all forms of algae, iron bacteria, protozoa, diatoms, and slime bacteria have been claimed for bromine.[47,48] Much lower concentrations of bromine than of chlorine resulted in maximal kill of chlorella pyrenoidosa.[49] Bromine and bromamine were found primarily algicidal, whereas chlorine and chloramines were mainly algistatic.

Fish Toxicity

Compared to chloramines, bromamines are much less stable in sewage or its receiving waters; they break down into harmless elements in less than one hour. Thus, bromine chloride is much less hazardous than chlorine to marine life. This is the basis for including BrCl with ozone and dechlorination in a bioassay project in Wyoming and Grandville, Michigan, supported by the Environmental Protection Agency (Research Grant No. 17060 H.J.B.), to evaluate safer alternatives to sewage chlorination.

Chloramines at concentrations below 0.1 ppm have produced fish kills in laboratory studies[4,50] and in stream trials with caged specimens.[51] In the latter, fish kills were observed at distances up to 0.8 mile below three of the four municipal outfalls tested. In two instances even after thorough mixing with the receiving water the chlorinated effluent was toxic up to 0.7 mile farther downstream. Under actual stream conditions the effect of chloramine toxicity may be indirect rather than direct. It has been observed that fish populations tend to avoid toxic materials, even at concentrations much less than necessary for toxic symptoms to occur. Therefore, the greatest effect of chlorinated-wastewater discharges probably is to render broad reaches of receiving water unavailable to many fish.[50] A side effect is the blockage of upstream migrations of certain fish during the spawning season.[52] Recent increases in chlorination along these waterways complicate the problem.

A parallel study was designed to determine the stability, biocidal activity, and acute fish toxicity of chlorine and bromine chloride in treated domestic sewage.[4] The appropriate halogen dose was added to one-liter aliquots of secondary (trickling filter) sewage effluent. After 15 minutes contact time (similar to that used in most sewage treatment plants) the halogen residual was measured; this liter of solution was then added to nine liters of Lake Huron water containing 10 fathead minnows. Fish survivals were determined over a 96-hour period. In the acute survival test all fish immersed in a diluted effluent treated with bromine chloride survived. By contrast all fish immersed in diluted effluent treated with equal-weight dosages (8 ppm) of chlorine were killed (Figure 6-5). After dosing the sewage, total bacteria or total coliforms were plated at 5-minute intervals over the 15-minute contact time. The results of these tests are also shown in Figure 6-5. Halogen residuals were also measured at identical times and after 24 hours. The 24-hr chloramine residuals showed an inverse effect on survival of fathead minnows. Only one of five chlorobrominated solutions contained a

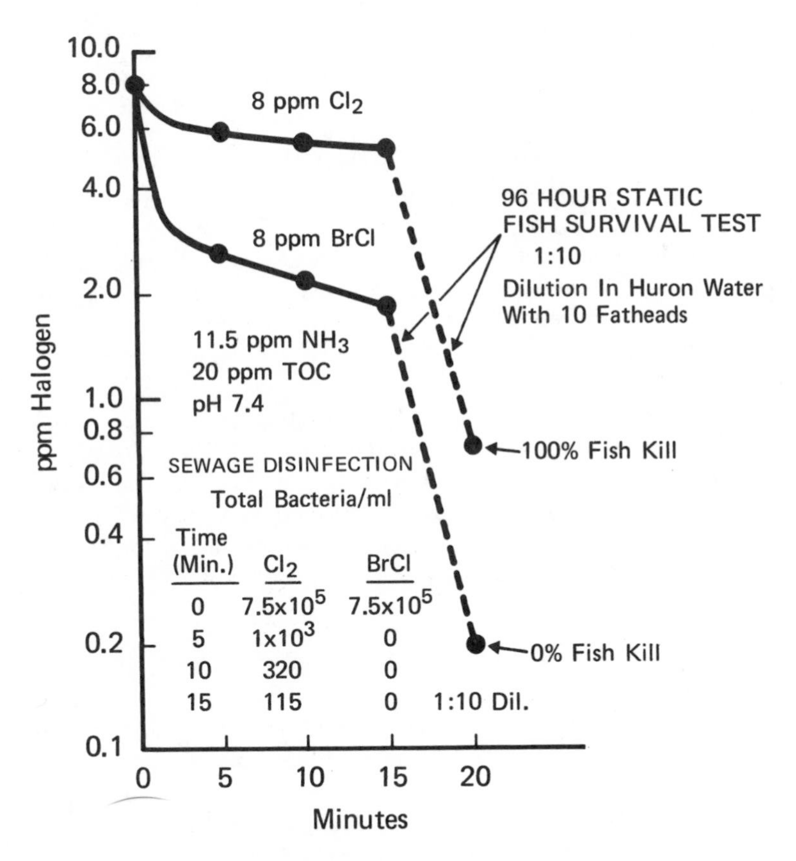

Figure 6-5. Disinfection of secondary sewage effluent.

residual after two hours, while chloramine residuals lasted as long as 48 hours. The results of these tests are summarized in Table 6-5.

The effects of lower concentrations of ammonia was also observed by studying chlorination and chlorobromination of nitrified secondary effluent.[4] The removal of most of the ammonia in sewage by biological oxidation or nitrification may eliminate the ecological hazards of chloramine residuals. In addition lower ammonia concentrations should decrease total oxygen demand as well as increase disinfection efficiency of chlorine.

Table 6-5. Effect of Halogen Residuals on Survival of Fathead Minnows

Dosage (ppm)	Initial Residual		24-hr Residual		% Fish Survival	
	Cl_2	BrCl	Cl_2	BrCl	Cl_2	BrCl
4	0.28	0.02	0.02	0	70	100
5	0.33	0.04	0	0	100	100
6	0.40	0.04	0.04	0	60	100
8	0.74	0.19	0.13	0	10	90
10	0.72	0.27	0.18	0	0	100
15	1.1	0.50	0.31	0	0	100

Using nitrification followed by high chlorine dosages for breakpoint chlorination, or nearly complete oxidation of ammonia to nitrogen, it is possible to form free chlorine residuals with great improvement in disinfection. At chlorine dosages of 25 ppm, the chloramine concentrations after breakpointing were too low to significantly affect biocidal activity or fish toxicity. While chlorine efficiency is increased BrCl performance is not improved significantly by nitrification. The high biocidal activity of bromamines makes it unnecessary to use high concentrations and breakpoint conditions to obtain active halogen residuals as is the case for chlorine. Breakpoint reaction with bromine chloride, similar to that for bromine,[53] is reached almost instantly in the presence of even a slight excess of bromine at pH 7-8.

$$2NH_3 + 3BrCl \rightarrow N_2 + 3HBr + 3HCl$$

By comparison, the theoretical amount of halogen required to breakpoint one part ammonia is 10.2 parts BrCl, 14.1 parts bromine and 6.26 parts chlorine.

According to Johannesson,[43] the very strong bactericidal properties of the bromamines can be exploited by using a small combined bromine residual since there is no need to reach breakpoint to achieve strong disinfectant properties. With chlorine it is necessary to add a quantity in excess of the breakpoint to obtain such sterilizing properties. This property of bromine chloride could make it more economical to use than chlorine, which requires the activity of free halogen residuals.

Table 6-6 compares the estimated costs of several alternative processes for disinfecting wastewater. Ozone and BrCl dosages of 5 ppm were assumed to be equivalent in performance to 8 ppm chlorine. Bromine chloride and dechlorination with sulfur dioxide appear to be the most economical alternatives not hazardous to marine life; both cost 0.80 cents per thousand gallons of sewage treated.

Table 6-6. Estimated Cost Comparisons (64 mgd Plant)

Disinfection Practice	Dosage (ppm)	Cost (incl capital)[a] ¢/M Gal
Fish Hazard		
Cl_2[b]	8	0.49
NaOCl[b]	8	0.90
Non-Hazardous		
BrCl	5	0.80
Cl_2/SO_2	8	0.80
$NaOCl/SO_2$	8	1.20
Ozone[b]	5	1.30
Cl_2 (break point)	25	1.30[c]

[a]Based on 1973 OPD prices and BrCl at 16¢/lb (Dow Chemical Co.).
[b]Yeo, *Water and Wastes Eng.* (January, 1972), p. 30.
[c]Does not include nitrification capital.

IODINE MONOCHLORIDE

Iodine monochloride is a heavy, reddish-brown, oily liquid with two solid forms, alpha (stable) m p 27°C and beta m p 14°C. It is soluble in water, alcohol and dilute hydrochloric acid. Iodine monochloride is stored in glass or noncorrosive metal containers and shipped under the classification of a corrosive liquid (white label). This interhalogen is used as an analytical reagent, iodinating agent, catalyst, and topical antiseptic.

Preparation of Iodine Monochloride

Continuous manufacture of ICl has been described[54] as well as laboratory methods.[55] Iodine monochloride can be conveniently prepared by direct combination of chlorine and iodine in or without a solvent.[56] Excess chlorine causes yellow crystals of iodine trichloride to form, but this reaction can be reversed by passing dry air through the solution to remove the excess chlorine.

$$I_2 + Cl_2 \rightleftharpoons 2ICl$$

$$ICl + Cl_2 \rightleftharpoons ICl_3$$

Another and common method for preparing ICl is oxidation of an iodide in hydrochloric acid solution:

$$2KI + KIO_3 + 6HCl \rightarrow 3ICl + 3KCl + 3H_2O$$

In addition to potassium iodate,[57] other oxidizing agents used in this method include potassium permanganate,[58] dichloramine-T,[59] 1,3-dichlorodimethylhydantoin,[60] trichlorocyanuric acid,[61] chlorine water,[62] and ozone.[63] Some of these combinations have been suggested or used for various disinfectant applications.

Chemical Properties of Iodine Monochloride

Iodine monochloride is a very reactive chemical which quickly attacks the eyes, nose and throat and causes bad skin burns. An effective counter measure for ICl burns is immediate washing with 20% HCl.[56] ICl also reacts quickly with cork, rubber and other organic substances. Although iodine monochloride reacts with many metals invariably yielding the metal chlorides, no reaction is observed with lead, zirconium, tantalum, chromium, molybdenum, tungsten or platinum under normal conditions.[64]

The reactions of iodine monochloride with organic compounds were summarized by Mellor.[2] Reactions with organic compounds can result in iodination, chlorination, decomposition, or the formation of a halogen addition compound. An example of iodination is the electrophilic substitution reaction between the amino acid tyrosine and iodine monochloride in acetic acid to give the di-iodo derivative.[65] In the reaction with 1,4-dioxane free-radical initiated chlorination prevails giving the 2,3-dichloro derivative.[66] Iodine monochloride readily adds to olefinic double bonds, the reaction being initiated by the electrophilic attack by the iodine end of the interhalogen.[67] Addition of iodine monochloride to unsaturated fatty acids was reported to be 33 times faster than addition of iodine bromide, but 40 times slower than the addition of bromine chloride (Table 6-7).[37]

In water, iodine monochloride rapidly hydrolyzes to hypoiodous and hydrochloric acids. In the absence of excess hydrochloric acid (or a

Table 6-7. Rate of Addition of Halogen to *cis*-Cinnamic Acid

Halogen	I_2	IBr	Br_2	ICl	BrCl
Relative addition rate	1	3×10^3	10^4	10^5	4×10^6

chloride), hypoiodous acid then disproportionates into iodic acid and iodine which is precipitated.

$$ICl + H_2O \rightarrow HOI + HCl$$

$$5HOI \rightarrow HIO_3 + 2I_2 + 2H_2O$$

Iodine monochloride forms a stable, yellow solution in aqueous hydrochloric acid or sodium chloride. This is due to the formation of the dichloroiodate iòn.[68]

$$ICl + Cl^- \rightleftharpoons ICl_2^-$$

The low dissociation of this polyhalide ion contributes to the stability and solubility of ICl in both aqueous and nonaqueous chloride solutions.[69] The equilibrium constant determined for the dissociation of the dichloroiodate (reverse of above equation) was 6.0×10^{-3}.[70] The ultraviolet spectrum of the yellow chloride solutions containing iodine monochloride has a pronounced absorption band at 227 nm associated with the dichloroiodate ion. Yellow solutions are also formed by solvents or solutes possessing an electron-donor group for complexing or solvation. On the other hand, noncomplexing solvents are red-brown with an absorption maximum near 470 nm for pure ICl gas (see Table 6-8).

Iodine monochloride has many uses in analytical chemistry. In Wijs' solution (ICl in glacial acetic acid), it is used to determine iodine values of fats and oils by a quantitative addition reaction. Reactive aromatics such as phenol, resorcinol, 2-naphthol, and 8-hydroxyquinoline may be directly titrated with iodine monochloride in 4*N* HCl using starch-iodide indicator.[70]

A solution of ICl in dilute hydrochloric acid cannot be titrated directly with sodium thiosulfate, but if sodium acetate and potassium cyanide are added first the reaction is quantitative.[71]

$$ICl + 2S_2O_3^= \rightarrow I^- + Cl^- + S_4O_6^=$$

Iodine monochloride will liberate free iodine from acidified potassium iodide solutions. The liberated iodine is then determined by the standard iodometric procedure of titration with a standard solution of sodium thiosulfate using starch as the indicator, or by the amperometric titration method.[72]

Disinfectant Properties of Iodine Monochloride

Iodine monochloride has been used as a topical antiseptic.[73] A solution similar to a "Tincture of Iodine" solution can be prepared using 2% ICl in aqueous sodium chloride. This solution is diluted about 1:1000 for use as an antiseptic. On dilution the iodine monochloride quickly hydrolyzes to produce free iodine and the result is essentially the same as with "Tincture of Iodine." A solution of 3% ICl in 15-20% sodium chloride is reported useful in controlling brucellosis (undulant fever).[74]

Iodine monochloride may be complexed with nonionic or anionic detergents to produce bactericides and fungicides useful in cleansing or sanitizing formulations.[75] Other ICl-polymer complexes similar to those described for iodine bromide and elemental iodine are claimed to be useful bactericides.[76-78] The polymer structure generally contributes greater stability, increased solubility and lower volatility; however, by reducing the free halogen concentration in use solutions, polymers reduce both the chemical and bactericidal activity. From a practical viewpoint, these complexes of the interhalogens are useful disinfectants which compromise lower bactericidal activity with increased stability.

IODINE BROMIDE

Iodine bromide is a crystalline, purple-black, heavy solid that melts at 36°C. It is soluble in water, alcohol, and dilute hydrobromic acid. Iodine bromide is best stored in either glass or noncorrosive metal containers.

Preparation of Iodine Bromide

Direct combination of iodine and bromine to form IBr is among the most rapid halogen-halogen reactions with a rate constant of 510 ℓ/mole sec.[79] The reaction can be carried out in or without a solvent.

$$I_2 + Br_2 \rightleftharpoons 2IBr$$

The preparation of iodine bromide in acetic acid has been used to make both Hanus and Waburn solutions to determine iodine numbers and total unsaturation, respectively.[80]

Another method for preparing IBr is by oxidation of an iodide in hydrobromic acid solution.

$$2KI + 2KIO_3 + 6HBr \rightarrow 3IBr + 3KCl + 3H_2O$$

The procedure used and the alternative oxidizing agents that may be substituted for iodate in this reaction are similar to those described for the preparation of iodine monochloride.

Chemical Properties of Iodine Bromide

The chemical reactions of iodine bromide (along with those of ICl and BrCl) have been reviewed by Burger.[81] Although less is known about the properties of IBr, its chemistry is similar to iodine monochloride.

The reaction of IBr with aromatic compounds generally tends to give iodination in polar solvents and bromination in non-polar solvents.[82] Reactions of IBr in aqueous solution at 25°C have been used to synthesize various iodine derivatives such as triiodophenol, triiodoaniline, triiodoresorcinol and diiodo-o-hydroxyquinoline from their parent aromatic compounds.[83] On the other hand, bromination of various aromatic compounds including phenol and aniline was accomplished using IBr in aqueous acetic acid.[84]

In electrophilic substitution reactions excess bromide ions weaken, and excess chloride ions strengthen, the dipolar character of IBr giving, respectively, less and more iodination than IBr alone.[85] Iodine bromide is complexed and its solubility is increased more effectively by bromide than by chloride ions.[70]

$$IBr_2^- \rightleftharpoons IBr + Br^- \ (K_D = 0.0027)$$

$$IBrCl^- \rightleftharpoons IBr + Cl^- \ (K_D = 0.023)$$

This effect is supported by the ten-fold difference in dissociation constants for the two polyhalide ions. Decrease in iodination by bromide ion might be due in part to the disproportionation of the bromoiodate ion:

$$3IBr_2^- \rightleftharpoons I_3^- + 2Br_3^-$$

and subsequently the lower reactivity of the triiodide ion.

Complexes of IBr with organic nitrogen compounds such as pyridene or acetamide have been shown to be $Py{\cdot}I^+IBr_2^-$ type ionization complexes.[86] These complexes are quite often more stable than the interhalogen alone, yet highly active in oxidation and halogenation reactions.

The degree of hydrolysis by iodine bromide has been discussed by Faull.[70] Primary hydrolysis occurs in the presence of hydrobromic acid,

$$IBr + H_2O \rightleftharpoons HOI + HBr$$

$$K_H = \text{ca. } 1.5 \times 10^{-7}$$

whereas in the absence of acid,

$$5IBr + 3H_2O \rightleftharpoons 2I_2 + HIO_3^- + 5HBr$$

$$K_H = \text{ca. } 10^{-21}$$

At equilibrium calculations show that in the latter case the concentration of HOI is about 1% of the HIO_3—a negligible amount. The proportion is about the same in the case of iodine monochloride.

Dissociation of IBr and ICl in aqueous solution was calculated by Faull.[70] In aqueous 1*N* hydrobromic acid the dissociation constant for IBr was 1.18×10^{-5}, compared with 1.62×10^{-10} for ICl in 2*N* hydrochloric acid. The low dissociation in these solutions is presumed due partly to formation of the very stable polyhalide ions. Degree of dissociation of IBr in carbon tetrachloride solution is 9.5%[87] which is close to the 8.9% measured in the vapor state at 25°C.[88]

The interhalogens and their polyhalide anions have distinct visible and ultraviolet spectra with maximum absorption as shown in Table 6-8.[89–91]

Table 6-8. Absorption Spectra of the Interhalogens

	CCl_4		Aq HCl		Aq HBr	
	$\lambda_{m\mu}$	e_{max}	$\lambda_{m\mu}$	e_{max}	$\lambda_{m\mu}$	e_{max}
BrCl	380	97	–	–	–	–
ICl	460	152	343	303	361	320
			227	53,000		
IBr	490	390	361	320	375	590
	266	200	232	38,700	253	50,400

The interhalogens are similar to the elemental halogens in displaying dichromism in various solvents. In going from a solution in carbon tetrachloride to alcohol a greater shift in maximum absorption to lower wavelengths is observed for the interhalogens, although it is not as apparent in the visible color range as for iodine solutions.[90] Although spectrophotometric analysis is useful to identify various halogen species, the interhalogens are most often determined quantitatively by iodometric analysis.

Disinfectant Properties of Iodine Bromide

Most reports of iodine bromide use as a disinfectant deal with complexed or stabilized forms. Although IBr may have some potential advantages over elemental iodine as a disinfectant, its hydrolysis and dissociation reactions in aqueous solutions are obvious limitations. Its properties are similar to those of iodine monochloride. Like bromine chloride, the germicidal activity of IBr should not be reduced by haloamine formation since bromamines are very active and iodoamines are not produced.

The Shelanski and Winicov patent claims that preparations of iodine bromide complexed with nonionic or anionic surface-active iodine carriers in place of iodine alone exhibit markedly enhanced germicidal activity and performance.[77] Some of the more stable nonionic iodine carriers are alkyl phenol ethylene oxide condensates used in sanitizing dairy equipment. An effective iodine carrier will dissolve and complex with at least 10% of iodine or iodine bromide, and dilute clearly in water without precipitation or crystallization of iodine. When use solutions are applied to control microorganisms the complex releases the IBr gradually, forming free iodine upon decomposition of the iodine bromide, whose decomposition takes place readily as the iodine bromide is released.

Other polymer complexes comprising a polyacrylic compound[76] and polyvinyl-2-oxazolidone[78] with IBr, ICl or I_2 are claimed to be useful bactericides. Strong base quaternary ammonium anion-exchange resins in the polyhalide form containing IBr_2^-, ICl_2^-, $BrCl_2^-$ ions have been suggested for use in a water treatment process for rapid control of microorganisms.[92] The polyhalide resins dissociate giving the free interhalogens in equilibrium concentrations depending on the loading or amount of halogen remaining complexed with the resin.

HALOGEN MIXTURES AS DISINFECTANTS

Bromine and Chlorine Mixtures

Mixtures of bromine and chlorine have been proposed for use as germicides superior to either halogen alone. A comparison of activities in killing *B. coli* was reported by Kamlet (Table 6-9).[27] Using 0.25 ppm total halogen dosage the percentage kill of *B. coli* after 5 minutes contact was 10, 80, and 100 for chlorine, bromine and bromine chloride, respectively.

Table 6-9. Comparison of Activities in Killing *B. coli*

	Dosage for 100% Kill, ppm		
Minutes	Cl_2	Br_2	BrCl
5	2.0	0.5	0.02
10	1.5	0.5	0.25
30	1.5	0.25	0.25

A French patent claims that drinking water sterilized with chlorine together with bromine or iodine, in the proportion of 10% to 50% of the chlorine used is most efficient and prevents formation of unpleasant chloramines.[93]

A continuous process for purifying water with a mixture of bromine and chlorine utilizes the electrochemical oxidation of bromide and chloride ions.[94] The halogens oxidize organic material in the water and are at the same time reduced to halides then recycled by electrolysis as free halogens. More recent work by Farkas-Himsley[95] confirmed the advantages of using mixtures of the two halogens. A proposed explanation for the increase in bactericidal activity of these mixtures is attack by bromine on sites other than those affected by chlorine. This synergistic effect could also hold for the equilibrium mixture bromine chloride in which, under some conditions, hypobromous and hypochlorous acids are present after hydrolysis.

Mixtures Containing Bromide or Iodide Salts

Another method for producing essentially the same results as those obtained using the interhalogen compound or mixtures of two halogens, is oxidation of bromide or iodide salts. This reaction can be used to prepare the interhalogen compound (see ICl preparation, p. 129) or the hypohalous acid:

$$HOCl + NaBr \rightarrow HOBr + HCl$$

At pH 7-9, the formation of hypobromite from bromide and hypochlorite is extremely rapid. However, if the hypochlorite is in excess, additional reactions may result in decomposition of the hypochlorite-hypobromite mixture.[96]

Houghton[97] reported that the rate of bacterial sterilization by chlorine in water containing ammonia is greatly accelerated in the presence of small amounts of bromides (Table 6-10). The enhanced sterilization effect is substantial with 1 ppm bromamines, but as little as 0.25 ppm bromamines may be significant under some conditions. If chloramines are formed prior to contact with bromide ion, the reaction and effect are diminished.

Even in the absence of ammonia, bromide at concentrations of 0.25 ppm clearly reduced the survival bacteria count. This enhanced disinfectant activity of bromides[98-101] and iodides[101] admixed to hypochlorite or chlorine has been recommended for commercial use. Improved germicidal action is also claimed for mixtures containing bromides[102,103] and iodides[104,105] with various organic chlorine releasing compounds including N-chloro-4-toluene sulfonamide, 1,3-dichloro-5,5-dimethyl-hydantoin, sodium N-chlorobenzenesulfonamide, chloromelamine, and chlorinated cyanurates. Some of these compositions have been used to disinfect

Table 6-10. Effect of Bromide on Chlorine Sterilization of Bacti. coli. (at 0.05 ppm Cl_2, pH 7.9, and 16°C)

NH_3 ppm	Br ppm	Bacti Number			Residual Cl_2, ppm
		Initial	10 min	30 min	
0.30	0	6800	2400	406	0.45
0.30	2.0	6800	20	3	0.42
0.60	0	1530	1130	105	0.35
0.60	1.0	1530	870	1	0.32

swimming pools.[104] Shere and coworkers[106] demonstrated that bromide improves the bactericidal properties of dichloroisocyanuric acid and hypochlorite against several bacteria.

Some organic halogen-releasing compounds have been used to directly release both·bromine and chlorine without using halide salts. Paterson has patented the composition of N-bromo-N-chloro-5,5-dimethyl-hydantoin as a swimming pool disinfectant.[107] Another bromine- and chlorine-releasing compound, N-bromo-N-chlorosulfamyl benzoic acid, is claimed to provide good biocidal properties with sufficient water solubility.[108] Several other similar compounds are claimed to be active disinfectants for water.[109]

The chemistry of the bromide effect on the germicidal activity of chlorine solutions is probably similar to that of the enhanced bleaching activity described by Lewin[110] and Bloch.[111] An acceleration of up to 80% of the bleaching of cotton and pulp by adding bromide was observed. Thus, hypobromite is observed to oxidize organic matter faster than hypochlorite in bleaching as well as disinfecting operations. In fact, bromine containing compositions are useful for their combined bleaching and disinfectant properties.

ENVIRONMENTAL EFFECTS OF BROMINE AND IODINE USE

The use of interhalogen compounds to disinfect wastewater might result in many unknown organic and inorganic halogen-containing products.[20,21] Although some wastes containing reactive organics may produce significant amounts of organic halogen compounds, in most cases the major products are the inorganic iodides and bromides (see Table 6-2). In view of possible recycling of wastewaters, we should consider the potential toxicity hazards of all halogen-containing products of disinfection.

Toxicity of Iodides

Iodine has long been recognized as a disinfectant for providing potable water. During World War II water purification tablets containing iodine were developed for the military. A tablet containing tetraglycine hydroperiodide was used to liberate 8 mg of elemental iodine in one quart of water.

Public Health Officials have been concerned with certain physiological aspects of iodine use in water supplies because of the extensive knowledge of the role played by iodine and iodides in the thyroid gland of animals and man. Several short-term studies have been conducted on the tolerance of small groups of people to iodinated water.[112,113] A six-month feeding study using iodine dosages of 12 mg to 19.2 mg per man per day showed no evidence of altered thyroid activity, anemia, cardiovascular damage, bone marrow depression, vision failure or loss of weight. Acute inhibition of thyroid hormone formation by large amounts of iodide is well known.[114] However, no evidence of hypersensitivity or other adverse effects were detected when iodine was used at 1 ppm to disinfect the water supply of an entire prison community for 9 months.[115] In 1962 the U.S. Public Health Service tentatively approved iodine for swimming pool disinfection, provided the maximum concentration of iodine in all forms does not exceed 5 ppm. When iodine was used as a disinfecting agent in three outdoor swimming pools at Stanford University, there was no evidence of inhalation, ingestion, or absorption of iodine by 30 male students who swam in the pools for one month.[116]

Toxicity of Bromides

Acute bromide intoxication is rare because amounts sufficient to elevate the serum bromide to a toxic level can hardly be ingested and retained without vomiting. Chronic bromide intoxication from the continued use of large doses (> 3-5 g daily) is known as bromism. The symptoms are skin rash, glandular excretions, gastrointestinal disturbances, and neurological disturbances. Bromide is rapidly absorbed from the intestinal tract and distributed in the body in a manner almost exactly similar to that for chloride. Also, bromide is excreted mainly by the kidney, in the same manner as chloride.

Brominated drinking water does not significantly increase the amount of bromine taken internally, and the amount is definitely well below the national or world standards of accepted safe levels of bromine. Assuming a daily water intake of undiluted sewage effluents treated with 5 ppm BrCl as shown in Table 6-11, the 3.5 mg/day bromine, added to a normal diet already containing 10 mg/day, is an approximate 30% increase in total

Table 6-11. Margin of Safety of Bromides in Recycled Water

BrCl Dose in Sewage Disinfection (ppm)	Environmental Dilution Factor	Maximum Residual Bromide (ppm)	Daily Intake of Bromide[a] (mg/day)	Margin of Safety[b]	1 mg/kg/day (WHO)[c]
10	None	7.00	17.50	180	4
5	None	3.50	8.75	350	8
5	5	0.70	1.75	1700	40
5	10	0.34	0.88	3400	80
5	20	0.17	0.44	6800	160

[a]Data represents man's complete daily water intake (2500 g) with maximum residual bromide.

[b]Safe assumed to be 3000 mg/day

[c]World Health Organization recommendations

bromide consumption. Using the safe therapeutic dose (STD) of 3000 mg/day, we still have a safety factor of 222 as opposed to a safety factor of 300 previously. According to Jolles[3] the bromine content of the body is between 1.7 and 30 mg/liter and blood bromide values reported by different observers are between 5 and 15 ppm. Considering the lowest value of 1.7 ppm bromide in the body, percent increase of bromide intake will be 0.0029%. The body is quite capable of excreting this increase without detrimental effects. It is concluded then that the amount of additional bromine in chlorobrominated water will not significantly increase the bromine content of the body nor cause bromism.[5]

One final concern might be the metabolism of organic bromine compounds. Comar and Bonner have stated "Organic bromine compounds are debrominated by the body. After the administration of brominated aliphatic hydrocarbons or dibromotyrosine there is a rise in inorganic bromide in the blood and the urine which indicates that organic bromine is not used but that it is actually debrominated by the body and used in its elemental form."[117]

REFERENCES

1. Greenwood, N. N., ed. *A Comprehensive Treatise on Inorganic and Theoretical Chemistry*, Suppl. II, Vol. 1 (London: Longmans Green and Co., 1956), p. 476.
2. Greenwood, N. N. *A Comprehensive Treatise on Inorganic and Theoretical Chemistry*, Suppl. I, Vol. 2 (London: Longmans Green and Co., 1956).
3. Jolles, Z. E. *Bromine and its Compounds* (London: Ernest Benn Ltd., 1966), Chapt. 4.
4. Mills, J. F. *Div. Water Air Waste Chem., Proc. Amer. Chem. Soc.* **13** (1), 65 (1973).
5. Mills, J. F. *Div. Water Air Waste Chem., Proc. Amer. Chem. Soc.* **13** (2), 137 (1973).
6. Ksenzenko, V. I. and D. S. Stasinevitch. *Technology of Bromine and Iodine* (Moscow: State Publishing House of Chemical-Scientific-Technical Literature, 1960).
7. *Chem. Eng. News* (February 25, 1974), pp. 11-12.
8. Mills, J. F. "Iodine," In *The Encyclopedia of the Chemical Elements*, C. A. Hampel, ed. (New York: Reinhold Book Corp., 1968), p. 290.
9. *Chemical Marketing Reporter, Oil Paint and Drug Reporter* (New York: Schell Publishing Co., 1973), pp. 3, 23, 24.
10. Emeleus, H. F. and J. S. Anderson. *Modern Aspects of Inorganic Chemistry*, 3rd ed. (New York: Van Nostrand Reinhold, 1960), p. 452.
11. Jost, W. *Z. Phys. Chem.* **B14**, 413-420 (1931).
12. Forbes, G. S. and R. M. Fuoss. *J. Amer. Chem. Soc.* **49**, 142-156 (1927).

13. Burger, K. and E. Schulek. *Ann. Univ. Sci. Budapest Rolando Eotvos Nominatae, Sect. Chem.* **2**, 133-138 (1960).
14. Barrott, S. and C. P. Stein. *Proc. Roy. Soc., Ser. A.* **122**, 582 (1929).
15. Vesper, H. G. and G. K. Rollefson. *J. Amer. Chem. Soc.*,**56**, 620 (1934).
16. Popov, A. I. and J. J. Mannion. *J. Amer. Chem. Soc.* **74**, 222 (1952).
17. Mills, J. F. and J. A. Schneider. *Ind. Eng. Chem. Prod. Res. Develop.* **12** (3), 160-165 (1973).
18. Cole, L. G. and G. W. Elverum. *J. Chem. Phys.* **20**, 1543-1551 (1952).
19. Glew, D. N. and D. A. Hames, private communication, 1970.
20. *The Effect of Chlorination on Selected Organic Chemicals.* (Washington, D. C.: The Manufacturing Chemists Association, 1972).
21. Ingols, R. S. and P. E. Gaffney. *Proc. 14th S.W.R. P.C.C.*, 175-181 (1965).
22. White, G. C. *Handbook of Chlorination.* (New York: Van Nostrand Reinhold Co., 1972), p. 139.
23. Kanyaev, N. P. and E. A. Shilov. *Trans. Inst. Chem. Tech. Ivanovo, (USSR)* (3), 69 (1940).
24. Liebhafsky, H. A. *J. Amer. Chem. Soc.* **56**, 1500 (1934).
25. Connick, R. E. *J. Amer. Chem. Soc.* **69**, 1509 (1947).
26. Farkas, L. and M. Lewin. *J. Amer. Chem. Soc.* **72**, 5766 (1950).
27. Kamlet, J. U.S. Patent 2,662,885 (1953).
28. Goodenough, R. D. *Swimming Pool Age* (April, 1964), pp. 25-30.
29. Galal-Gorchev, H. and J. C. Morris. *Inorg. Chem.* **4** (6), 899-905 (1965).
30. Johnson, J. D. and R. Overby. *J. San. Eng. Div.* 8425 (October, 1971).
31. Plonka, J. private communication, 1973.
32. Schulek, E. and K. Burger. *Talanta* **1**, 224 (1958).
33. Schulek, E. and K. Burger. *Ann. Univ. Sci. Budapest, Rolando Eotvos Nominatae, Sect. Chim.* **2**, 139 (1960).
34. Walling, C. *Free Radicals in Solution.* (New York: John Wiley and Sons, Inc., 1957), p. 379.
35. Schulek, E. and K. Burger. *Talanta* **1**, 147 (1958).
36. *Bromine Chloride Handbook.* Brochure Form No. 101-15-72 (Midland, Mich.: The Dow Chemical Company, 1972), p. 6.
37. White, E. P. and P. W. Robertson. *J. Chem. Soc.*, 1509 (1939).
38. Konishi, R., *et al. Anal. Chem.* **60**, 2198 (1968).
39. Mills, J. F. and B. D. Oakes. *Chem. Eng.*, 102-106 (1973).
40. Kohler, P. W. *J. Amer. Water Works Assoc.* **43**, 553 (1953).
41. Johannesson, J. K. *Nature* **181**, 1799-1800 (1958).
42. Schaffer, R. B. and J. F. Mills. *Proceedings of the National Symposium on Quality Standards for Natural Waters.* (Ann Arbor, Mich.: The University of Michigan Press, 1966), pp. 158-159.
43. Johannesson, J. K. *Amer. J. Publ. Health* **50**, 1731-1736 (1960).
44. Krusé, C. W. and Yu-Chih Hsu. U.S. Army Contract DA-49-163-MD-2314, Johns Hopkins University (August 1967).
45. Kott, Y. and J. Edlis. "Effect of Halogens on Algae-I. Chlorella Sorokiniana," *Water Res.* **3**, 251-256 (1969).
46. Kott, Y. "Effect of Halogens on Algae-III. Field Experiment," *Water Res.* **3**, 265-271 (1969).

47. Albright, J. C. *Petrol. Process.* **3**, 421 (1948).
48. The Dow Chemical Company, Inorganic Chemicals Department, Midland, Mich. Brochure Form No. 101-19-73.
49. Kott, Y., *et al.* *Appl. Microbiol.* **14** (1), 8-11 (1966).
50. Zillich, J. A. *J. Water Pollution Control Fed.* **44** (2), 212 (1971).
51. Basch, R. E., M. R. Newton, J. G. Truchan, and C. M. Fetterolf. *Water Pollution Control Research Series*, EPA, WQO Grant No. 18050 GZZ (1971).
52. Tsai, Chu-Fa. *Chesapeake Sci.* **9** (2), 83 (1968), and **11** (1), 34 (1970).
53. Brooke, M. *J. Amer. Water Works Assoc.* **43**, 847-848 (1951).
54. Dagra, N. V. *Neth. Appl.* 6510508 (1965).
55. *Org. Syn., Coll.* **II**, 197,344 (1943).
56. Cornog, J. and R. A. Karges. *Inorg. Syn.* **1**, 165 (1939).
57. Philbrick, F. A. *J. Chem. Soc.*, 2254-2260 (1930).
58. Lang, R. *Z. Anorg. Chem.* **122**, 332-348 (1922).
59. Bradfield, A. E., K. J. P. Orton and I. C. Roberts. *J. Chem. Soc.*, 782 (1928).
60. "HIO-DYNE" product bulletin, Nease Chemical Company, State College, Penn. U.S. Patent 3,136,716 (1964).
61. Ellis, J. G. and V. Dvorkovitz. U.S. Patent 2,815,311 (1957).
62. Fourneau, E. and E. Donard. *Bull. Sci. Pharmacol.* **27**, 629-634 (1920).
63. Marks, H. C., *et al.* U.S. Patent 2,580,809 (1952).
64. Gutmann, V. *Z. Anorg. Chem.* **264**, 169 (1951).
65. Block, P. and G. Powell. *J. Amer. Chem. Soc.* **65**, 1430 (1943).
66. Kucera, J. J. and D. C. Carpenter. *J. Amer. Chem. Soc.* **57**, 2346 (1935).
67. Ingold, C. K. and H. G. Smith. *J. Chem. Soc.* 2742 (1931).
68. Faull, J. H. and S. Baeckstrom. *J. Amer. Chem. Soc.* **54**, 620 (1932).
69. Buckles, R. E. and J. F. Mills. *J. Amer. Chem. Soc.* **76**, 4846 (1954).
70. Faull, J. H. *J. Amer. Chem. Soc.* **56**, 522 (1934).
71. Gengrinovich, A. I. *Farmatsiya* **10** (2), 23-28 (1947).
72. *Standard Methods*, 13th ed. (Washington, D.C.: American Public Health Association, 1971), pp. 110-116.
73. *The Merck Index*, 7th ed. (Rayway, N.J.: Merck & Co., Inc., 1960), p. 559.
74. Zharov, V. G. *Probl. Zoogigieny Vet. Sanit. Zhivotnovod. Fermakh, Mater. Plenuma*, 210-217 (1970).
75. Shelanski, M. V. and M. W. Winicov. U.S. Patent 2,863,798 (1958).
76. Shelanski, M. V. and M. W. Winicov. U.S. Patent 2,826,528 (1958).
77. Shelanski, M. V. and M. W. Winicov. U.S. Patent 2,868,686 (1959).
78. Werner, J. U.S. Patent 2,987,505 (1961).
79. Walton, P. R. and R. M. Noyes. *J. Amer. Chem. Soc.* **88**, 4324-4325 (1966).
80. Kartha, A. R. S. *J. Sci. Ind. Res.* **21B**, 555-556 (1962).
81. Burger, K. *Magy. Tud. Akod. Kem. Orzt. Kozlemeny* **21** (3), 313-338 (1964).
82. Jargiello, P. A. *Gannon Coll. Chem. J.*, 21-27 (1964).
83. Nazrullaev, S. N., *et al.* *Tr. Tashkentsk. Farmatsevt. Inst.* **3**, 406-410 (1962).

84. Schlessinger, G. C. *Gannon Coll. Chem. J.* **26** (2) (1965).
85. Schulek, E. and K. Burger. *Magy. Tud. Akod. Kem. Orzt. Kozlemeny* **11**, 315-318 (1959).
86. Fialkov, Y. A. *Izvest. Akad. Nauk. S.S.S.R.*, Otdel. Khim. Nauk 972 (1954).
87. Yost, D. M., T. F. Anderson and F. Skoog. *J. Amer. Chem. Soc.* **55**, 552-555 (1933).
88. McMorris, J. and D. M. Yost. *J. Amer. Chem. Soc.* **53**, 2625 (1931).
89. Gillam, A. E. and R. A. Morton. *Proc. Roy. Soc.* **A124**, 604-616 (1929).
90. Buckles, R. F. and J. F. Mills. *J. Amer. Chem. Soc.* **76**, 4845 (1954).
91. Gillam, A. E. *Trans. Faraday Soc.* **29**, 1132 (1933).
92. Mills, J. F. U.S. Patent 3,462,363 (1969).
93. Derreumaux, A. French Patent 2,171,890 (1973).
94. Guter, G. A. and L. M. Tint. U.S. Patent 3,582,485 (1971).
95. Farkas-Himsley, H. *Appl. Microbiol.* **12**, 1 (1964).
96. Lewin, M. and M. Avrahami. *J. Amer. Chem. Soc.* **77**, 4491 (1955).
97. Houghton, G. U. *J. Soc. Chem. Ind.* **65**, 324 (1946).
98. Diversey Corp., Chicago. British Patent 781,708 (1955).
99. Diversey Corp., Chicago. British Patent 781,703 (1955).
100. Diversey Corp., Chicago. U.S. Patent 2,815,311 (1958).
101. Marks, H. C. and E. B. Strandskov. U.S. Patent 2,443,429 (1948).
102. Diversey Corp., Chicago. British Patent 781,730 (1957).
103. McClain, H. K. and L. E. Meyer. U.S. Patent 3,583,922 (1971).
104. Kitter, V. U.S. Patent 3,136,716 (1964).
105. Zsoldos, F. J., Jr. U.S. Patent 3,161,588 (1964).
106. Shere, L., M. J. Kelly, and J. H. Richardson. *Appl. Microbiol.* **10**, 537 (1962).
107. Paterson, L. O. U.S. Patent 2,779,764 (1957).
108. Paterson, L. O. U.S. Patent 3,647,836 (1972).
109. Paterson, L. O. U.S. Patent 3,147,219 (1964).
110. Lewin, M. *Bull. Res. Council Israel* **2**, 101 (1952).
111. Bloch, R., *et al.* U.S. Patent 2,461,105 (1949); British Patents 596,192 (1948), 615,604 (1949).
112. Nelson, N.E.D., *et al. J. Clin. Invest.* **26**, 301 (1947).
113. Morgan, D. P. and R. J. Karpen. *U.S. Armed Forces Med. J.* **4**, 725 (1953).
114. Childs, D. S., Jr., *et al. J. Clin. Invest.* **29**, 726 (1950).
115. Freund, G., W. C. Thomas, Jr., E. D. Bird, R. N. Kinman, and A. P. Block. *J. Clin. Endocrinol. Metab.* **26** (6), 619-624 (1966).
116. Byrd, O. E., H. M. Malkin, G. B. Reed and H. W. Wilson. *U.S. Publ. Health Rep.* U.S. Department Health, Education, Welfare, **78** (5), 393-397 (1963).
117. Comar, C. L. and F. Bronner. *Mineral Metabolism, An Advanced Treatise,* Vol. II, Part B. (New York and London: Academic Press, 1964), pp. 266-268.

CHAPTER 7

THE COMPARATIVE MODE OF ACTION OF CHLORINE, BROMINE, AND IODINE ON f_2 BACTERIAL VIRUS

V. P. Olivieri, C. W. Krusé, Y. C. Hsu, A. C. Griffiths and K. Kawata

The Johns Hopkins University
School of Hygiene and Public Health
Department of Environmental Health
Baltimore, Maryland

INTRODUCTION

The virucidal activity of the halogens under a wide variety of environmental and experimental test conditions has been well documented. Reactions of the halogens with compounds of biological importance have also been reported. By contrast, little information has been available on the fundamental mechanism and reactions responsible for the inactivation of viruses. Little attention has been directed to correlate the many possible reactions of the halogens with the viral protein and nucleic acid to inhibit their biological functions.

The conspicuous discontinuity in the understanding of halogen disinfection can be explained, in part, by the very effectiveness of the halogens as disinfectants. To date the halogens, particularly chlorine, have been used almost exclusively to disinfect public water supplies, treated wastewater, and swimming pools. Halogen disinfection has been our primary line of defense against waterborne disease. Virtual absence of waterborne bacterial diseases and the rare waterborne outbreaks of infectious hepatitis testify to the effectiveness of halogens as disinfectants. As a result, there has been little impetus to evaluate and understand the fundamental mechanisms producing non-infectious microbial agents. Expanding urbanization and increasing use and reuse of water for domestic and recreational purposes have caused the quality of raw water for public use to deteriorate and made the safety of shellfish-growing waters tenuous.

Evidence is accumulating on the mode of action of the halogens. The explanation for inadequate poliovirus inactivation[1] with a high concentration commercial iodine preparation was, in part, the inhibitory effect of high concentration iodide which is formed as a product of reaction of iodine and organic substances.[2]

The notorious waterborne epidemic of infectious hepatitis in Delhi, India, was associated with heavy sewage contamination of the raw water source. Standard treatment employing coagulation, filtration and post-chlorination was adequate only against bacterial, not viral, pathogens. Later observations indicated that, while post-chlorination with combined halogens inactivated bacterial pathogens, these chemical species are quite unreliable as virucides compared to the free halogen.[3] Alternative disinfectants to halogens are currently gaining attention, particularly for wastewater disinfection. Comparative modes of action and the relative efficiency of the candidate disinfectant must be known if future Delhi-like epidemics are to be avoided.

The security afforded by the predictable behavior of bacteria to halogen disinfection can be adequately explained by the sulfhydryl hypothesis advanced by Knox *et al.*[4] The large number of enzymes in bacteria that depend upon the integrity of the SH group provide many possible sites for inactivation. However, several workers[2,5,6] have demonstrated a marked resistance of some viruses to sulfhydryl reagents which suggests that halogen inactivation of viruses may not involve the SH group.

This study attempts to further elucidate the possible modes of viral inactivation with the halogens through knowledge of the effect on the more fundamental components of the virus. The study concentrates on the free halogen species since these are the principal virucidal agents.

In 1963 the bacterial virus f_2 was chosen as a model virus for studies of the mechanism of inactivation with iodine because f_2 was chemically and physically similar to the enteroviruses. Subsequent experiments have shown that f_2 responds to chlorine and iodine in a manner similar to the enteric viruses and, based on the limited data available, also responds to chlorine in a manner similar to the infectious hepatitis agent. The bacterial virus f_2 can be readily prepared and purified to yield stock virus suspension with a total of approximately 10^{15} PFU (plaque-forming units). Input virus titers of about 10^{10} PFU/ml are necessary to provide adequate viral material for chemical and biological assay of viral components. Large numbers of samples can be assayed with the bacterial virus plaquing techniques, less fastidious than those for animal viruses. Literature on a variety of biochemical procedures and assay systems with direct application to studies on the mode of action of halogens on virus has expanded rapidly.

Use of f_2 virus as a model for animal virus disinfection studies requires some clarification; f_2 is *not* being proposed as an indicator for evaluating the enteric viral quality of water and wastewater systems. Sufficient epidemiological and laboratory evidence is not available to support this use of f_2. Rather, f_2 was used as a classical model. A simpler, technically more workable system, as similar to the more complex system as possible, is used to obtain basic information concerning fundamental parameters and reactions. The information gained in the simple system can then be used to limit and design key experiments in the more difficult system.

A comparison of the known physical and chemical characteristics of f_2, poliovirus, T even, ϕx 174, and the agent of infectious hepatitis is given in Table 7-1. As can be seen, f_2 and poliovirus are strikingly similar with the exception of the host organism. Both f_2 and poliovirus are small icosahedral virions without an envelope and contain a single strand of RNA. According to the virus classification suggested by Lwoff, Horne and Tournier (LHT), as of 1971 both poliovirus and f_2 belong to the type-genus (family) Napoviridae.[7]

Hsu and co-workers[2,8] first reported the use of f_2 as a model virus for disinfection studies with iodine. They showed that poliovirus 1 RNA like f_2 RNA was resistant to iodination and that inactivation of both f_2 and poliovirus 1 were inhibited by increasing iodide ion concentration. Dahling[9] compared two enteric viruses, poliovirus 1 and Coxackievirus A9; two DNA phages, T2 and T5; two RNA phages, f_2 and MS2; and *E. coli* ATCC 11229 under demand-free conditions with free chlorine at pH 6.0. The enteric viruses were found to be most resistant to free chlorine followed by the RNA phages, *E. coli* and the T phages. Shah and McCamish[10] compared the resistance of poliovirus 1 and the coliphages f_2 and T2 to 4 mg/l combined chloride. f_2 was shown to be more resistant to this form of chlorine than poliovirus 1 and T2 coliphage. Cramer and Kruse[11] compared the inactivation of poliovirus 3 (Leon) and f_2 with chlorine and iodine in buffered sewage. Both viruses were treated together in the same reaction flask thereby eliminating any inherent differences due to virus preparations and replicate systems. In sewage effluent at pH 6.0 and 10.0 with a 30 mg/l dosage of halogen under prereacted (halogen added to sewage, allowed to react, viruses added at 0 time) and dynamic (virus added to sewage, halogen added at 0 time) conditions, f_2 in each case was at least as, or more resistant to, chlorine and iodine than poliovirus 1. The f_2 appears to be more sensitive to free chlorine but more resistant to combined chlorine than poliovirus 1. The agent of infectious hepatitis was found by Stokes and Neefe to be inactivated by breakpoint chlorination (free chlorine) but not completely inactivated by combined chlorine.[12]

Table 7-1. Characteristics of Some Animal and Bacterial Viruses[13-18]

Criteria	Animal Virus		Bacterial Virus		
	Poliovirus	Infectious Hepatitis Agent	f_2	ϕx 174	T even
Nucleic acid	RNA		RNA	DNA	DNA
Capsid symmetry	Cubic		Cubic	Cubic	Binal
Envelope	None	None?	None	None	None
Capsid diameter in Å	270-300	<1000	200-250	250	Head - 650x950 Tail - 20x950
LHT[a] family	Napoviridae		Napoviridae	Microviridae	Urovirales[b]
Virion wt x 10^6 daltons	5.5-6.8		3.0	6.2	220
Nature of nucleic acid					
No. strands	1		1	1	2
Shape	Linear		Linear	Circular L	Linear
Molecular wt x 10^6 daltons	2.5-2.7		0.7-1.2	1.7	130
Nucleotides	6000		3300	5500	200,000 pairs
Nature of protein					
No. different polypeptides	4		2	4	8
Amino acid not found			Histidine		
pH Stability	3-10		3-10		4-10
Stability at 50°, 30 min	Cationic stabilized	Stable	Stable	Stable	Stable

Chlorine susceptibility					
Cl_2	++++	++++	++++	++++	++++
HOCl	++++	++++	++++	++++	++++
OCl^-	+++	+++	+++		+++
Combined chlorine	+	–	–		+
Biological properties					
Host	Primate cells	Man	Male *E. coli*	*E. coli*	*E. coli*
Adsorption	Cell surface		F pili-sides	Cell wall	Cell wall
Entrance into cell	Virion		RNA	DNA	DNA
Release	Rupture of surface vacuoles		Lysis	Unstable spheroplasts	Lysis

[a]Lwoff, Horne and Tournier classification system for viruses.

[b]Order

MATERIALS AND METHODS

Methods and procedures for the iodine and bromine experiments were previously teported by Hsu *et al.*[2,8] and Griffiths.[19]

All chlorine experiments were performed in organic-free, distilled water prepared according to the method of Israel[20] containing 0.002M phosphate. The preparation and determination of chlorine and the reaction system employed were described earlier by Olivieri *et al.*[21] f_2 bacterial virus was prepared according to the methods described by Loeb and Zinder[22] with some modification. Tritium-labeled uridine was incorporated into f_2 virus using TPG medium described by Oeschger and Nathans[23] with a uridine-requiring male strain of *Escherichia coli* (K-12 C 3000-38) as host. Sulfur-35-labeled f_2 was prepared in a similar manner except the TPG medium was prepared at a minimum sulfur concentration. RNA was prepared and assayed according to the methods described by Hofschneider and Delius,[24] with some modifications. f_2 attachment experiments to bacteria and pili were performed according to the methods described by Brinton and Beer[25] with some modification, using *E. coli* K-13 ATCC No. 15766 and *E. coli* K-15 T^-.

RESULTS AND DISCUSSION

The effect of p-chloromercuribenzoate (PCMB), a sulfhydryl reagent, on f_2 and *E. coli* is shown in Figure 7-1. *E. coli* is inactivated rapidly when exposed to $10^{-3}M$ PCMB at pH 8.0 while f_2 is noticeably resistant. Table 7-2 shows the effect of $10^{-4}M$ PCMB on f_2 at three pH values, 5.5, 7.5 and 10.5. At each pH little inactivation is observed. The role of the sulfhydryl group in the inactivation of f_2 virus appears limited. Whether the SH groups are buried deep in the protein and are unavailable for reaction or the integrity of the SH group is not necessary for viral function, has not been determined.

The effect of 50 mg/l iodine at pH 7.0 and 27°C on intact f_2 virus, RNA prepared from iodine-treated f_2 virus (RNA′) and f_2 virus RNA is shown in Figure 7-2a. Both the RNA prepared from iodine-treated virus and the naked RNA are resistant to iodination despite the inactivation of the intact virus. Poliovirus RNA was found resistant to iodine at concentrations as high as 200 mg/l and contact time as long as 60 minutes.[8] Tobacco mosaic virus RNA was previously reported[26] to be resistant to iodine. These results clearly suggest that inactivation of virus by iodine may be attributed to reactions with the viral protein.

Figure 7-2b shows a similar series of experiments with bromine at 15 mg/l, pH 7.5 and 0°C. RNA prepared from bromine-treated virus (RNA′)

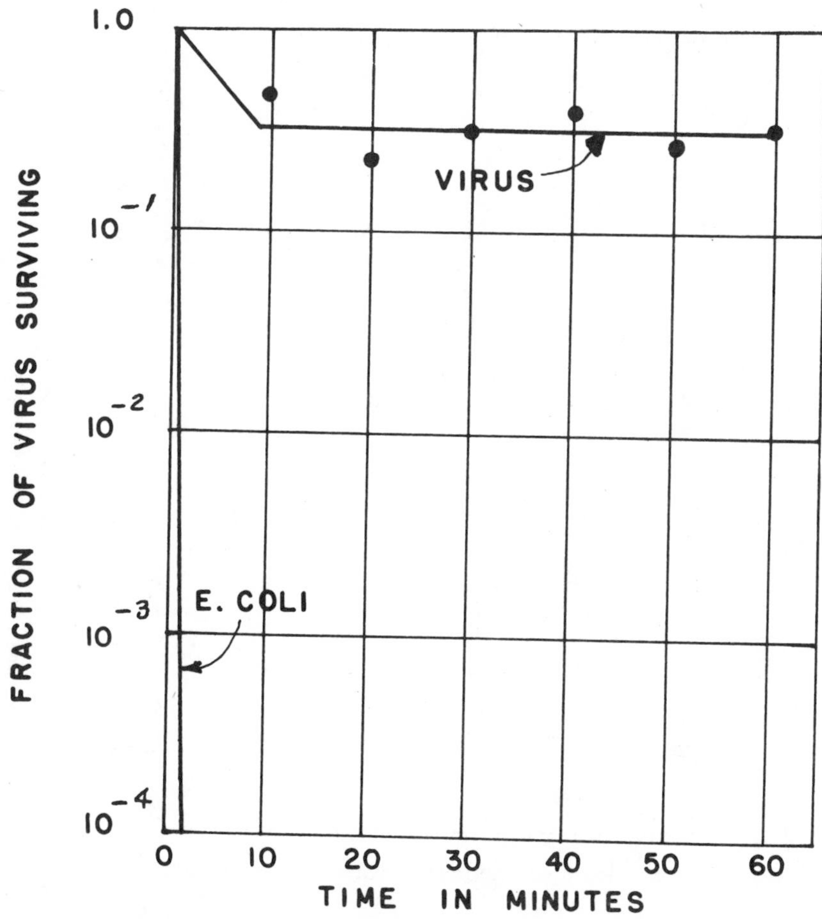

Figure 7-1. Effect of p-chloromercuribenzoate (sulfhydryl reacting agent) on f_2 virus and *E. coli* 10^{-3} *M* PCMB, pH 8.0 and room temperature.

Table 7-2. Effect of p-chloromercuribenzoate on f_2 Virus at 10°C

	Survival Fraction N/No		
	pH 5.5	pH 7.5	pH 10.5
1 minute	6.7×10^{-1}	5.8×10^{-1}	9.2×10^{-1}
10 minutes	6.7×10^{-1}	6.6×10^{-1}	7.5×10^{-1}
30 minutes	4.1×10^{-1}	5.8×10^{-1}	6.7×10^{-1}
PCMB concentration	$7.8 \times 10^{-5} M$	$8.5 \times 10^{-5} M$	$9.5 \times 10^{-5} M$

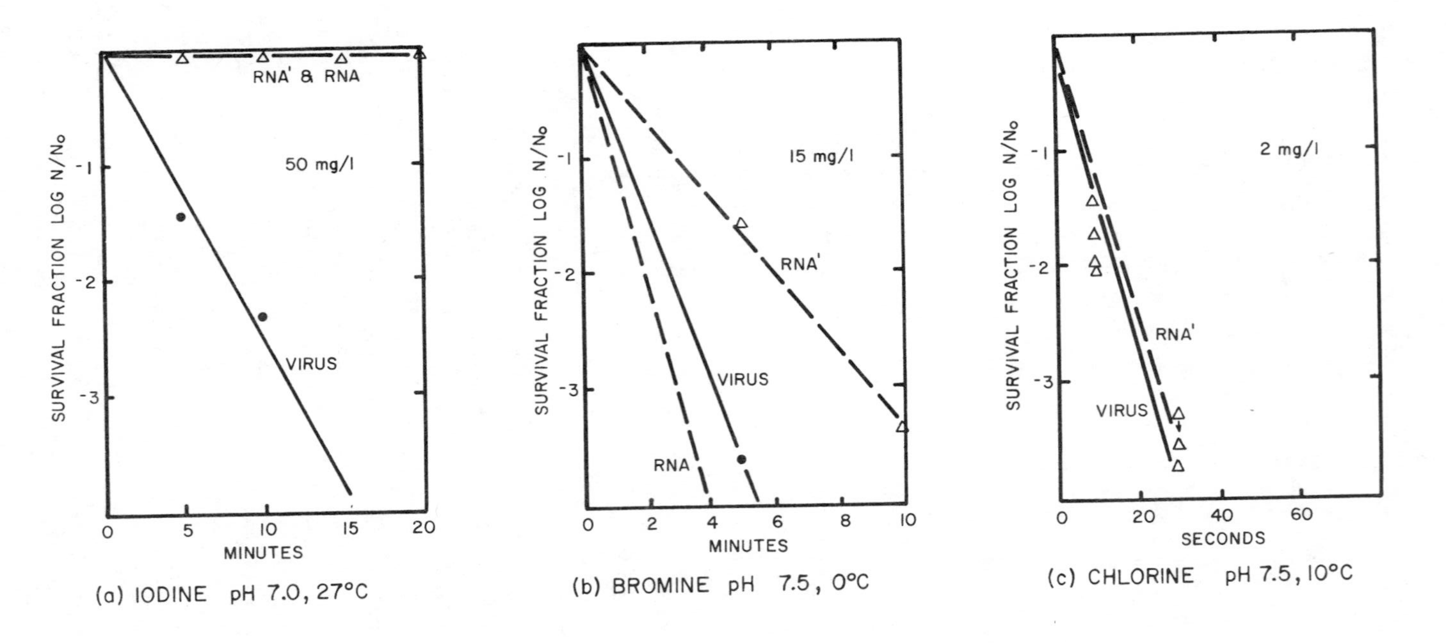

Figure 7-2. The effect of iodine, bromine and chlorine on f_2 virus, RNA prepared from halogen-treated f_2 virus (RNA$'$) and f_2 RNA. Halogen dosages, pH, temperature and contact time are as indicated.

lags significantly behind the inactivation of the intact virus. However, the naked f_2 virus RNA appears to be at least as sensitive to bromine as the intact virus. Unlike iodine, bromine does inactivate the naked f_2 viral RNA and the RNA prepared from bromine-treated intact f_2 virus. The primary site of inactivation, however, does not appear to be associated with the viral RNA under the experimental test conditions, and more likely involves the reaction of bromine with the protein moiety of the virus.

A similar series of experiments for chlorine is shown in Figure 7-2c. The chlorine dose was 2 mg/l at pH 7.5 and 10°C. The RNA extracted from chlorine-treated virus (RNA′) during the initial seconds of reaction was inactivated at a rate similar to that observed for the intact virus. The effect of 2 mg/l chlorine on intact f_2 and RNA prepared from the same chlorine-treated virus (RNA′) at pH 5.5, pH 7.5, and pH 10.5 during the initial seconds of reaction can be seen in Figure 7-3a. Similar rates of inactivation were observed for the intact virus and the RNA′ preparations at each pH providing strong evidence for the mechanism of chlorine action to be associated with the viral RNA. Figure 7-3b shows the effect of chlorine (2.0 mg/l dose) on f_2 viral RNA at pH 6.1, pH 7.7 and pH 10.6 at 10°C. The rate of inactivation of the naked f_2 RNA at pH 6.1 and 7.7 appears to be at least that observed for the intact virus and RNA extracted from chlorine-treated f_2. However, at pH 10.6 the naked viral RNA was markedly more susceptible to chlorine. More than 99.99% (4 logs) inactivation occurred in less than 60 seconds compared with less than 90% inactivation of the intact virus and RNA prepared from chlorine-treated virus. At pH 10.5 more than 99% of the chlorine present exists as hypochlorite ion (OCl^-). The noticeable resistance of f_2 virus at pH 10.5 does not appear to result from the difference in the chemical reactivity of HOCl and OCl^- toward viral RNA but, more likely, is due to the inability of chlorine as OCl^- to breech the viral coat protein.

The data presented for the chlorine, bromine, and iodine were obtained over a ten-year period by different investigators in our laboratory. The experiments were designed to elucidate modes of action and not intended to evaluate the relative efficacy of chlorine, bromine and iodine as virucides.

Li[27] observed that increased iodide ion and hydrogen ion concentrations each inhibited the iodination of the amino acid tyrosine. Figure 7-4 compares the effect of increasing amounts of potassium iodide (KI) on the iodination of the amino acid tyrosine and the inactivation of f_2 virus with 10 mg/l of iodine. The degree of inhibition of both the iodination of tyrosine and f_2 inactivation was observed to be a function of iodide ion concentration. Figure 7-5 shows a similar experiment conducted with poliovirus type 1 with 20 mg/l iodine dosage. The inactivation of poliovirus was also inhibited by increasing iodide ion concentration. The effect of

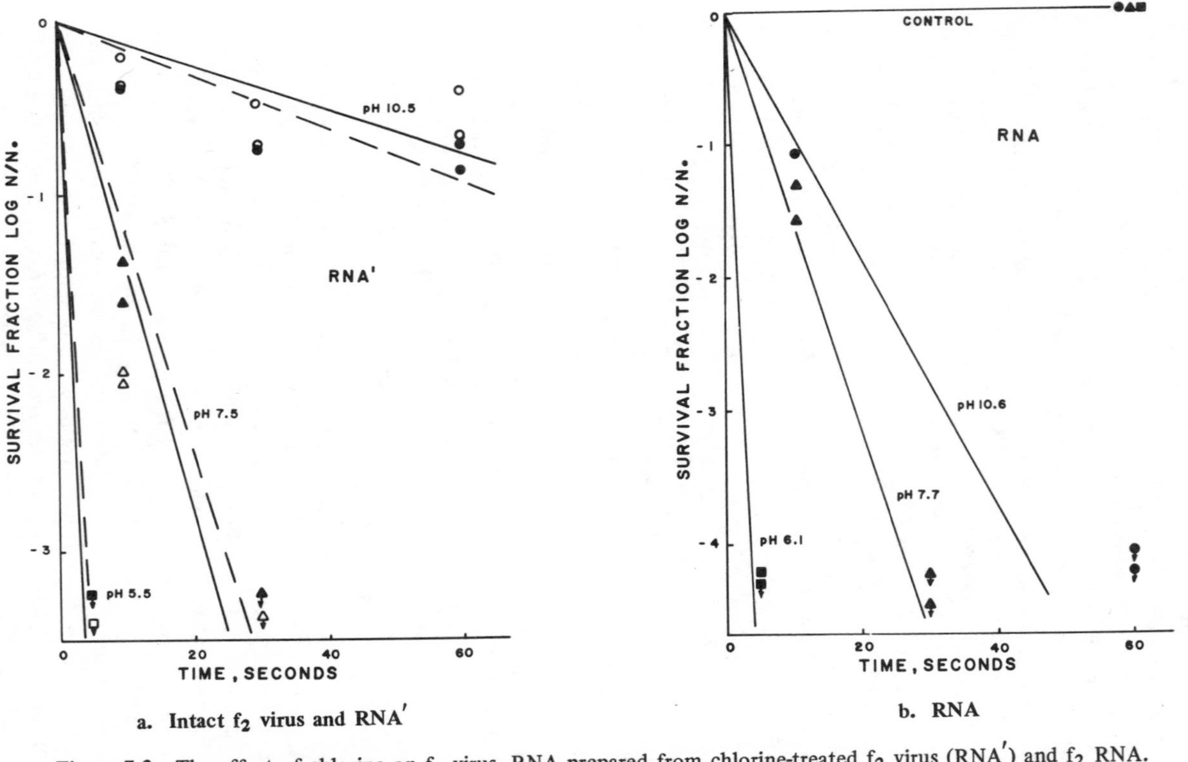

Figure 7-3. The effect of chlorine on f_2 virus, RNA prepared from chlorine-treated f_2 virus (RNA$'$) and f_2 RNA. 2 mg/l chlorine, 10°C, pH and contact time as indicated.

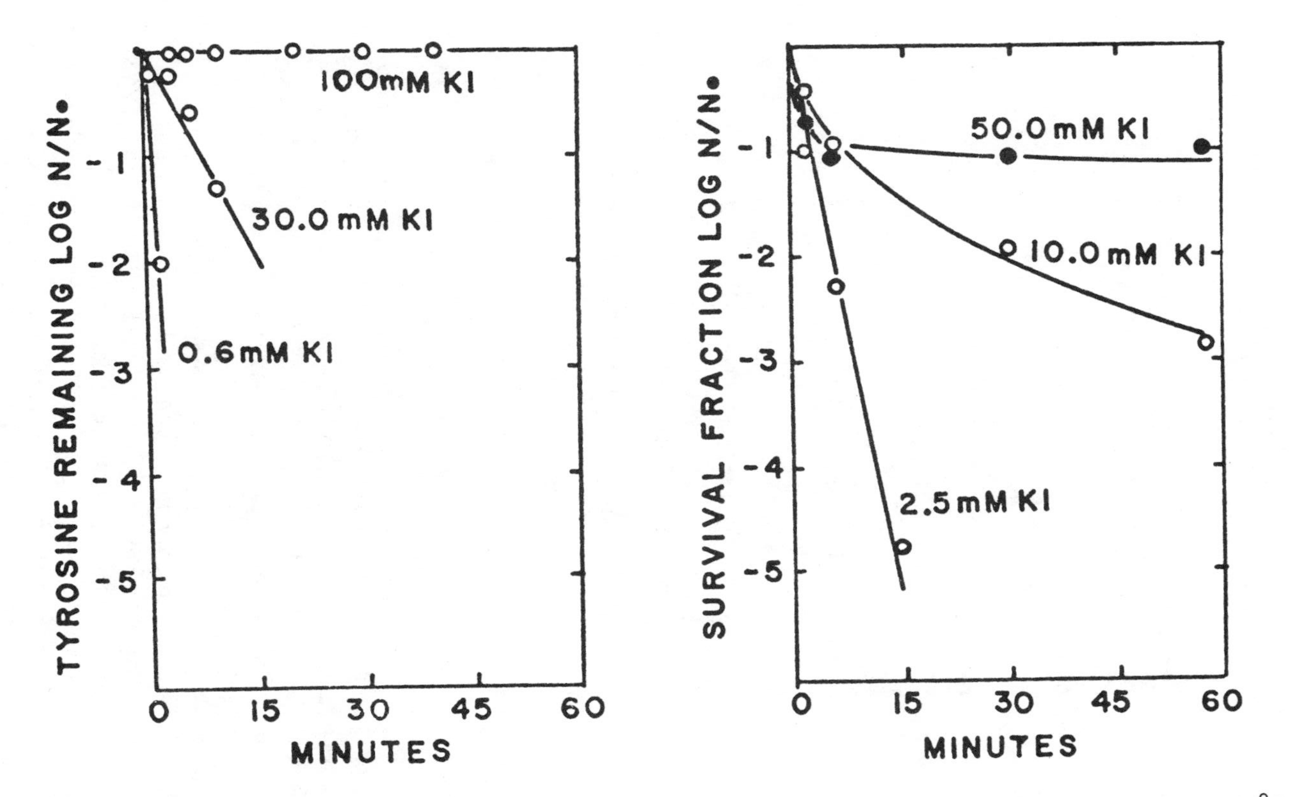

Figure 7-4. Effect of varying amounts of KI on the iodination of tyrosine and inactivation of virus with 10 mg/l iodine at pH 7.0 and 26°C.

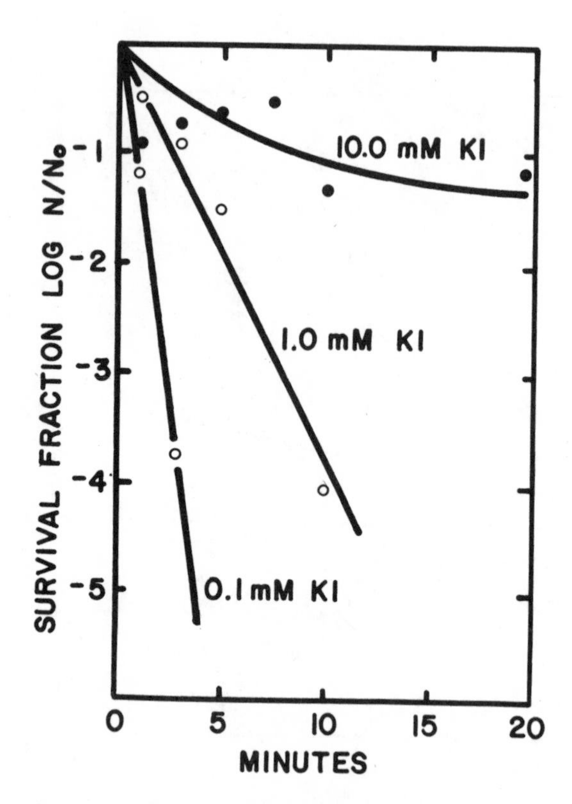

Figure 7-5. Effect of varying amounts of KI on the iodination of Poliovirus Type 1, with 20 mg/l iodine at 26°C and pH 7.0.

hydrogen ion concentration on the inactivation of f_2 with 10 mg/l iodine containing 0.048 m*M* KI at 5°C is shown in Figure 7-6. The inactivation of f_2 decreased as the hydrogen ion increased and is similar to the observations of Li for tyrosine. Thus, the same conditions of iodide ion concentration and pH which inhibit the iodination of tyrosine also inhibit the inactivation of f_2 virus.

The ability of f_2 virus, following chlorine disinfection, to initiate the viral productive cycle was studied to evaluate the role of the protein portion of f_2 virus. Sulfur-35-labeled f_2 was treated with chlorine at pH 5.5, 7.5 and 10.5 and samples were removed with time for virus assay and attachment to male and female strains of *E. coli*. Figure 7-7 compares the attachment to host bacteria, corrected for specificity, of chlorine-treated virus to viral inactivation. At each disinfection pH (10.5, 7.5, 5.5) the inhibition of the specific attachment of virus to bacteria host cells lags

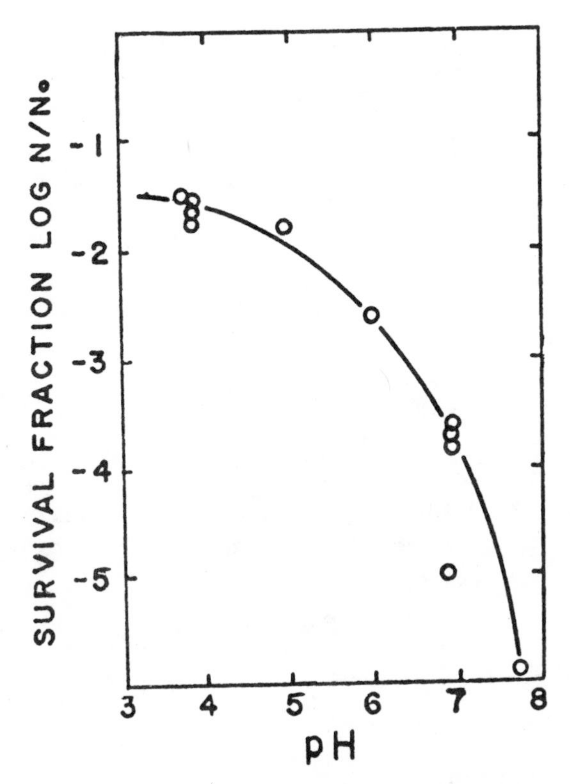

Figure 7-6. The effect of pH on the inactivation of f_2 virus with 10 mg/l iodine, 0.048 m*M* KI at 5°C.

noticeably behind the inactivation of the virus as shown in Figures 7-7a, b, and c respectively. Upon extended contact with chlorine, however, the virus adsorption becomes nonspecific. Figure 7-7d shows the attachment of chlorine treated f_2 for varying disinfection contact times and pH to non-host strain of *E. coli* (non-specific attachment). At pH 5.5 six logs of virus were inactivated in 10 seconds. The corresponding non-specific attachment of f_2 virus at 10 seconds was similar to untreated controls. Upon prolonged contact with chlorine, after the virus had been inactivated, the non-specific attachment markedly increased. At pH 7.5 six logs of virus were inactivated in 1.0 minute and again the corresponding non-specific attachment was similar to untreated f_2 controls. At pH 10.5 six logs of virus were inactivated in 5.0 minutes. The corresponding non-specific attachment was slightly higher than untreated f_2 controls and only 20-25% of the 10-minute value. Therefore it should be noted that the non-specific

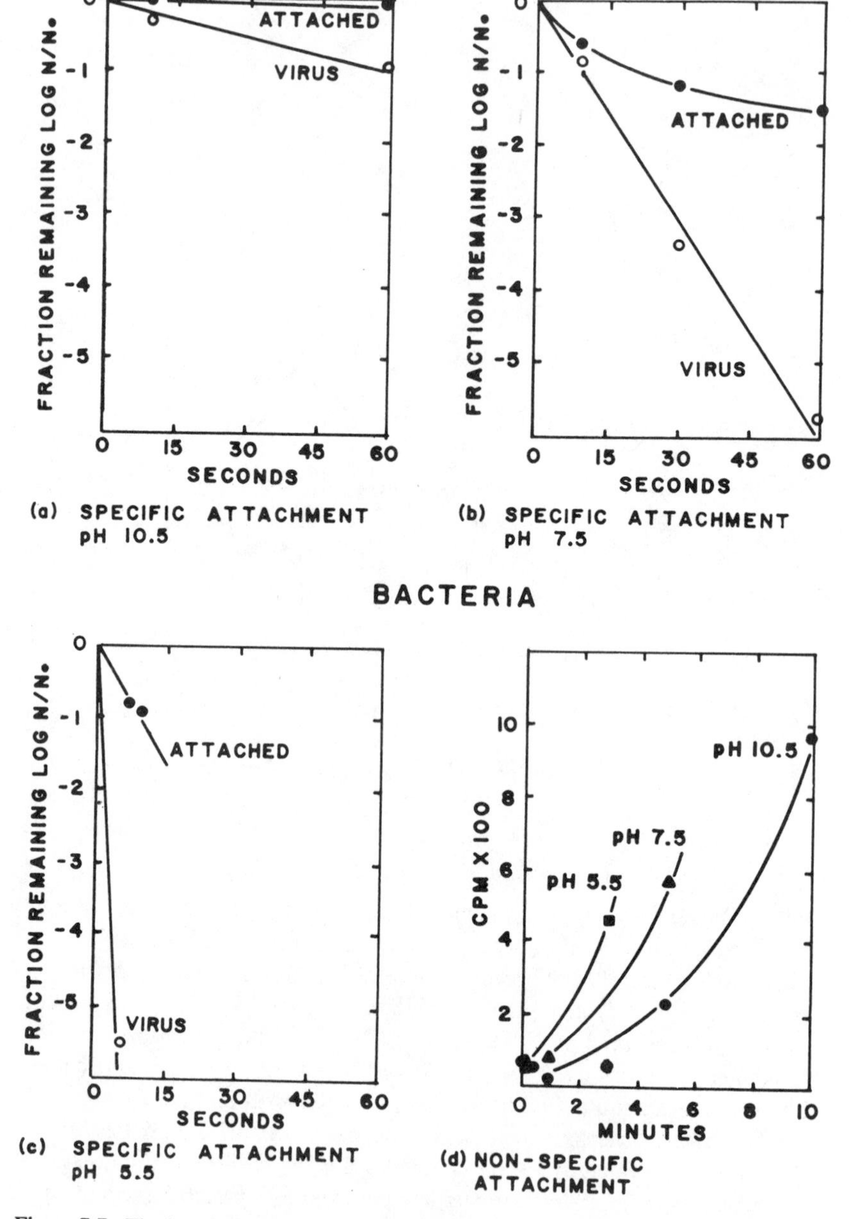

Figure 7-7. The inactivation of f_2 with chlorine and the adsorption of chlorine-treated f_2 to *E. coli* K-13 ATCC No. 15766 and *E. coli* K-15 T^-, with 3 mg/l chlorine at 10°C, contact time and pH as indicated.

attachment occurred long after the virus had already been inactivated. Preliminary results from modified Hershey-Chase blender experiments suggest that a portion of the RNA from the adsorbed virus enters the host bacteria.

In an effort to eliminate possible interference of the bacterial surface in the virus adsorption, similar attachment experiments were conducted with crude preparations of pili from *E. coli* K-13 (host) and *E. coli* K-15T$^-$ (non-host). Similar to the results obtained with whole bacteria, the inhibition of the attachment of virus to pili lags behind the inactivation of the virus at each pH as shown in Figures 7-8a, b, and c. Upon extended contact with chlorine, non-specific attachment is observed after viral inactivation has occurred (Figure 7-8d).

The protein apparently still retained a portion of its functional ability even after the virus was inactivated. The reaction of chlorine with the viral protein resulting in non-specific attachment to non-host bacteria occurs too slowly to explain the viral inactivation. Thus, the primary site of inactivation of virus with chlorine does not appear to be associated with the viral protein.

SUMMARY AND CONCLUSIONS

The halogens belong to a class of general protoplasmic poisons and are effective disinfectants for a wide spectrum of microorganisms. Their mode of action on viruses, however, appears to differ. Iodine functions through iodination of the amino acid tyrosine in the protein moiety of the virus with almost no effect on the nucleic acid, even at extremely high concentrations. Bromine, although capable of inactivating the nucleic acid does not appear able to penetrate the protein. The primary mechanism is more likely involved with the viral protein. Chlorine readily inactivates the RNA in intact virus in a manner similar to that observed for the whole virus with respect to contact time and pH. The protein is apparently still able to function in adsorption to host cells after the virus has been inactivated. The subsequent alteration of the protein by chlorine does not appear to explain the inactivation of f_2. The relative resistance of f_2 virus to hypochlorite ion appears to be due to an inability of this species of free chlorine to penetrate the coat protein to the RNA core.

ACKNOWLEDGMENT

This investigation was supported by the U.S. Army Medical Research and Development Command and U.S. Public Health Service General Research Support Grant.

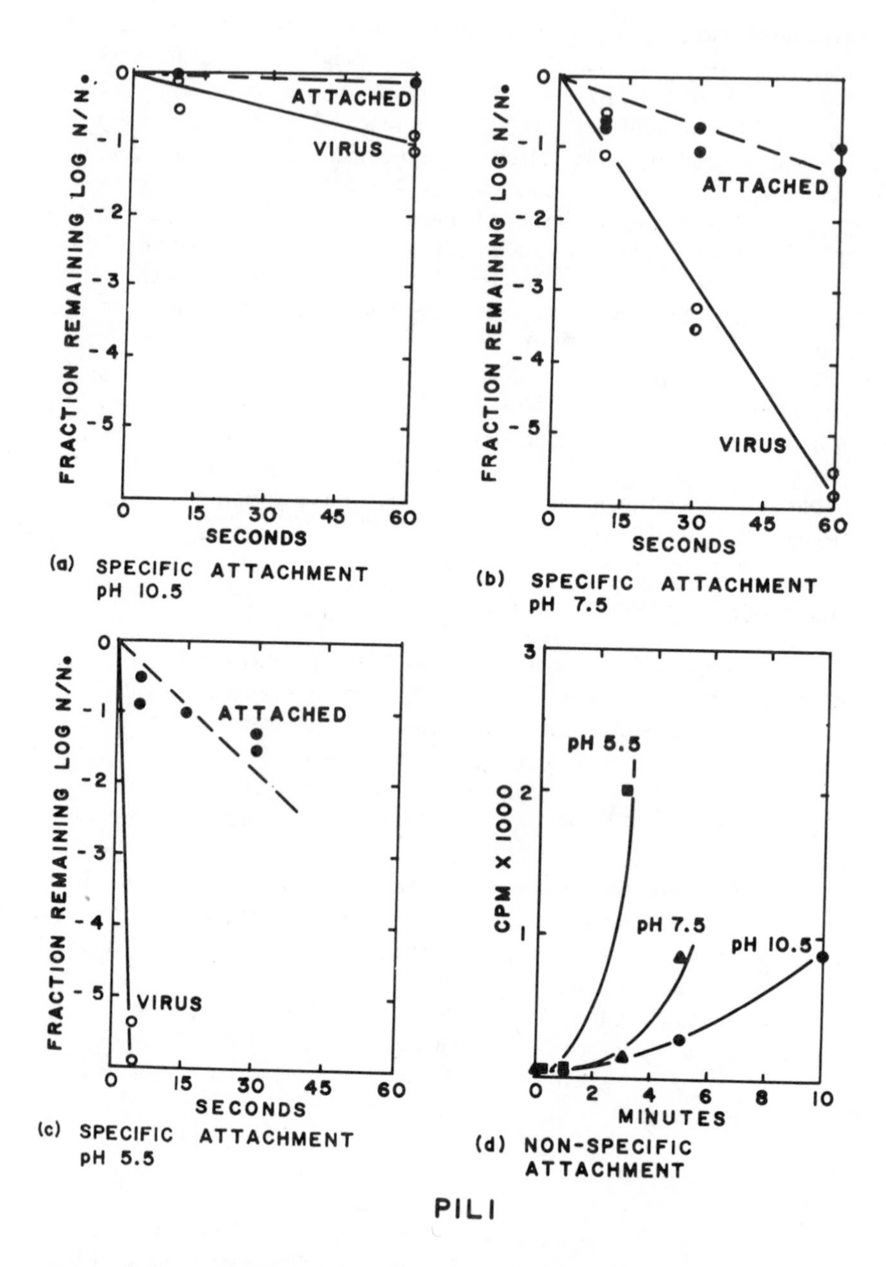

Figure 7-8. The inactivation of f_2 with chlorine and the adsorption of chlorine-treated f_2 to crude pili preparations from *E. coli* K-13 ATCC 15766 and *E. coli* K-15 T^-. 3 mg/l chlorine, 10°C, contact time and pH as indicated.

REFERENCES

1. Wallis, C., *et al.* "The Ineffectiveness of Organic Iodine (Wescodyne) as a Viral Disinfectant," *Amer. J. Hyg.* **78**, 325 (1963).
2. Hsu, Y., S. Nomura, and C. W. Krusé. "Some Bactericidal and Virucidal Properties of Iodine not Affecting Infectious RNA and DNA," *Amer. J. Epidemiol.* **82**, 3 (1966).
3. Krusé, C. W. "Mode of Action on Bacteria, Virus and Protozoa in Water Systems," Final Technical Rep., U.S. Army Med. R&D Comman Command, Contract No. DA-49-193-MD 2314 (September, 1969).
4. Knox, W. E., *et al.* "The Inhibition of Sulfhydryl Enzymes as the Basis of the Bactericidal Action of Chlorine," *J. Bacteriol.* **55**, 451 (1948).
5. Alison, A. C. "Observation on the Inactivation of Viruses by Sulfhydryl Reagent," *Virology* **17**, 176 (1962).
6. Choppin, P. W. and L. Philipson. "The Inactivation of Enterovirus Infectivity by the Sulfhydryl Reagent p-Chloromercuribenzoate," *J. Exper. Med.* **113**, 713 (1961).
7. Lwoff, A. and P. Tournier. "Remarks on the Classification of Viruses," In *Comparative Virology,* K. Maramorosch and E. Kurstak, Eds. (New York and London: Academic Press, 1971).
8. Hsu, Y. "Resistance of Infectious RNA and Transforming DNA to Iodine which Inactivates f_2 Phage and Cells," *Nature* **203**, 152 (1964).
9. Dahling, D. R., *et al.* "Destruction of Viruses and Bacteria in Water by Chlorine," presented at the Annual Meeting of the American Society for Microbiology, 1972.
10. Shah, P. and J. McCamish. "Relative Resistance of Poliovirus 1 and Coliphage f_2 and T2 in Water," *Appl. Microbiol.* **24**, 658 (1972).
11. Cramer, W. and C. W. Krusé. "Disinfection of f_2 Bacteriavirus and Poliovirus by Chlorine and Iodine," presented at the Annual Meeting of the American Society for Microbiology, 1973.
12. Neefe, J., *et al.* "Disinfection of Water Containing the Causative Agent of Infectious (Epidemic) Hepatitis," *J. Amer. Med. Assoc.* **128**, 1076 (1945).
13. Lauria, S. and J. Darnell. *General Virology.* (New York: John Wiley and Sons, Inc., 1967).
14. Maramorosch, K. and E. Kurstak. *Comparative Virology.* (New York and London: Academic Press, 1971).
15. Fraenkel-Conrat, H. *Molecular Basis of Virology*, ACS Monograph 164 (New York: Reinhold Book Corp., 1968).
16. Mathews, C. *Bacteriophage Biochemistry,* ACS Monograph 166 (New York: Van Nostrand Reinhold Co., 1971).
17. Horsfall, F and I. Tamm. *Viral and Rikettsial Infections of Man,* 4th ed. (Philadelphia, Pa.: J.B. Lippincott Co., 1965).
18. Davis, B., *et al. Microbiology* (New York: Harper & Row, 1968).
19. Griffiths, A. C. "The Activity of Bromine Against f_2 Bacteriophage and its Host *Escherichia coli* K-12 Under Various Aqueous Conditions," Master's Thesis, The Johns Hopkins University, Baltimore, Md., 1966.
20. Israel, B. M. "Hydrolysis of Organo-halogenating Agents," Ph.D. Thesis, University of Wisconsin, 1961.

21. Olivieri, V. P., T. K. Donovan and K. Kawata. "Inactivation of Virus in Sewage," *J. San. Eng. Div. ASCE* **97** (SA5), 661 (1971).
22. Loeb, T. and N. Zinder. "A Bacteriophage Containing RNA," *Proc. Natl. Acad. Sci., U.S.* **47**, 282 (1961).
23. Oeschger, M. P. and D. Nathans. "Differential Synthesis of Bacteriophage Specific Proteins in MS2 Infected *Escherichia coli* Treated with Actinomycin," *J. Mol. Biol.* **22**, 235 (1966).
24. Hofschneider, P. H. and H. Delius. "Assay of M12 Phage RNA Infectivity in Spheroplasts," In *Methods in Enzymology,* L. Grossman and K. Moldarc, Eds. (New York and London: Academic Press, 1968), Vol. XII, Part B.
25. Brinton, C. and H. Beer. "The Interaction of Male-Specific Bacteriophages with F. Pili," In *The Molecular Biology of Viruses*, J. S. Coulter and W. Paranchych, Eds. (New York: Academic Press, Inc., 1967).
26. Fraenkel-Conrat, H. "The Reaction of Tobacco Mosaic Virus with Iodine," *J. Biol. Chem.* **217**, 373 (1955).
27. Li, C. H. "Kinetics and Mechanism of 2,6-Diiodotyrosine Formation," *J. Amer. Chem. Soc.* **64**, 1147 (1942).

CHAPTER 8

BROMINE DISINFECTION OF WASTEWATER EFFLUENTS

F. W. Sollo, H. F. Mueller, T. E. Larson

Illinois State Water Survey
Urbana, Illinois

J. Donald Johnson

Department of Environmental Sciences
and Engineering
University of North Carolina
Chapel Hill, North Carolina

INTRODUCTION

To protect the sanitary quality of surface waters, which serve as sources of public water supply or are used for recreational purposes, disinfection of major waste discharges to these waters has been found necessary. Although there are other bactericidal agents that may be used for the disinfection of waste effluents, chlorine is almost invariably the method of choice. The effectiveness of chlorine as a disinfectant depends upon several factors, including pH, temperature, and the chemical state of the chlorine. Sewage effluents usually contain sufficient ammonia to convert the chlorine added to chloramines. The chloramines formed have only a fraction of the disinfection potential of free chlorine, especially at high pH where they form rapidly and hydrolyze poorly. These chloramines also have the disadvantage of being relatively stable and extremely toxic to fish and aquatic life when discharged to streams.[1-4]

Bromine, another of the halogens, has been compared to chlorine in a number of investigations.[5-7] Like chlorine, bromine reacts with ammonia, forming bromamines, but the bromamines are better disinfectants and approach free chlorine and free bromine in this respect.[8]

However, the bromamines are somewhat unstable, especially dibromamine.[5,9] Due to this low stability of dibromamine and also because of rapid demand reactions from reducing agents, a portion of the bromine added is lost almost immediately and contributes nothing to the desired disinfection. Thus, an evaluation of bromine as a disinfectant must include not only data relating residual bromine concentrations to the degree of disinfection, but also data regarding the quantities of bromine required to produce those concentrations or, more simply, both dosage and residual concentrations must be considered.

The primary aims of this study were to compare the efficiency of chlorine and bromine as disinfectants in wastewater, and to determine whether or not bromine might have advantages over chlorine under certain conditions. Since the degree of disinfection varies with temperature and pH as well as concentration of the disinfectant, time of contact, and other factors, it was necessary to compare these agents under a variety of conditions.

Chlorine and bromine were added to wastewater effluents, in varying amounts, and their effectiveness determined using the reduction in number of coliforms, fecal coliforms, and total bacteria as measures of disinfection. Chlorine and bromine have been compared at pH levels of 6.0, 7.5, and 9.0, and at temperatures of 10, 20 and 30°C.

In addition to the comparisons made with chlorine and bromine, a few preliminary tests were made on combinations of chlorine and bromine.

EXPERIMENTAL

Reagents

Standard Bromine Solutions

Standard bromine solutions were prepared by appropriate dilutions of a stock bromine solution with glass-distilled deionized water. The solutions were standardized by iodometric titration with sodium thiosulfate. The stock bromine solution was prepared by adding 2 ml of liquid bromine to 1 liter of 0.15*M* NaOH.

Standard Chlorine Solutions

Standard solutions of chlorine were prepared from dilutions of chlorox (5.25% NaOCl) with glass-distilled deionized water and standardized iodometrically against sodium thiosulfate.

N,N-Diethyl-p-Phenylenediamine Oxalate Indicator Solution (DPD)

One gram of DPD oxalate was dissolved in a volume of water (approximately 100 ml) containing 8 ml of sulfuric acid (1:3) and 0.2 g disodium ethylenediamine tetraacetate dihydrate (EDTA). This solution was diluted to one liter and stored in a dark bottle in the cold.

Phosphate Buffer, pH 6.4

Dibasic sodium phosphate, 24 g and 46 g of monobasic potassium phosphate were dissolved in 500 ml of water and added to a solution that contained 0.8 mg of EDTA. The final solution was brought up to one liter and 20 mg of mercuric chloride was added as a preservative.

Methods

Residual Chlorine or Bromine

The DPD colorimetric procedure[10] was used for total residual analyses. Total chlorine or total halogen was determined on 100-ml samples added to a beaker containing 5 ml of DPD indicator and 5 ml of phosphate buffer. Approximately 1 g of potassium iodide was added to the sample with thorough mixing. After five minutes the absorbance was measured at 552 nm. To determine total bromine alone, the same procedure was followed without the addition of potassium iodide. In the combination studies where both chlorine and bromine were added to the samples, total halogen was measured on a 100-ml sample by the above procedure, and bromine was separately determined on a parallel sample without the addition of potassium iodide. Total chlorine was then calculated from the difference in the two determinations.

Total Coliform Counts

The coliforms were measured by the MPN test in accordance with the "Standard Methods" procedure,[11] using serial dilutions of the effluent samples. Lactose broth was used as the presumptive test media and brilliant green bile lactose broth was used as the confirmatory media. Incubation was at 35°C. The results were reported as the most probable number (MPN).

In some studies, coliform analyses were made by the two-step millipore filter procedure, using M-Endo LES agar as the final medium. Lauryl tryptose broth was used as the enrichment medium for a pre-incubation period of two hours. Incubation of the millipore filters was at 35°C for a 24-hr period.

Fecal Coliform Counts

Fermentation tubes of EC medium were inoculated with a loopful of broth from the presumptive tubes that gave a positive test. The tubes were then incubated at 44.5°C for 24 hours in a water bath. Gas production within 24 hours or less was considered to be a positive reaction. The results for fecal coliforms were also reported as MPN.

Total Bacterial Counts

TGE agar plates were poured from the serial dilutions of the effluent samples and incubated at 35°C. Total counts were made after 24 hours.

Procedures

Disinfection-pH and Temperature

Grab samples of activated sludge effluent taken from the Urbana-Champaign treatment plant were used for the chlorination and bromination experiments. After collection, the samples were filtered through loosely packed cotton, adjusted to the desired temperature and pH, and thoroughly mixed before being divided into aliquots for treatment with varying amounts of chlorine and bromine. The quality of this effluent was variable but was generally a good secondary sewage. The BOD was 10-20 mg/l, COD 40-80 mg/l and ammonia 10-20 mg/l.

In the studies relating the effect of pH and temperature to disinfection, solutions of sodium hypochlorite or sodium hypobromite were added to the aliquots to give the calculated halogen dosages. In each experiment four levels of disinfectant were studied, and control samples were assayed in duplicated. To insure thorough mixing after addition of the halogens, the samples were gently stirred throughout the test period. After a 10-minute interval, at the midpoint of the contact period, an aliquot from each of the treated effluents was removed, and the chlorine or bromine residual determined by the DPD colorimetric procedure. At the end of the 20-minute contact period an aliquot taken from each of the treated effluents was dehalogenated with an excess of sodium thiosulfate, treated in a Waring blender to break up the particulate matter, and assayed for coliforms, fecal coliforms, and total bacterial numbers. Degree of disinfection was determined on the basis of the percent survival of the organisms in these tests.

Disinfection–Pre-chlorination

The effect of pre-chlorination upon the efficiency of disinfection with bromine was studied on effluent samples prepared as described above. In one series of tests, three samples of effluent were first treated with a low dosage of chlorine, which had been shown in earlier work to be essentially ineffective. After a 10-minute contact period, bromine and a molar equivalent of chlorine were added to two of the individual samples. The remaining sample served as a control. An additional sample of the effluent, not prechlorinated, was treated with an equivalent amount of bromine alone. Chlorine and bromine residuals were determined on aliquots at 10-minute intervals over a contact period of 30 minutes. Total plate count values were made on aliquots that had been dehalogenated with sodium thiosulfate, and from these values, the degree of disinfection was determined.

In a second series of tests, chlorine was added to samples of effluent for a contact period of 10 minutes, after which time bromine was added in varying concentrations. Samples having comparable amounts of bromine added, but no chlorine, served as controls in comparing the relative effectiveness of the combined agents. Four levels of bromine were studied. As in all tests, the samples were stirred throughout the contact period. Residuals and total plate count values were determined by the procedures previously described.

RESULTS

The results of studies showing the effect of pH and temperature upon the disinfection of activated sludge effluents by chlorine and bromine dosage and residual are summarized in Figures 8-1 and 8-2, respectively. The curves as presented cover only survivals of 1 to 100%, but can be extrapolated to lower survivals since the original data included many points with survival levels of 0.1% or less. The curves were developed from a number of determinations and located by the least squares method. Individual data points for each curve are not shown but correlation coefficients are given.

The data curves show that the usual trend of increasing effectiveness in disinfection with increasing temperature was generally observed. However, the two halogens did not follow the same pattern with respect to pH. Chlorine followed the pattern which was expected, that is, increasing effectiveness with decreasing pH. Bromine effectiveness, however, increased with increasing pH and was considerably more effective at pH 9.0 than 7.5 and 6.0.

In Table 8-1 a summary is given of the chlorine and bromine residuals required to obtain 90% and 99% reduction in the total coliforms by the MPN test. Comparing chlorine and bromine on the basis of resulduals required for equal disinfection, these data indicate that bromine was three times as effective as chlorine at pH 6,0, eight times as effective at pH 7.5, and 30 times as effective as chlorine at pH 9.0. These differences are due, in part at least, to the relative decay patterns of the chlorine and bromine residuals. When chlorine is added to the effluent, a small fraction of the

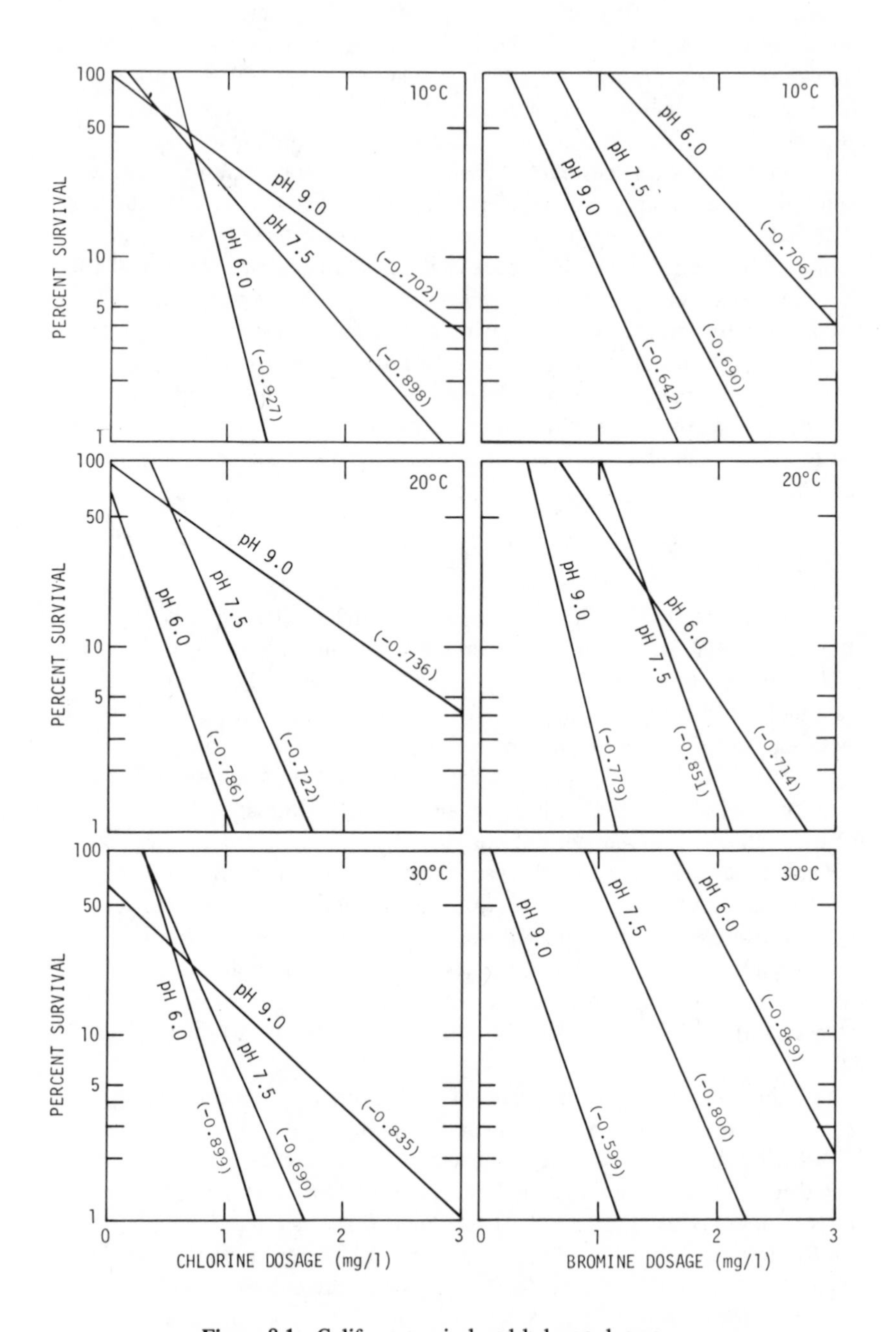

Figure 8-1. Coliform survival and halogen dosage.

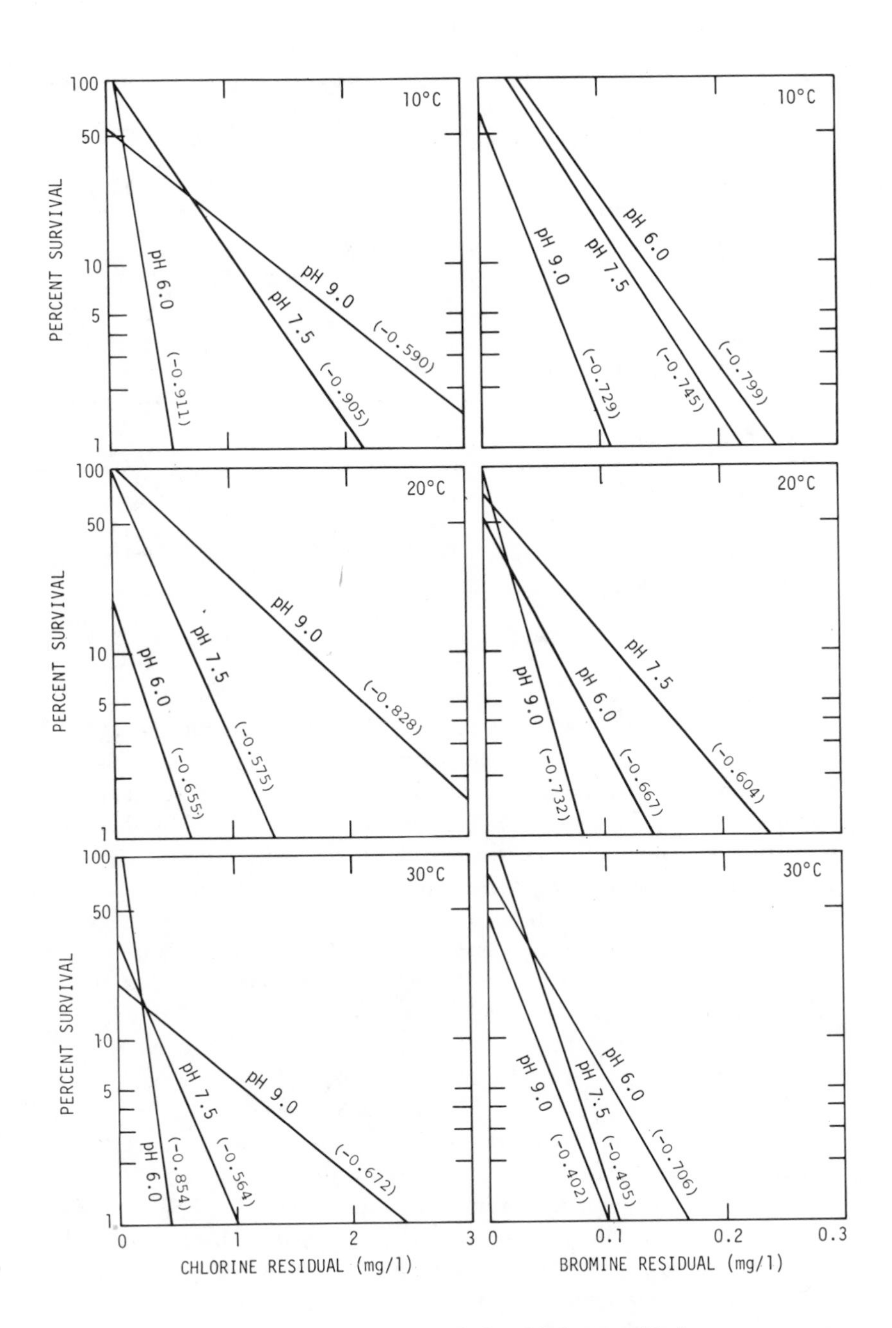

Figure 8-2. Coliform survival and halogen residual.

Table 8-1. Required Residuals[a] of Chlorine and Bromine for 90% and 99% Disinfection as Measured by Coliforms

	Chlorine - mg/l			Bromine - mg/l		
	Temperature - °C			Temperature - °C		
pH	10	20	30	10	20	30
90% Disinfection						
6.0	0.29	0.12	0.23	0.14	0.06	0.08
7.5	1.09	0.68	0.36	0.12	0.11	0.06
9.0	1.44	1.67	0.60	0.05	0.04	0.04
99% Disinfection						
6.0	0.55	0.67	0.43	0.25	0.14	0.17
7.5	2.16	1.37	1.02	0.22	0.24	0.11
9.0	3.40	3.30	2.47	0.11	0.08	0.10

[a]Residuals measured at 10 minutes

amount added is immediately consumed by demand from reducing agents. After the first minute or two, the residual has formed monochloramine and is fairly constant, and the residual at 10 minutes is a reasonable measure of the average residual through the 20-minute contact period. Only the changing proportion of chlorine to chloramine is unknown. With bromine the situation is quite different. A portion of the added bromine is consumed by demand from reducing agents as in the case of chlorine. The remainder forms dibromamine which disappears rather rapidly because it is more reactive than monochloramine. Thus, the measured bromine residual at 10 minutes is only a small fraction of that added, or of the initial bromine concentration after demand is satisfied. Another factor that must be considered in interpreting these differences is the limited accuracy of the DPD test for bromine at the very low residuals encountered.

The ratios of the required dosages of the two halogens for equal disinfection are somewhat different from the ratios of the required residuals. In Table 8-2 the chlorine and bromine dosages required for 90% and 99% disinfection are presented. Compared on the basis of dosage required for 99% disinfection, the two halogens were approximately equivalent at pH 7.5. At pH 6.0 chlorine was almost three times as effective as bromine, but at pH 9.0 bromine was approximately three times as effective as chlorine.

Table 8-2. Dosage of Chlorine and Bromine Required to Obtain 90% and 99% Disinfection as Measured by Coliforms

	Chlorine - mg/l			Bromine - mg/l		
	Temperature - °C			Temperature - °C		
pH	10	20	30	10	20	30
90% Disinfection						
6.0	0.94	0.49	0.78	2.45	1.69	2.46
7.5	1.49	1.04	0.98	1.47	1.57	1.57
9.0	2.08	2.22	1.34	0.94	0.78	0.63
99% Disinfection						
6.0	1.32	1.07	1.26	3.87	2.74	3.28
7.5	2.82	1.72	1.66	2.28	2.14	2.25
9.0	4.25	4.44	3.05	1.64	1.14	1.16

This comparison of the effectiveness of chlorine and bromine on the basis of required dosages should not be extended to other wastes, since variations in the concentration of halogen-reactive substances could materially change this comparison.

The per cent survival of both the fecal coliforms and the total bacterial numbers followed essentially the same disinfection patterns as was observed for the MPN data, so the results of these studies will not be included in this report.

In a number of the studies made, the coliform survival data obtained by the MPN procedure were compared with data obtained by the Millipore filter method. The values obtained by the Millipore filter method, with preliminary enrichment, were slightly lower than the values obtained by the MPN procedure, but were within the confidence limits for this test. Survival curves based on the Millipore filter results gave correlation coefficients of -0.92 to -0.97, which were considerably better than most of the MPN data.

The effect of pre-chlorination upon the efficiency of bromine as a disinfectant has been studied using plate count values as a measure of disinfection. In one series, chlorine (0.15 mg/l) in excess of the immediate demand was added to three samples of effluent. After a 10-minute interval, molar equivalents of chlorine (0.67 mg/l) and bromine (1.50 mg/l) were added to two of the samples. The remaining sample served as the pre-chlorinated control. To a fourth sample, bromine (1.50 mg/l) alone was added. The results of this study are given in Table 8-3. Sample 1, 2 and 3 showed essentially the same degree of disinfection after 10 minutes by the added chlorine dosage of 0.15 mg/l. The degree of disinfection for sample 1 remained about the same at the 20- and 30-minute intervals. In sample 2, the additional chlorine dosage of 0.67 mg/l reduced the per cent survival to 48% after 20 minutes and to 19.2% after 30 minutes. The addition of bromine to sample 3 was quite effective and, in a 10-minute interval, reduced the survival to 2.0%. A comparison of this figure with the 33.7% survival observed for bromine alone (sample 4), after a 10-minute interval, is indicative of the increased efficiency of disinfection with bromine by pre-treatment with chlorine.

Table 8-3. Effect of Pre-Chlorination on the Efficiency of Bromine as a Disinfectant – pH 7.5

	Cl_2 Residual - mg/l			Br_2 Residual - mg/l			Plate Count–% Survival		
	Minutes			Minutes			Minutes		
Treatment	10	20	30	10	20	30	10	20	30
Cl_2 0.15 mg/l	0.06	0.06	0.06				80.0	80.0	78.8
Cl_2 0.15 mg/l [a]Cl_2 0.67 mg/l	0.08	0.60	0.60				80.0	48.0	19.2
Cl_2 0.15 mg/l [a]Br_2 1.5 mg/l	0.08	0.08	0.08		0.18	0.22	83.0	2.0	3.0
Br_2 1.5 mg/l				0.09	0.09	0.09	33.7	18.6	17.9

[a]Added after 10 minutes.

In a second series of tests, bromine alone was compared to bromine following pre-chlorination at a low dosage. The results were like the first series of tests, in that pre-chlorination reduced the required bromine dosage necessary to obtain a particular degree of disinfection. In this series of tests the chlorine dosage ranged from 0.3 to 1.0 mg/l. Table 8-4 presents data representative of the tests in this series. A comparison of the survival data at all levels of bromine addition, shows an increase in the disinfection by bromine for the samples that were pretreated with chlorine. From the few studies we have made, the data would indicate that pre-chlorination conserves bromine by satisfying the immediate demand of the effluent.

Table 8-4. Effect of Pre-Chlorination (0.3 and 1.0 mg/l) of Secondary Effluents on the Efficiency of Bromine as a Disinfectant - pH 9.0

	Cl_2 Residual (mg/l)		Br_2 Residual (mg/l)		Plate Count (% Survival)	
	Minutes		Minutes		Minutes	
Treatment	10	20	10	20	10	20
No Cl_2						
Br_2 0.75 mg/l			0.04	0.03	100.0	96.0
Br_2 1.00 mg/l			0.09	0.03	100.0	94.0
Br_2 1.25 mg/l			0.11	0.08	98.0	27.5
Br_2 1.50 mg/l			0.11	0.06	15.0	8.5
Cl_2 Added			(0.3 mg/l)			
Br_2 0.75 mg/l	0.20	0.22		0.15	100.0	71.5
Br_2 1.00 mg/l	0.20	0.23		0.10	100.0	35.0
Br_2 1.25 mg/l	0.21	0.26		0.18	100.0	7.1
Br_2 1.50 mg/l	0.20	0.26		0.19	100.0	4.3
No Cl_2						
Br_2 0.75 mg/l			0.02	0.01	100.0	100.0
Br_2 1.00 mg/l			0.05	0	97.8	97.8
Br_2 1.25 mg/l			0.04	0	97.8	94.8
Br_2 1.50 mg/l			0	0	32.0	31.2
Cl_2 Added			(1.0 mg/l)			
Br_2 0.75 mg/l	0.66	0.58		0.21	100.0	69.0
Br_2 1.00 mg/l	0.65	0.61		0.28	100.0	33.7
Br_2 1.25 mg/l	0.72	0.66		0.33	100.0	7.6
Br_2 1.50 mg/l	0.69	0.67		0.33	100.0	3.0

DISCUSSION

The pH effect demonstrated by the results of this work correlates well with the chemistry of the bromamines and chloramines. For the chlorine experiments, the ammonia concentration of the wastewater was between 10-20 ppm, with chlorine dosages between 1-4 ppm. If these figures are converted to molar concentrations and the high and low values are taken, the range of NH_3/Cl mole ratios becomes 2.6-42. This is certainly a wide range; however, for the different pH values investigated (6, 7.5 and 9), monochloramine is seen to be the principal species formed in all cases.[12] Therefore the biocidal effect should be related to the chemistry of the stable specie monochloramine. Weil and Morris[12] showed that the rate of formation of monochloramine is a function of pH with the maximum rate occurring at pH 8.4 where the concentration of the molecular species HOCl and NH_3 is maximum. At pHs greater or less than 8.4, the rate of formation is reduced considerably. Thus, at pH 6 the formation of monochloramine is slower than at either pH 7.5 or pH 9. Because hypochlorous acid (HOCl) is the predominant specie at this pH, and is a stronger disinfectant than either monochloramine or hypochlorite ion (OCl^-),[13] it follows that under these experimental conditions, disinfection efficiency will be greatest because of the slow rate of the chlorine-ammonia reaction. At pHs 7.5 and 9 the formation rate of monochloramine is approximately the same, perhaps somewhat faster at pH 9. The difference in disinfection efficiency is then related to the two different free chlorine species present. At 25°C the transition pH for hypochlorous acid to hypochlorite ion is 7.6. Thus in these experiments, with monochloramine formation relatively the same in both cases, disinfection would be expected to be more rapid at pH 7.5 where hypochlorous acid is predominant than at pH 9.0 where hypochlorite ion (with much weaker disinfection power) comprises more than 90% of the free chlorine present.

A reverse effect was noted in the case of the bromine experiments, with higher pH yielding an increase in the efficiency of the disinfectant. An analysis similar to the one just presented, using an ammonia concentration range of 10-20 ppm and a bromine dosage, as cited, of 1-4 ppm, gives an ammonia to bromine molar ratio range of from 12-95. If this region is located on the bromine distribution diagram (Figure 9-6, Chapter 9) for the pH region 6-9 (the logarithm range of the N/Br ratio being 1.07-1.98), dibromamine and monobromamine are the principal species. The bromamines are known to form instantly, so the disinfection powers of the free species are unimportant. For any given N/Br ratio the amount of monobromamine will increase as pH increases (see distribution diagram), thus as the pH goes from 6 to 7.5 to 9, increasing amounts of the more stable monobromamine are formed. Thus the life of active bromine (*i.e.,* in the

+1 oxidation state) will be longer for the high pH experiments where monobromamine predominates than at low pH where the unstable dibromamine is the main specie. Actually for the ammonia concentration range used and the bromine dosage employed, the pH points were almost perfectly chosen to illustrate this effect. At pH 6 for all dosages dibromamine comprises 90-99% of the total bromine species; at pH 9 the case is similar for monobromamine; pH 7.5 lies right between the two and on the distribution diagram represents a transition from monobromamine to dibromamine. Assuming a constant ammonia concentration, an increasing dose of bromine will decrease the N/Br ratio causing a greater percentage of dibromamine to form during these experiments. While the bromine dosage increases the bromine concentration in solution, integrated over time to give the total time-concentration, actual dose might not increase as rapidly, or might actually decrease. This, of course would significantly affect the survival curves.

The final products of the halogen dosage also explain the difference in scale of the residual survival values in Table 8-1. In the case of chlorine monochloramine is formed and is stable, thus residual measurements are on the same scale as dosage, the only reduction in concentration being demand reactions. Bromine, however, is reduced to bromide both by demand and the monobromamine and dibromamine decomposition, thus the 10-minute residual measurements are more than 10 times lower than the corresponding dosage.

CONCLUSIONS

The disinfection of activated sludge effluents by chlorine followed the expected trend of decreased efficiency with increasing pH. By contrast, bromine was the most effective at the high pH level where monobromamine is the more stable bromamine formed. The dibromamine formed at lower pH was rather unstable. The decreasing disinfection efficiency at lower pH is a function of dibromamine instability. At higher pH, the bromamines are more stable and, thus, more effective in disinfection. The chloramines have been reported to be very ineffective as disinfectants,[6,8,12] whereas bromine is reported to be equally effective as the free agent or as bromamine.[5,6,9] In view of these results, it would appear that bromine could be useful in the disinfection of certain wastes at high pH and low reducing demand. Pre-chlorination was found to help reduce the inorganic demand and thus improved the efficiency of bromination. There are two kinds of demands:

1. demand from organic and inorganic reducing agents
2. demand from reactions with ammonia.

Chlorine and bromine demand of the first type consumes halogen with bromine reacting more rapidly, but to a lesser extent, since it is a slightly weaker oxidizing agent. However, bromine–because of its greater molecular weight and higher cost–is a poor way to reduce demand and is not efficiently used in a high-demand waste. The reactions with ammonia do not produce a problem for bromine. The bromamines are good disinfectants as shown by this work and previous studies. Also, the bromamine formed in wastewater is not stable and thus does not present the fish toxicity problems of the stable chloramines.

ACKNOWLEDGMENT

This research was supported, in part, by the U.S. Environmental Protection Agency, under Grant No. 17060 DNU.

REFERENCES

1. Zillich, J. A. "Toxicity of Combined Chlorine Residuals to Freshwater Fish," *J. Water Pollution Control Federation* **44**, 212 (1972).
2. Mills, J. F. "The Disinfection of Sewage by Chlorobromination," *Amer. Chem. Soc. Div. Water, Air, Waste Chem.* **13** (1), 65. Preprints of papers presented at 165th National Meeting, Dallas, Texas (April, 1973).
3. Esvelt, L. A., W. J. Kaufman, and R. E. Selleck. "Toxicity Assessment of Treated Municipal Wastewater," *J. Water Pollution Control Federation* **45**, 1558 (1973).
4. Brungs, W. A. "Effects of Residual Chlorine on Aquatic Life," *J. Water Pollution Control Federation* **45**, 2180-2193 (1971).
5. Johnson, J. D. and R. Overby. "Bromine and Bromamine Disinfection Chemistry," *J. San. Eng. Div. ASCE* **97** (SA5), 617-628 (1971).
6. Krusé, C. W., V. P. Olivieri and K. Kawata. "The Enhancement of Viral Inactivation by Halogens," *Water Sewage Works* **118**, 187-193 (1971).
7. McKee, J. E., C. J. Brokaw and R. T. McLaughlin. "Chemical and Colicidal Effects of Halogens in Sewage," *J. Water Pollution Control Federation* **32**, 795-819 (1960).
8. Johannesson, J. K. "Anomalous Bactericidal Action of Bromamine," *Nature* **181**, 1799-1800 (1958).
9. Galal-Gorchev, H. and J. C. Morris. "Formation and Stability of Bromamide, Bromimide and Nitrogen Tribromide in Aqueous Solution," *J. Inorg. Chem.* **4**, 899-905 (1965).
10. Palin, A. T. "Determination of Free Residual Bromine in Water," *Water Sewage Works* **108**, 461 (1961).
11. *Standard Methods for the Examination of Water and Wastewater*, 13th ed. (Washington, D.C.: American Public Health Association, Inc., 1971).

12. Weil, I. and J. C. Morris. "Kinetic Studies of the Chloramines, I," *J. Amer. Chem. Soc.* **71**, 1664-1671 (1948).
13. Butterfield, C. T. "Bactericidal Properties of Chloramines and Free Chlorine in Water," *Public Health Rep.* **63**, 934-940 (1948).

CHAPTER 9

BROMINE DISINFECTION OF WASTEWATER

J. Donald Johnson and William Sun

Department of Environmental Sciences & Engineering
School of Public Health
University of North Carolina
Chapel Hill, North Carolina 27514

INTRODUCTION

Wastewaters are disinfected in U.S. practice, primarily because of the growing probability of water reuse either as directly planned water reclamation projects, such as those in California and South Africa, or after the discharge of human waste into waters subsequently used for water supplies. The objectives of wastewater disinfection as discussed by Heukelekian and Faust[1] are primarily to destroy Salmonella and Shigella bacteria and the virus of infectious hepatitis. Chlorination has been the universally used method for disinfection of wastewaters, mainly because it is economical. In their comparison of chlorine, bromine and iodine, McKee *et al.*[2] concluded that of the halogens, chlorine was the most suitable for untreated wastewater. Cost and the presence of measurable chlorine residual at relatively low concentration inputs of chlorine were two key factors in their conclusion. Other halogens and ozone have recently received considerable interest for wastewater disinfection. The lesser effectiveness of the chloramines and hypochlorite ion as disinfectants especially for bacterial spores, protozoan cysts and viruses compared to their low but better reactivity against vegetative bacteria is the primary reason for considering other disinfectants.[3] The long persistance and high fish toxicity of chloramine residuals has been another reason for the consideration of some other wastewater disinfectants. Brungs[4] recommends that chlorine not exceed 0.002 mg/l because of the lethality of higher concentrations to trout, salmon and sensitive life stages of other fish. Marks and Strandskov[5] and many others have shown the importance of the chemical species of disinfectant present, on the rate of kill of enteric pathogens in wastewater. This is especially true for bromine compounds compared with chlorine because of the much greater effectiveness of the bromamines as disinfectants.

Wastewater bromination chemistry and its effect on the disinfection process is quite different than with chlorination. Free bromine, when present, exists almost entirely as HOBr since the dissociation of hypobromous acid to hypobromite ion occurs near pH 9.0 or 1.5 pH units higher than the similar reaction for the formation of hypochlorite ion. Since HOBr and HOCl are the most effective disinfectant forms of bromine and chlorine, this is a significant advantage of bromine. In advanced waste treatment processes, which might employ lime for phosphorus removal, bromine would have a significant advantage because of the higher than normal pH of the water to be disinfected. Here, free chlorine would exist as the ineffective virucide and cysticide, hypochlorite ion, OCl^-, while bromine remains predominately effective HOBr to pH values greater than 9. All halogens and ozone are effective, wide spectrum disinfectants when present in their free molecular, chemical forms of HOCl, HOBr, I_2/HOI and O_3. Because of their high chemical reactivity, however, they rarely exist for significant time in these forms in wastewater. Chlorine and bromine form chloramines and bromamines, chlorine forms hypochlorite, ozone forms ozonides and peroxides, all of which are measured as "residual" and are generally weak disinfectants. All these disinfectants also can be reduced to completely ineffective forms for disinfection: chloride, bromide, iodide and oxygen.

Olivieri *et al.*[6] have shown that the disinfection of virus in sewage is due primarily to short-lived virucidal intermediates, presumably free chlorine which is initially present on the addition of chlorine to wastewater. The presence of the long-lived chloramine species does not appear significant in their study on the disinfection of viruses. Extending this same evidence to bacteria would help explain the lack of correlation between combined chlorine or chloramine residuals and the disinfection efficiency of chlorine in wastewater treatment as measured by coliform.[1]

The chemistry of the reactions of chlorine and bromine with ammonia and other organic amine compounds present in wastewater is thus extremely important in determining the effectiveness of the disinfectant. A knowledge of these reactions and their rates as well as the disinfection strength of the products formed and their stability is also necessary to control wastewater disinfection. Morris[7] has shown for chlorine that the rates of formation of chloramines depend on how strong a base the nitrogen is in the various nitrogen compounds which can be present in wastewater. Feng[8] and others have shown that organic chloramines are present in wastewater and represent a significant source of the "chloramine residual," normally used to control wastewater disinfection. The ineffectiveness of chloramines as disinfectants and the universal usage of total chlorine residual to control wastewater disinfection also helps explain why very little correlation has been found between the bactericidal efficiency of chlorination and the chloramine residual measured.[1]

The poor disinfection efficiency of the chloramines, especially organic, contrasts with the disinfection efficiencies of the bromamines. Johannesson[9-12] has shown that the bromamines are nearly as effective bactericides as free bromines. Stringer, *et al.* (see Chapter 10) have shown that in the presence of ammonia, bromine is also nearly as effective a virucide and cysticide as free bromine.

The purpose of this study was to analyze the efficiency of bromine as a wastewater disinfectant in relation to the chemical species present. Bromine and chlorine were compared for the disinfection of *E. coli* in an alum-coagulated trickling filter effluent from municipal sewage. Alum coagulation and settling was done because of the growing need to remove phosphorus from wastewater and the relatively low cost of alum compared with that of the disinfectants. Bromine is more expensive than chlorine but less costly than ozone in the relatively convenient and less expensive and less corrosive bromine chloride form.[13] Because of the greater economy and efficiency of all the disinfectants in tertiary effluent and the growing need to do better wastewater treatment, the disinfection of a tertiary effluent appears to be a much more logical approach than that taken by McKee *et al.*[2] who compared the halogens in settled raw sewage.

METHODS

Sampling

Twenty-four hour composites of the secondary effluent from the University of North Carolina Wastewater Research Center at Chapel Hill's 3.5 million gal/day Mason Farm Treatment Plant were obtained immediately before use in a refrigerated sampler. Samples were brought up to 15°C by holding in a water bath for 30 minutes. After thorough mixing, the sample was dosed with 200 mg/l of alum with rapid mix for 30 seconds. The sample was flocculated for 30 minutes at 30 rpm, at the completion of which it was allowed to settle quietly for an additional 30 minutes. The supernatant was then decanted and held at a 15°C constant temperature for use in the halogenation experiments.

Reagents and Solutions

All solutions of halogen and buffers were prepared in chlorine demand-free water. This dilution water was filtered through a 1-μ depth filter, activated carbon and a highly regenerated and aged mixed-bed ion exchange resin. Any residual chlorine demand of this water was satisfied by adding approximately 3 mg/l free chlorine and letting it stand for three days. After checking that significant chlorine had not been lost during the three-day period, dechlorination was achieved by inserting a quartz tube ultraviolet lamp. When the residual chlorine had been removed, the water was stored in sealed carboys drained by syphon and protected from the atmosphere by sulfuric acid traps. Reagent grade chemicals were used throughout. Both chlorine and bromine stock solutions were standardized at pH 4.7 in acetate buffer in the presence of excess potassium iodide using thiosulfate standardized against iodate and iodine.[14] These solutions were approximately 1 m*M* each in concentration.

Procedures

The Stabilized Neutral Othotolidine test, SNORT, was used to determine free chlorine.[15] Combined chlorine and total bromine were both determined as those compounds which oxidize iodide to iodine at pH 7. In the determination of total bromine, the sample was first added to excess iodide. This iodine solution was then added to the SNORT reagents.

E. coli were determined using the Most Probable Number (MPN) presumptive test with lauryl tryptose broth (Difco Laboratory, Detroit, Michigan). Five tubes with five decimal dilutions were used in the appropriate ranges determined by trial runs. Sterilized water blanks were used for addition of 1-ml samples into 9 ml of sterile water for dilution except when a 10-ml sample was taken. Sterile water and unhalogenated control were also measured for comparison with the halogenated samples. All samples were subjected to a 50% dilution by the halogen solutions, so the control MPN number is divided by two for comparison and computation of survival.

Experimental

The pH of the sewage was first adjusted to 4 with 1*M* HCl or with 1*M* NaOH for pH 10, and Na_2HPO_4 and KH_2PO_4 from pH 6 to 8, after initially bringing the sample to the desired pH with NaOH or HCl. After completing the experiments, a final pH reading was taken to ensure that the final pH was within 0.5 units of the desired value.

Fifty ml of alum coagulated sewage at 15°C was placed in a beaker also maintained at 15°C. With rapid mixing 50 ml of halogen solution at the same temperature was added at time 0. After one minute, excess sodium thiosulfate was added to stop the action of the halogen. The same procedures were repeated for 15 minutes of contact time. Replicates of the above were performed for the analysis of halogen residual using the SNORT methods and omitting the sodium thiosulfate.

RESULTS AND DISCUSSION

Dosage Effect

The composite samples taken on two different days were buffered with phosphate at pH 7 and halogenated with equal weight dosages of 2, 5, 10 and 20 mg/l of chlorine or bromine. Figure 9-1 shows the free chlorine and total bromine residuals present after 1 and 15 minute contacts for the first sample. The corresponding coliform survivals after 1- and 15-minutes of contact on sample one are shown in Figure 9-2. The control MPN indices on unhalogenated samples, with the 50% dilution factor discussed above, were 6500/100 ml for both the 1- and 15-minute contact samples.

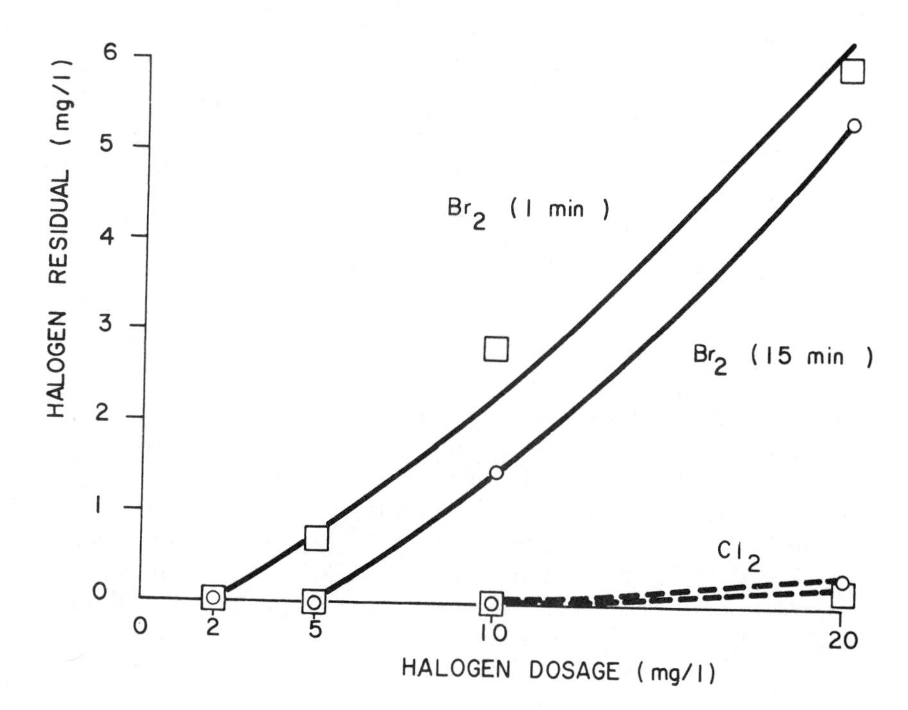

Figure 9-1. Free chlorine (- - -) and total bromine (——) after 1-minute (□) and 15-minute (○) contact

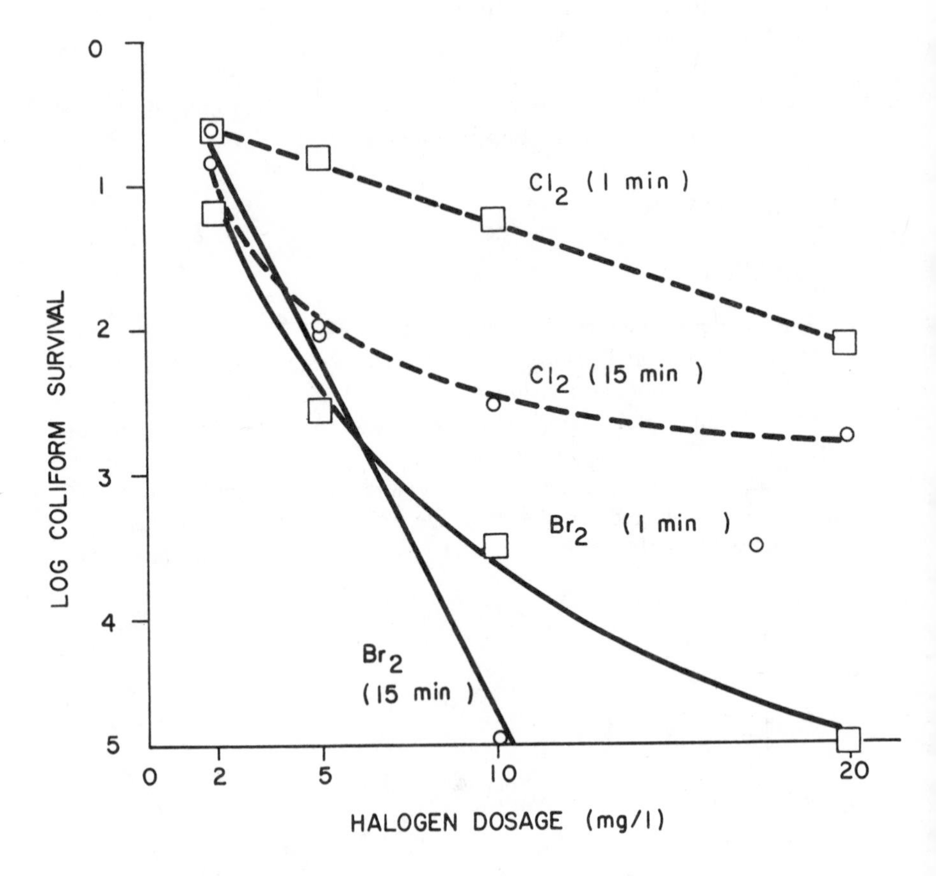

Figure 9-2. Coliform survival after 1 minute (□) and 15 minutes (○) of contact with chlorine (- - -) and bromine (——).

Chemical analysis including the dilution factor gave 15 mg/l total Kjeldahl nitrogen and 13 mg/l as ammonia nitrogen.

Figure 9-1 shows that even in one minute, the entire 2 mg/l bromine has been consumed by fast-acting demand. In spite of this, approximately 90% reduction in coliform bacteria occurred for both the bromine and the chlorine. Only slightly better disinfection is apparent after 15 minutes, probably because all the halogen has been consumed by fast-acting demand after the first minute. The differences in the survival fractions for the 1- and 15-minute samples at 2 mg/l show the magnitude of the experimental error is approximately one-half a log unit of survival. The 15-minute samples were done on a second aliquot of sewage which had the same chemistry and control MPN after alum coagulation and settling, as compared to the one-minute sample.

At the 5-mg/l dosage, less than 1 mg/l of total bromine residual remained after one minute and no bromine residual remained after 15 minutes. There was no free chlorine residual at this dosage; however, chloramine residual was present as shown by approximately a one-log greater chlorine disinfection after 15 minutes compared to the one-minute samples. For this 5-mg/l dosage, bromine residual disappeared and was clearly unstable in the 15-minute sample. Bromine produced, therefore, no additional increment of disinfection in the 15-minute sample. Chlorine, however, continued to produce an additional log of disinfection probably because of a more stable chloramine residual compared to the less stable bromamine residual. In all samples, where the initial demand of the sample has been satisfied, the same weight dosage of bromine and chlorine acting on the same sample for the same length of time showed bromine disinfection equal to or better than the chlorine disinfection. The second set of samples (Figures 9-3 and 9-4) contained 9-10 mg/l of ammonia and 12 mg/l total Kjeldahl nitrogen. Although

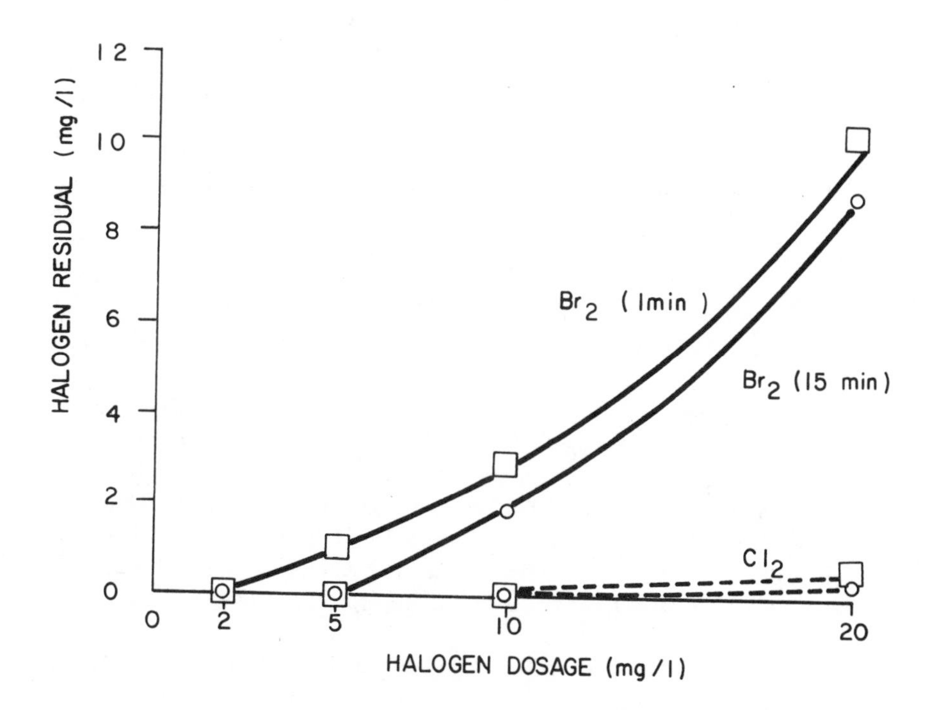

Figure 9-3. Free chlorine (- - -) and total bromine (——) after 1-minute (□) and 15-minutes (○) contact.

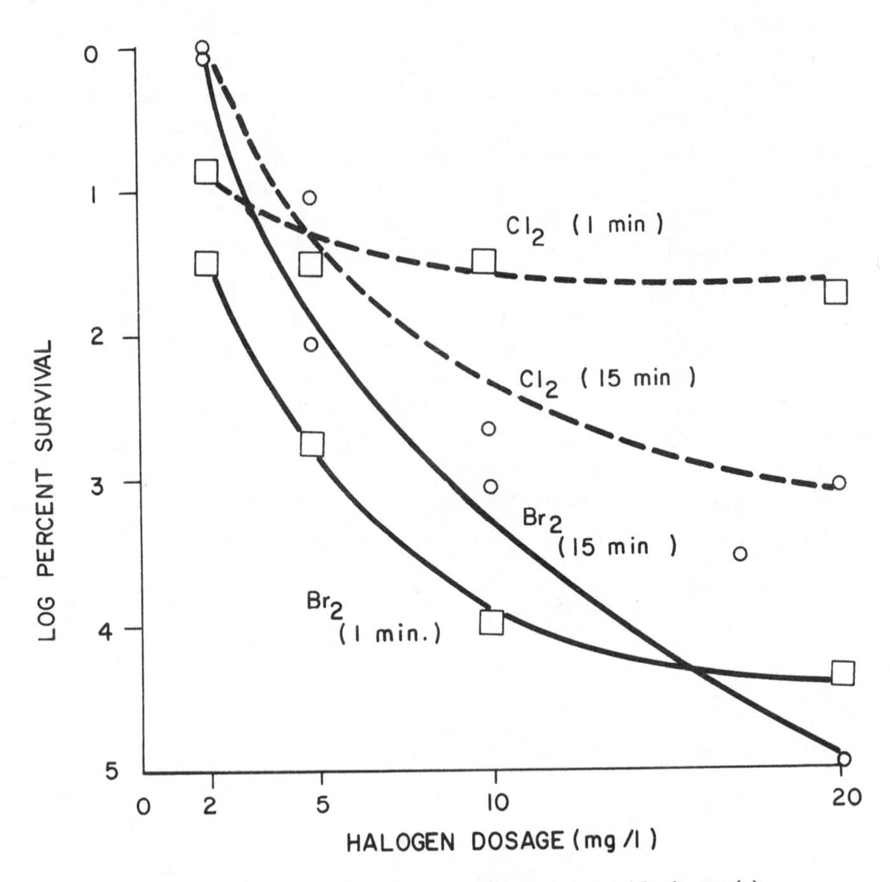

Figure 9-4. Coliform survival after 1-minute (□) and 15-minutes (○) contact with bromine (——) and chlorine (- - -).

the chemistry of these samples showed lower ammonia concentration, the initial MPN indices were significantly higher than the first set of samples. The control MPN's were 1.2 x 10^5/100 ml and 2.45 x 10^3/ml for 1- and 15-minute samples. A similar pattern of residual and disinfection for these samples (Figures 9-3 and 9-4) is present in spite of the variation in the sewage composition compared to the first set of samples (Figures 9-1 and 9-2). Bromine produces an approximately one-log greater kill of coliform as the same weight dosage of chlorine as in the first set of samples.

The greater effectiveness of bromine compared to chlorine is more apparent at the higher dosages of 10 and 20 mg/l than at the lower dosages. This is probably due to the fact that bromine, because of its higher molecular weight, produces fewer oxidizing equivalents per milligram than does chlorine.

Thus, at low concentrations of halogens a larger fraction of the bromine is consumed by fast-acting reducing demand than is chlorine. As soon as the fast-acting, reducing demand of the sample has been satisfied, chlorine will react with the ammonia and the other amines present in the sample to produce monochloramine, NH_2Cl, with ammonia and other principally monosubstituted chloramines with the organic amines present.[7] For all dosages and all samples, the ammonia was always in at least two-fold excess over the amount of chlorine added. At these neutral pHs, this excess of ammonia limits the compounds formed to monochloramine, NH_2Cl, as given in reaction (1).

$$NH_4^+ + HOCl = NH_2Cl + H_3O^+ \quad (1)$$

The second-order rate constant for this reaction is $3.5 \times 10^{-3}/H^+$ 1 $mole^{-1} sec^{-1}$ at 25°C.[7] At pH 7, 2 mg/l chlorine and 10 mg/l ammonia nitrogen, the half-life of free chlorine is only about 0.03 sec. Thus monochloramine, NH_2Cl, is formed very rapidly at this pH and concentration of ammonia. No significant concentration of dichloramine, $NHCl_2$ or trichloramine, NCl_3, is present in the sample. Organic chloramines can also be formed. From the data of Morris[7] aliphatic and aromatic amines such as methylamine and amino acids like glycine which can be present because of the incomplete biological oxidation of organics such as proteins, can also form chloramines. The majority of these compounds react even more rapidly than ammonia with the chlorine added but are present in much lower concentrations and will be formed competitively with monochloramine. Amides such as urea and the unhydrolized peptide linkages in proteins would not be chlorinated because of their generally slower rates of reaction with chlorine as compared with ammonia and their much lower concentrations. These organic chloramines are often measured by the analytical methods as if they were mono- or dichloramine. To the extent they are present, they represent generally poor biocides, and false residual halogen disinfectants.[5,8]

Bromine also forms bromamines.[16] The residuals measured in Figures 9-1 and 9-3 are entirely bromamine and, as is the case with chloramine, no significant free bromine was present even at dosages as high as 20 mg/l of bromine. Because of its higher molecular weight, the bromine molar equivalent added to the samples was approximately half the concentration of chlorine molar equivalents added. In relation to the ammonia present, a 5- to 100-fold molar excess of ammonia was present compared to the concentration of bromine added. This would result in the formation of dibromamine as shown in equation 2.[17]

$$NH_4^+ + 2HOBr = NHBr_2 + 2H_2O + H^+ \quad (2)$$

pH Effect

At neutral pH and this excess of ammonia, no significant concentrations of ammonia monobromamine, NH_2Br, or tribromamine, NBr_3, would be present in the solution. Dibromamine, $NHBr_2$, is rapidly formed and as shown by the 15-minute residual in Figures 9-2 and 9-3 is reasonably stable compared to the period of disinfection normally associated with wastewater treatment. Figures 9-2 and 9-4 show dibromamine is at least twice as effective as monochloramine as a coliform bactericide compared to the same weight of chlorine. This agrees with the previous work on bactericidal properties of bromine and the bromamines.[3,9–12,18]

Figure 9-5 shows the effect of coliform survival after one minute of contact with 5- and 20-mg/l of chlorine and bromine dosage as a function of pH. A control on the MPN index, N_0, for the 5-mg/l sample was 6500 at

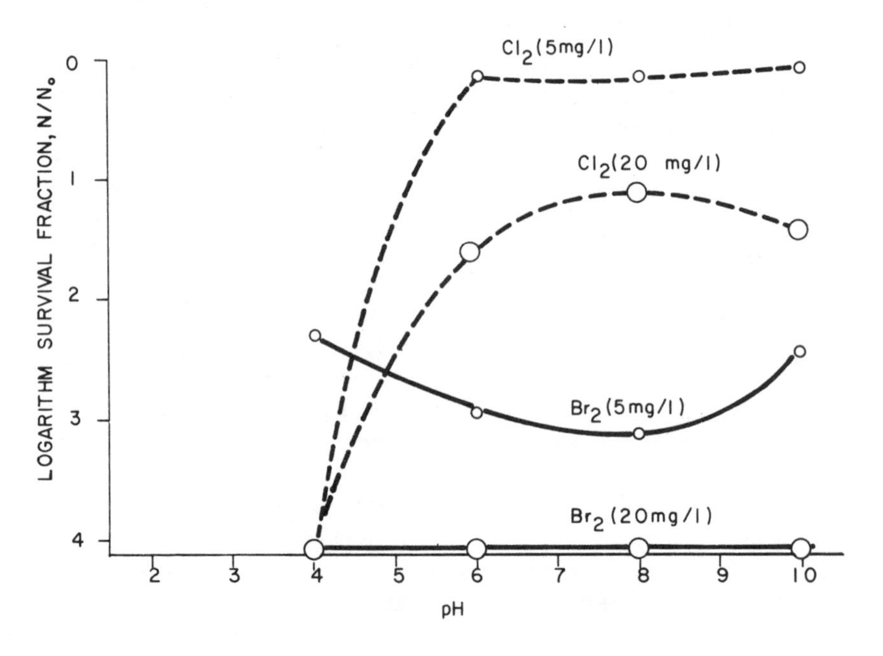

Figure 9-5. Coliform survival after 1 minute of contact with 5 mg/l and 20 mg/l, chlorine (- - -) or bromine (——).

pH 6 and 8 decreasing to 3950 at pH 10 and 1750 at pH 4. The same sample was used for chlorine and bromine at these four pH values. The negative logarithm of the survival fraction, N/N_0, is plotted as a function of chlorine as the dotted lines and bromine as the solid lines. The 20-mg/l sample had a control MPN of 6500 for pH 6, 8 and 10 dropping to 3500 at pH 4. At all pH values the bromine MPN index for the 20-mg/l sample was 0. This is plotted as a survival fraction of 10^{-4} by comparison with the control index. Although no chemistry is available on these samples, it is clear from the survival fraction that effective bromine residual was present at both 5- and 20-mg/l dosage at all pHs. Chlorine by contrast is most effective at pH 4 and least effective at pH 8. Other samples studied as a function of pH showed the same trend at dosages of 2, 5, 10 and 20 mg/l of bromine and chlorine. The chlorine consistently gave better disinfection at pH 4 while bromine produced better disinfection between pHs 6 and 10.

Figure 9-6 shows the principal species of bromine and the bromamines present after dosing with given mole ratios of bromine to ammonia as a function of pH. From this diagram, it can be seen that the principal species

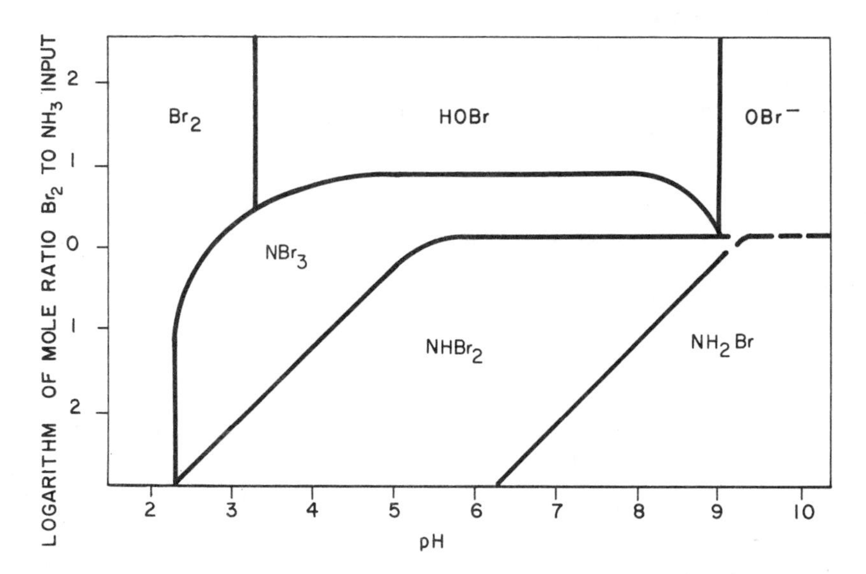

Figure 9-6. Principal species of bromine and bromamine predominating after dosing with given mole ratio of Br_2 to NH_3. Lines represent equal equivalent concentration.

of bromine present at pH 4 is NBr_3 at the mole ratio of 5-100 excess ammonia present in these samples. At pH 6 dibromamine, $NHBr_2$, predominates, at pH 8 a mixture of dibromamine and monobromamine are present and monobromamine predominates at pH 10. At these short contact times, chlorine is a mixture of free chlorine, HOCl and OCl^-, and monochloramine, NH_2Cl. Insignificant concentrations of the other chloramines are present at these large excesses of ammonia and short contact times. Even at pH 4 no significant concentration of dichloramine will be present although this compound does form at this low pH after long contact. From the work of Weil and Morris[19] the rate of monochloramine formation as a function of pH is strikingly similar to the rate of disinfection as a function of pH. The rate of formation of monochloramine reaches a maximum value at pH 8.2, decreasing linearly with pH at lower and higher pH values. Thus reaction (1) with a half-life of only a few hundredths of a second in the neutral pH range, has an approximate half life of 30 seconds in free chlorine at pH 4 and a concentration of 10 mg/l ammonia nitrogen. This persistence of free chlorine for a short time after an addition of chlorine, explains the much greater disinfection efficiency of chlorine at low pH. In a neutral pH range, however, the chlorine added reacts rapidly to form monochloramine which is a relatively poor disinfectant. By contrast, bromine reacts rapidly in all ranges of pH to form the bromamines. The difference in the reactions of bromine are in the products formed rather than their rate of formation as is the case with chlorine. In the low pH range tribromamine (NBr_3) is formed, in the neutral pH range dibromamine ($NHBr_2$) is the predominate form, while at high pH monobromamine (NH_2Br) predominates. Unlike monochloramine, however, all three of these forms of the bromamines appear to be effective disinfectants. Recorded lack of disinfection efficiency[2] is probably due either to high chemical demand reducing the bromine added to bromide, or to the long term instability of dibromamine, $NHBr_2$.[17]

ACKNOWLEDGMENT

The assistance of Dr. Linda Little in the bacteriological analyses and Messrs. Tom Bates and Tony Owen in the chemical analyses is gratefully acknowledged.

REFERENCES

1. Heukelekian, H. and S. D. Faust. "Compatability of Wastewater Disinfection by Chlorination," *J. Water Pollution Control Federation* **33**, 932 (1961).
2. McKee, J. E., C. J. Brokaw and R. T. McLaughlin. "Chemical and Colicidal Effects of Halogens in Sewage," *J. Water Pollution Control Federation* **32**, 795 (1960).
3. Krusé, C. W., Y. Hsu, A. C. Griffiths and R. Stringer. "Halogen Action on Bacteria, Viruses, and Protozoa," In *Proc. Natl. Specialty Conf. on Disinfection, ASCE,* (New York: ASCE, 1970), pp. 113-136.
4. Brungs, W. A. "Effects of Residual Chlorine on Aquatic Life," *J. Water Pollution Control Federation* **45**, 2180-2193 (1973).
5. Marks, H. C. and F. B. Strandskov. "Halogens and Their Mode of Action," *Ann. NY Acad. Sci.* **53**, 163-171 (1950).

6. Olivieri, V. P., T. K. Donovan and K. Kawata. "Inactivation of Virus in Sewage," *J. San. Eng. Div., ASCE* **97**, 365-384 (1970).
7. Morris, J. C. "Kinetics of Reaction Between Aqueous Chlorine and Nitrogen Compounds," In *Principles and Applications of Water Chemistry*, S. D. Faust and J. Hunter, Eds. (New York: John Wiley & Sons, 1967), pp. 23-51.
8. Feng, T. H. "Behavior of Organic Chloramines in Disinfection," *J. Water Pollution Control Federation* **38**, 614-628 (1966).
9. Johannesson, J. K. "Anomalous Bactericidal Action of Bromamine," *Nature* **181**, 1799-1800 (1958).
10. Johannesson, J. K. "The Determination of Monobromamine and Monochloramine in Water," *Analyst* **83**, 155-159 (1958).
11. Johannesson, J. K. "Bromamines. Part I. Mono- and Di-bromamine," *J. Chem. Soc., London*, 2998-3001 (1959).
12. Johannesson, J. K. "The Bromination of Swimming Pools," *Amer. J. Pub. Health* **50**, 1731-1736 (1960).
13. Mills, J. F. "The Chemistry of Bromine Chloride in Waste Water Disinfection," In *Reprints of Papers Presented at 166th National Meeting, American Chemical Society, Division of Environmental Chemistry*, Mellon Institute, Pittsburgh, Pennsylvania (1973), pp. 137-143.
14. *Standard Methods for the Examination of Water and Wastewater,* 13th ed. (Washington, D.C.: American Public Health Association, 1971).
15. Johnson, J. D. and R. Overby. "Stabilized Neutral Orthotolidine, SNORT, Colorimetric Methods for Chlorine," *Anal. Chem.* **41**, 1744-1749 (1969).
16. Galal-Gorchev, H. and J. C. Morris. "Formation and Stability of Bromamide, Bromimide and Nitrogen Tribromide in Aqueous Solution," *Inorgan. Chem.* **4**, 899-905 (1965).
17. Johnson, J. D. and R. Overby. "Bromine and Bromamine Disinfection Chemistry," *J. San. Eng. Div., ASCE* **97**, 617-628 and **99**, 371-373 (1971).
18. Wyss, O. and J. R. Stockton. "The Germicidal Action of Bromine," *Arch. Biochem.* **12**, 267 (1947).
19. Weil, I. and J. C. Morris. "Kinetic Studies on the Chloramines I. The Rates of Formation of Monochloramine, N-chlormethylamine," *J. Amer. Chem. Soc.* **71**, 1664-1671 (1949).

CHAPTER 10

COMPARISON OF BROMINE, CHLORINE AND IODINE AS DISINFECTANTS FOR AMOEBIC CYSTS

Richard P. Stringer

Gilbert Associates Inc.
Environmental Regulatory Department
Reading, Pennsylvania

William N. Cramer, Cornelius W. Krusé

The Johns Hopkins University
School of Hygiene and Public Health
Baltimore, Maryland

INTRODUCTION

Cysts of amoeba are an important consideration in water and wastewater disinfection for two reasons: (1) at least two species of amoebae are pathogenic to man, and the waterborne route of transmission for both species dictates a need to establish and maintain cysticidal doses of halogen disinfectants in water, and (2) amoebic cysts serve as good test organisms to evaluate the universality of disinfectants, because they can survive under environmental conditions that would be lethal to most bacteria and viruses.

Entamoeba histolytica causes amoebic dysentery, a disease that can result when the susceptable individual ingests food or drink contaminated with sewage containing viable cysts. Chapters in the public health history of Chicago dramatize the epidemiological significance of the waterborne route of infection for amebiasis caused by *Entamoeba histolytica.* An outbreak of amebiasis during the 1934 Chicago World's Fair caused 98 deaths and 1409 infections.[1] The misfortune was the consequence of cross connections and sewage dripping into the ice water supply of a complex of hotels. The next year during a three-day fire in the Chicago stockyards, 52% of the firemen developed amebiasis after quenching their thirsts with water contaminated with sewage containing amoebic cysts. These and other waterborne epidemics of amebiasis have occurred in the United States[2] and establish the pattern of direct contamination of drinking water by fresh sewage through defects in plumbing or cross connections.

Amebiasis is much more prevalent in the tropics. In Columbia, South America, 20 to 60% of the population is infected.[3] The frequent travel of Americans to and from the tropics is a continuing threat that an infected traveler returning to the United States could be a carrier. During the Vietnam war, thousands of troops suffered from amoebic dysentery and hepatic complications caused by amoeba.[4] Returning soldiers are potential carriers.

The infected person produces up to 14,000,000 cysts per day,[5] and each cyst is a potential source of infection if it is deposited in water or food ingested by another person. Because cysts die quickly when they are out of water, there seems to be little doubt that the waterborne route of infection is the one that occurs most frequently and, in fact, all documented epidemics of amoebiasis caused by *E. histolytica* have been traced to water contaminated with cysts.[6]

Viable cysts of *E. histolytica* have been found in the chlorinated effluent of a sewage treatment plant. If viable cysts are in raw water entering a drinking water treatment plant, the combination of coagulation, sedimentation, filtration and chlorination to a free residual will remove or destroy the cysts from treated water leaving the plant. However, cross connections and back siphonage possibilities in transmission lines have been sources of contamination for treated drinking water that might not have been adequately neutralized, even when a good chlorine residual was maintained throughout the distribution system.

Naegleria gruberi, the causative agent of amoebic meningoencephalitis, is the other amoeba considered important in disinfection studies. Although epidemiological evidence regarding this central nervous system disease suggests a waterborne route such as swimming, this route has not been proved.[7]

The literature on cyst disinfection shows inconsistencies regarding the relative effectiveness of the halogens. This might be due to differences in the concentrations and species of amoebic cyst and the bioassay system employed; difficulties in identifying and measuring the free or combined chemical form of the halogen; or unfamiliarity with the effect of rapid-mixing on halogen in test suspensions, especially in water of high pH or, as in sewage, containing interfering nitrogenous substances. Chlorine is the disinfectant most extensively used for public water and wastewater treatment. The Globaline tablet is the iodine-containing disinfectant used by the military for canteen treatment under emergency conditions in the field. Bromine, as a substitute for chlorine, is being proposed in special cases for water disinfection such as aboard ship or for transport aircraft.

Although a large number of studies have been concerned with the cysticidal properties of chlorine and iodine, only one included bromine as a candidate cysticide.[8] Table 10-1 documents results from previous studies selected for the complete description of experimental conditions. The divergence of results is apparent. This report summarizes a study in which chlorine, bromine, and iodine are compared as cysticides under the same test conditions utilizing the same bioassay technique. The cysticidal efficiency was determined for each halogen in buffered test cyst suspensions in (1) distilled water, (2) distilled water containing organic nitrogen, (3) distilled water containing organic nitrogen as glycine, and (4) in filtered autoclaved secondary sewage effluent. All tests were conducted at room temperature. Disinfection was determined from either prereacted or dynamic experiments. The latter is more representative of field conditions where the halogen is rapidly mixed with cyst suspension at time zero and reflects the dynamic combination and equilibration of reactants at the buffered pH. Prereaction experiments, however, allow the buffered water with all its constituents to react fully with the halogen, after which the cysts are mixed in at time zero. This permits a "freezing" of the halogen chemical species of interest and determines its particular property.

MATERIALS AND METHODS

Amoebic cysts were obtained from monkeys that became infected under natural circumstances. They consisted predominantly of *Entamoeba histolytica* and *Entamoeba coli.* Mixed species of cysts were found to be equally sensitive to the halogens as *E. histolytica* cysts from a human carrier, but much less sensitive than cultured *E. histolytica* cysts.[9]

All tests were performed in duplicate. The reaction flasks were thoroughly mixed throughout the experiments, and temperature was maintained at 27° to 30°C. Samples were taken from the reaction flasks at designated time intervals, and halogen was quickly neutralized by dumping into sodium thiosulfate. The halogen concentrations were determined before, after, and frequently during each experiment. Control cysts in buffered test suspensions without halogen were withdrawn and dumped into sodium thiosulfate in the same manner. Cyst density varied from 200 to 2000 cysts/ml. The greater cyst density admittedly increased the halogen demand, but also increased the sensitivity of the test to approximately 4 logs of inactivation (99.99%).

Cysts were concentrated from aliquots on 0.8μ Millipore filters by the procedure described in a previous publication.[10] In this procedure, a syringe equipped with a two-way valve was used to pump the sample through a Swinnex-25 Millipore filter holder containing a 22 mm MF, Type AA, 0.8μ

Table 10-1. Selected Results and Experimental Parameters from Previous Studies of the Amoebic Cysticidal Properties of Halogens

Strain Source of *E. histolytica* Cysts	Halogen	ppm Halogen Dose	ppm Halogen Residuals	ppm Nitrogeous Materials	Contact Time in Minutes	pH	Temperature (C)	Cysts/ml	% Kill	Method Used to Measure Cyst Viability	Method Used to Measure Halogen Concentration	Author
						Chlorine						
Amebiasis patient	chlorine	10	6.5	n.g.[a]	10	7	23-26°	113	13.2	Eosin stain	Starch iodide	11
"	"	50	41.7	"	"	"	27°	113	92.5	"	"	"
"	"	100	78.1	"	"	"	27°	220	97.6	"	"	"
"	"	100	76.3	"	"	"	1°	220	25	"	"	"
?	chlorine	500 (free)	n.g.	n.g.	10	n.g.	n.g.	n.g.	100	n.g.	n.g.	12
NRS	chlorine	4	2	0.12	20	n.g.	22°	1819	100	Culture	Orthotolidine	13
"	"	"	2.2	"	10	"	"	"	90	Observation	"	"
"	"	"	2	"	20	"	"	"	100	Observation	"	"
"	"	"	2.2	"	10	"	"	"	80	Eosin stain	"	"
"	"	"	2	"	20	"	"	"	92	Eosin stain	"	"
"	"	10	5.5	"	20	"	"	"	100	Culture	"	"
"	"	"	5.5	"	20	"	"	"	100	Observation	"	"
"	"	"	5.5	"	20	"	"	"	94	Eosin stain	"	"
NRS or P	chlorine	n.g.	2.2	0.09	15	5.6	10°	60-80	100	Culture	Starch iodide	14
"	"	"	5.4	"	"	7.0	"	"	"	"	"	"
"	"	"	6.0	"	"	8.2	"	"	"	"	"	"
"	"	"	7.0	"	"	9.0	"	"	"	"	"	"
"	"	1.75	1.0	n.g.	"	7.0	30°	60-100	"	"	"	"
Amebiasis patient	chlorine	18	12.7	High (in sewage)	30	n.g.	n.g.	n.g.	Less than 100	n.g.	n.g.	15

NRS	chlorine	n.g.	1.0	0.5-2.0	15	5.0	27°	100	100	n.g.	n.g.	11
NRS or P	chlorine	3.8	2.4	0.1-0.2	15	6.8-7.2	18°	30-75	100	Culture	Starch iodide	10
″	″	4.0	3.0	0.1	15	6.0	10°	75	″	″	″	″
″	″	6.0	4.5	0.1	15	7.0	110°	75	″	″	″	″
″	″	10.0	n.g.	n.g.	10	5.0	n.g.	n.g.	20-30	Orthotolidine stain	Orthotolidine	″
NRS or P	chlorine	1.8-2.4 (free)	1.8-2.4 (free)	n.g.	100	8.3	Room temp.	3-10	Less than 100	Culture	Orthotolidine	16
NRS	HTH	1.0	n.g.	5-2	15	5.0	27°	100	100	Culture	n.g.	17
NRS	HTH	n.g.	56.6	0.1-0.65	15	6.5-7.2	19-28°	20	Less than 100	Culture	Starch iodide	18
″	″	″	″	″	20	″	″	″	100	″	″	″
NRS or P	HTH	4.5	3.5	0.1-0.2	15	7	18°	30-62	100	Culture	Starch iodide	19
″	″	13	12.5	″	″	8.5	″	″	″	″	″	″
NRS or P	HTH	n.g.	22.9	n.g.	10	5.0	3°	30	100	Culture	Orthotolidine	10
″	″	″	25.1	″	″	7.0	″	″	″	″	″	″
″	″	″	144	″	″	9.0	″	″	″	″	″	″
″	″	″	806	″	″	9.9	″	″	″	″	″	″
″	″	″	2	″	″	5.5	28°	″	″	″	″	″
″	″	″	2.8	″	″	7.2	″	″	″	″	″	″
″	″	″	35	″	″	9.0	″	″	″	″	″	″
″	″	″	425	″	″	10.0	″	″	″	″	″	″
Amebiasis patient	HTH	n.g.	2.0	0.4	20	7	18°	25-40	100	Culture	Orthotolidine	20
Amebiasis patient	Chlorinated lime	100 (free)	n.g.	n.g.	Several hours	n.g.	n.g.	n.g.	Less than 100	Eosin stain	n.g.	21
?	Chlorinated lime	350	n.g.	n.g.	n.g.	n.g.	n.g.	n.g.	Less than 100	n.g.	n.g.	22
Amebiasis patient	Chlorinated lime	1750	n.g.	n.g.	10	n.g.	n.g.	n.g.	0	Eosin + I_2 stain & plasmolysis of cell	Weight	23

Table 10-1, continued.

Strain Source of *E. histolytica* Cysts	Halogen	ppm Halogen Dose	ppm Halogen Residuals	ppm Nitrogeous Materials	Contact Time in Minutes	pH	Temperature (C)	Cysts/ml	% Kill	Method Used to Measure Cyst Viability	Method Used to Measure Halogen Concentration	Author
Chlorine (continued)												
Amebiasis patient	Chlorinated lime	350	n.g.	n.g.	30	n.g.	37°	n.g.	100	Culture	Weight	24
Amebiasis patient	Chlorinated lime	3500	n.g.	n.g.	360	n.g.	n.g.	n.g.	Less than 100	Culture	Weight	25
Amebiasis patient	Chlorinated lime	3.5	n.g.	n.g.	30	n.g.	n.g.	45	0	Disintegration of nuclei	Weight	26
Amebiasis patient	Chlorine + NH_4 - (2:1)	9.5-10	n.g.	n.g.	10	7	21.5°	231	4.1	Eosin stain	Starch iodide	11
"	"	500	"	"	"	"	"	"	22.8	"	"	"
NRS or P	Chlorine + NH_4OH	6	5	0.13-0.2	15	6.8-7.4	18°	30-45	100	Culture	Starch iodide	19
"	"	14	11	"	"	8.6	"	"	"	"	"	"
"	Chlorine + peptone proteose	55	25	22.2-35.5	30	3.5-4.5	"	35-60	"	"	"	"
NRS or P	Chlorine + NH_4Cl (1:1)	n.g.	10.1	n.g.	10	6.9	23°	30	100	Culture	?	10
"	"	"	6	"	"	5.1	"	"	"	"	"	"

						Iodine						
NRS or P	iodine:I^- (1:0)	4	3.2	n.g.	10	5	23°	60	100	Culture	Starch iodide	10
〃	iodine:I^- (1:1)	5	3.8	〃	〃	〃	〃	〃	〃	〃	〃	〃
〃	iodine	〃	〃	〃	〃	〃	〃	30	〃	〃	〃	〃
?	iodine	8	+4	n.g.	10	?	+0°	30	100	Culture	Starch iodide	27
〃	〃	〃	n.g.	〃	〃	5	3°	〃	〃	〃	〃	〃
〃	〃	3	〃	〃	〃	〃	23°	〃	〃	〃	〃	〃
NRS	iodine	10	6.5-7.5	n.g.	6	6	23°	50-500	99.95	MPN	Acid starch iodide	28
〃	iodine + I^-	n.g.	I_2 15 I^- 5000	〃	10	6.3	35°	n.g.	100	Culture	〃	29
〃	〃	〃	I_2 20 I^- 5000	〃	〃	〃	25°	〃	〃	〃	〃	〃[b]
〃	〃	〃	I_2 60 I^- 5000	〃	〃	〃	6°	〃	〃	〃	〃	〃[b]
NRS	iodine	n.g.	7.5	n.g.	10	6.5	3°	30-60	100	Culture	Acid starch iodide	30
〃	〃	〃	2.5	〃	10	6.5	35°	〃	〃	〃	〃	〃
〃	〃	〃	1.75	〃	20	6.6	35°	〃	〃	〃	〃	〃
〃	〃	〃	13.7-14.5	〃	5	8	6°	〃	〃	〃	〃	〃
〃	〃	〃	6.9-7.2	〃	5	8	25°	〃	〃	〃	〃	〃
						Bromine						
NRS or P	bromine	+14	+3.6	n.g.	10	7.1	23°	30-60	100	Culture	n.g.	10

[a]n.g. – not given in reference.

filter. Cysts remained on the filter, and the liquid passed through it. Extra water was passed through the unit to wash cysts adhering to the filter. By disassembling the unit and turning the filter over, cysts could be backwashed into a 15-ml centrifuge tube and concentrated into 0.4 ml of liquid. To the concentrated cysts 2.6 ml of excystation medium was added, and cyst viability was determined.[10] The excystment culture was incubated for at least 10 hours at 37°C to allow viable amoebae to excyst and leave behind an empty cyst wall. Dead amoebae remained in their cysts. Under high dry magnification both full (dead) and empty (live) cysts could be counted and the per cent of live cysts calculated.

Bromine stock solutions were prepared by bubbling nitrogen gas through a train of three gas washing bottles containing elemental bromine, deionized distilled water, and neutralized serum solution. Stock bromine solution containing 10,000 to 15,000 mg/l of bromine was produced in the middle bottle. Bromine concentration was determined by the colorimetric bromcresol purple test of Larson and Sollo.[31] Iodine solution was made by dissolving elemental iodine in deionized distilled water. Stock chlorine solution was made by bubbling chlorine gas into 0.1*M*, pH 7.0 phosphate buffered distilled water. Black and Whittle[32] colorimetric tests were used to measure concentrations of iodine and chlorine. Colorimetric values were standardized against halogen concentrations, which were determined by means of standard titration procedures. Ammonium chloride or glycine was prereacted with the halogen to prepare the desired organic or inorganic halogen test solution. Secondary sewage effluent was filtered and autoclaved before it was used to limit the problem of millipore filtering and excessive bacterial growth during the incubation step of cyst bioassay.

RESULTS AND DISCUSSION

Summarized in Figure 10-1 are the results of many prereaction dose response experiments conducted in distilled water buffered over a pH range of 4 to 10. The cyst kill, corrected for controls at the end of 10 minutes of contact, are related to the 10-minute residual halogen. Compared with chlorine and iodine, bromine was a more effective cysticide throughout the pH range. Furthermore, the pH value influenced cysticidal action of bromine much less than that of chlorine and iodine. At pH 4, 1.5 ppm bromine residual was needed to produce 99.9% cyst mortality in 10 minutes. The same level of cyst mortality could be produced by 5 ppm iodine and 2 ppm chlorine residual respectively. Higher pH values weaken all three halogens as cysticides; however, at pH 10 a residual of only 4 ppm bromine is required to produce 99.9% mortality as compared with 12 ppm chlorine and more than 20 ppm iodine.

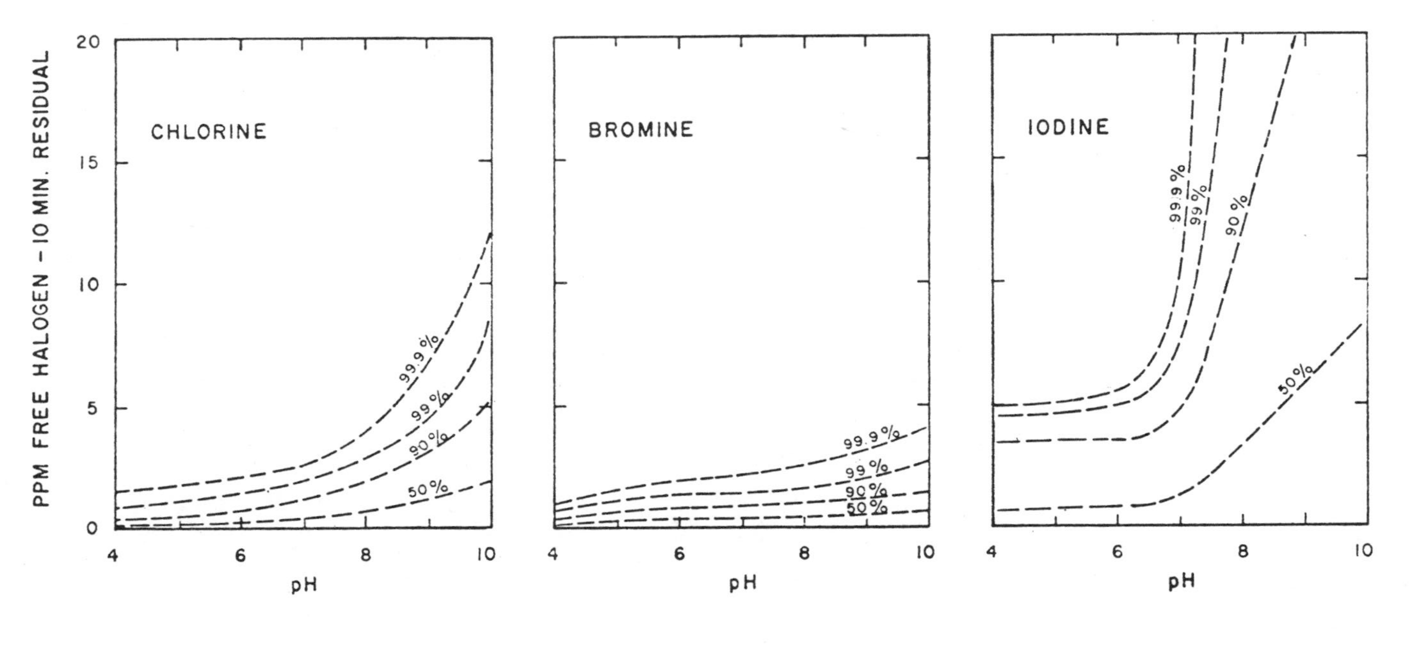

Figure 10-1. The residual free halogen concentration after 10 minutes of contact time required to obtain the indicated level of amoebic cyst mortality. Amoebic cysts from simian hosts mixed into buffered distilled water at 30°C to a final density of 1600-2000 cysts/ml.

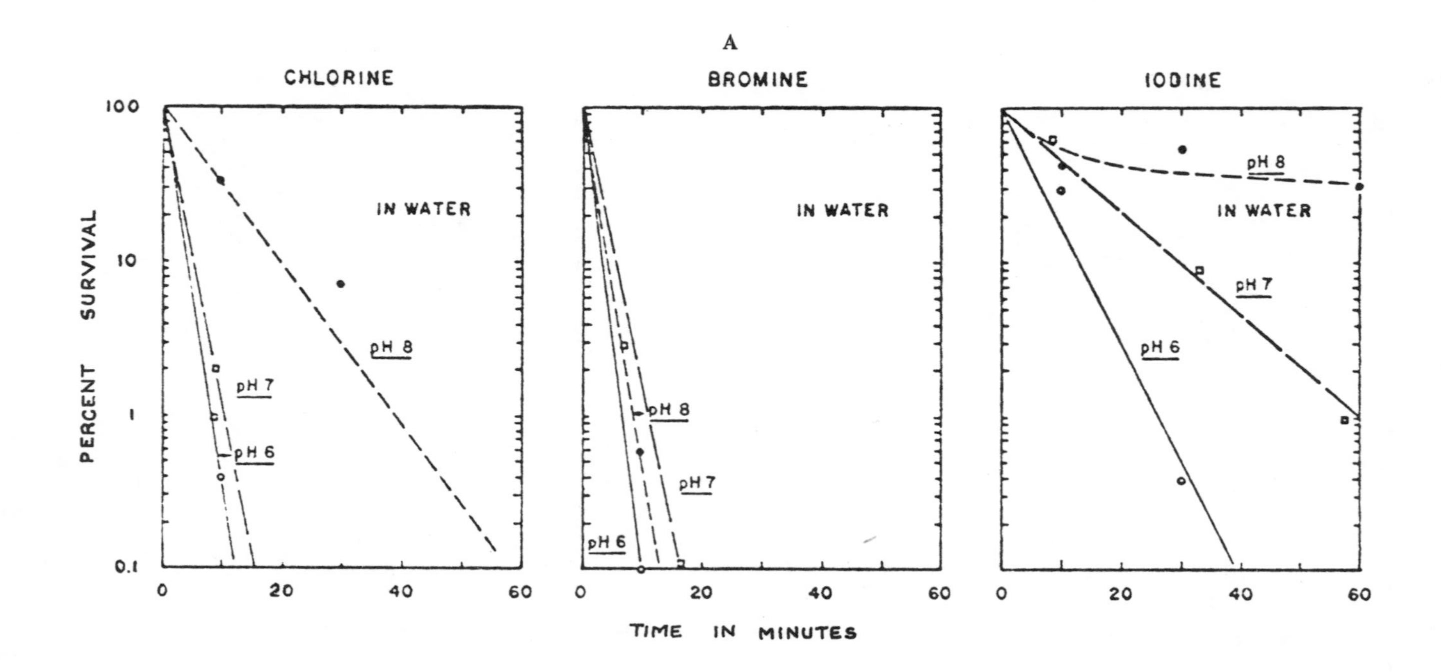
A
CHLORINE
BROMINE
IODINE
IN WATER
pH 8
pH 7
pH 6
IN WATER
pH 8
pH 7
pH 6
pH 8
IN WATER
pH 7
pH 6
PERCENT SURVIVAL
100
10
1
0.1
0
20
40
60
TIME IN MINUTES

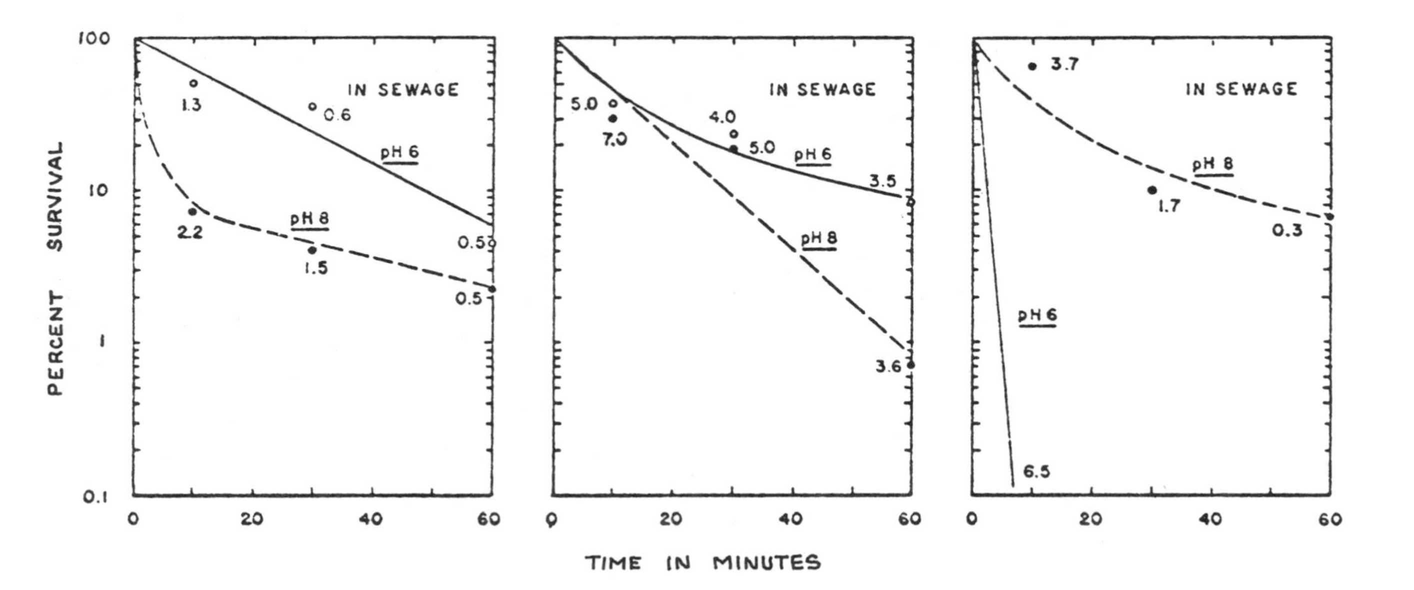

Figure 10-2. Comparison of the cysticidal properties of the halogens flash-mixed at time zero using:

A. 2 mg/l concentration in water buffered to pH 6, 7 and 8

B. 13.7 mg/l dose of halogens to secondary treated sewage effluent buffered to pH 6 and 8.

(Number given in the sewage disinfection experiments represent the combined chlorine and bromine residuals and the iodine free residual, all in mg/l at various time periods. Simian cyst density 200/ml. Temperature 27°C.)

Figure 10-2 compares the performance of the halogens reacted dynamically in clean and dirty water systems with constant dose of chlorine, bromine and iodine. These tests more closely approximate natural conditions with a cyst density of 200/ml, and the halogen flash mixed into the cyst test suspension, buffered to pH 6, 7 and 8, at time zero. Halogen dosages were selected to give a good representation of the difference in efficiency. In the clean water 2 mg/l (Figure 10-2A) and in the dirty (sewage effluent) water 13.7 mg/l (Figure 10-2B) were selected for comparison at pH 6 and 8. Results from the dynamic studies in distilled water (Figure 10-2A) reiterate what was shown in the prereaction tests (Figure 10-1)–namely, that free bromine is a more efficient cysticide than either chlorine or iodine, and that pH effect on its cysticidal potency is insignificant when compared with the effect of pH on chlorine and iodine. A 15-minute contact time resulted in death of 99.9% of the cysts exposed to bromine at both pH 6 and pH 8, but to chlorine at only pH 6. To achieve 99% mortality with chlorine at pH 8 required a contact time exceeding 60 minutes. Iodine at pH 8 was ineffectual as a cysticide, and at pH 6 more than 40 minutes contact time was required to produce 99.9% cyst mortality with 2 mg/l of iodine.

It might be concluded then that bromine would be the disinfectant of choice for amoebic cysts. However, results from the dynamic studies of the halogens and cysts in sewage effluent shown in Figure 10-2B demonstrate the major cysticidal strength of iodine and weakness of bromine and chlorine. It was observed in these tests that no free residual chlorine or bromine was detected after one second in sewage; however, free iodine could be detected even after one hour of contact time, when only combined residual chlorine and bromine could be detected. Bromine and chlorine react with compounds in the sewage to produce the haloamines that are shown to be poor cysticides; iodine does not react with the interfering compounds and remains in the elemental state. In sewage at pH 6, 13.7 mg/l iodine killed nearly 99.9% of the cysts in less than 10 minutes. In sewage at pH 8, iodine was as ineffectual a cysticide as were both chlorine or bromine.

Figure 10-3 summarizes tests comparing the cysticidal properties of 2 ppm dose of the three halogens added to a $10^{-3} M$ ammonium chloride solution, a $10^{-3} M$ glycine solution and in deionized distilled water. Cysticidal action of the organic haloamine is very similar to that of the halogen in sewage, suggesting that it is the organic haloamines which predominate in halogenated sewage effluent. Of the three forms of halogens (free, organic amino and inorganic amino), inorganic haloamines are more cysticidal than the organic haloamines and the free halogens are most cysticidal. Unlike bromine and chlorine, iodine does not react with ammonium

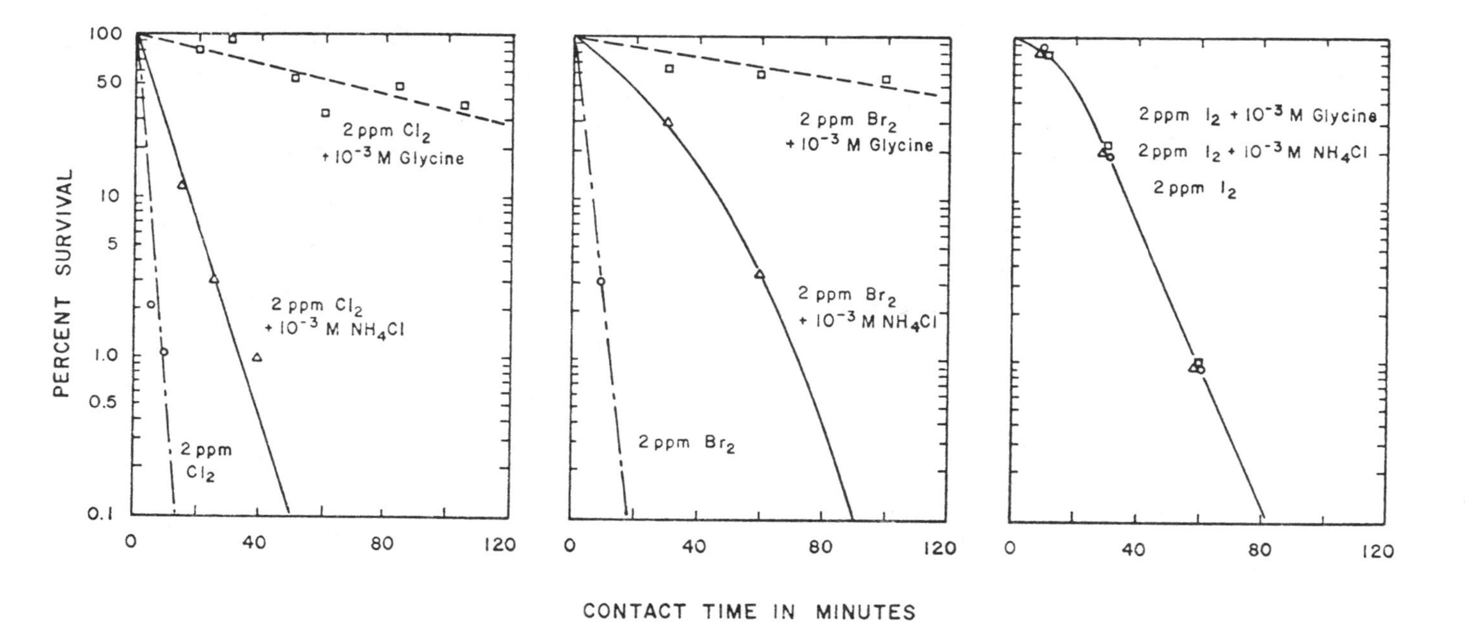

Figure 10-3. Comparative response of amoebic cysts after varying time of exposure to dosage of 2 mg/l of prereacted ammonia-haloamine, N-halo glycine and free halogen in water buffered to pH 7.0 at 30°C. Amoebic cysts from simian hosts at a density of 200 cysts/ml.

chloride and glycine compounds. As Figure 10-3 shows, iodine in distilled water is nearly as cysticidal as iodine in the ammonium chloride and glycine solutions.

Results from this study strongly reinforce the use of slightly acid-buffered iodine to disinfect water during such emergency situations as wars, national disasters when the water may be heavily contaminated, and on camping trips. The Globaline tablet (Van Brode Milling Company, Clinton, Mass.) is a standard issue of the U.S. Armed Forces for canteen water disinfection. The tablet is compounded with tetraglycine hydroperiodide that releases 8 mg iodine/l of water. An acid buffer (disodium dihydrogen pyrophosphate) lowers the water pH to 6.5 to 7.0 so that the predominant iodine species present is I_2 (diatomic iodine), which has been confirmed to be the most cysticidal of the halogen species tested in water with sewage. To disinfect larger volumes of water with iodine when a municipal supply is unavailable, the iodine saturator[33] maintains a reliably constant iodine residual in the water. With this system, a solution saturated with elemental iodine is pumped into the waterline at a rate proportional to the flow, which can be adjusted to maintain the desired residual concentration of iodine in the final effluent. The system's simplicity and efficiency make it ideal for emergency use or when halogenation to a free bromine or chlorine residual is not practical.

SUMMARY

Bromine is compared with iodine and chlorine as a disinfectant for the causative agent of amoebic dysentery (*Entamoeba histolytica*) in distilled water, in distilled water with ammonium chloride or glycine and in secondary sewage effluent. Variables include pH, contact time and halogen concentration. The following conclusions may be drawn:

(1) In distilled water, which is free of interfering substances, bromine is the superior cysticide; compared with chlorine and iodine at the same dosage, it is more cysticidal throughout the pH range 4 to 10 and its action is not as greatly influenced by pH.

(2) All free and combined species of halogen have better cysticidal properties in water at less than pH 7.0. Differences in the kill are attributed to the predominating halogen species at the given pH. The chemical species of free halogen in very low pH is the diatomic halogen and sometimes a mixture of the diatomic and hypohalous acid form. In the pH range of 4 to 7, I_2, HOCl, and mixtures of HOBr and Br_2 predominate. In the highest pH range of water the hypohalite ion will predominate along with mixtures with the hypohalous acid. In the pH range of 7 to 9, mixtures of I_2 and HOI, HOCl and OCl^- and HOBr predominate. In the very high pH range

9 to 12, HOI, HOBr, OBr^- and OCl^- will be the dominating species. Of these HOI, OBr^- and OCl^- are not very cysticidal agents.

(3) The combined forms of chlorine and bromine are slow reacting and much poorer cysticides compared to free halogens. The combined forms are more cysticidal in waters below pH 7.0. In sewage effluent at pH 6 only iodine has significant value as a cysticide. In sewage at pH 8 chlorine, bromine, and iodine are poor cysticides. The loss of cysticidal potency of both chlorine and bromine in sewage is attributed to its affinity to combine with interfering nitrogenous substances. Unlike bromine and chlorine, iodine does not have this affinity and remains uncombined in sewage.

(4) Organic haloamines have about the same cysticidal activity as the halogens in sewage. Inorganic (ammonia) haloamines are slightly more cysticidal than the organic haloamines (N-haloglycine).

(5) Results reinforce the position that iodine is the best disinfectant for amoebic cysts in heavily contaminated water where a free residual of chlorine or bromine is difficult to attain. The cysticidal efficiency of iodine is, however, a strong function of pH. Thus, the pH must be slightly acidic.

ACKNOWLEDGMENTS

This report was supported by the U.S. Army Medical Research and Development Command. However, the report's findings are not to be construed as an official Department of the Army position unless so designated by other authorized documents.

We would like to thank especially the New York City Health Department, particularly Dr. DiRemos of the Bronx Tropical Disease Clinic for generously supplying us with the human cysts used in comparative studies.

REFERENCES

1. American Public Health Association. "Waterborne Outbreaks in the United States and Canada (1930-36) and Their Significance," *Amer. Public Health Assoc. Yearbook,* (New York: American Public Health Association, 1937-38).
2. Maxcy, K. F. and M. J. Rosenau. In *Preventive Medicine and Public Health*. P. E. Sartwell, Ed. (New York: Appleton-Century-Crofts, 1956).
3. Botero, D., C. Bravo and O. Dunque. "Clinical, Pathological and Therapeutic Studies of Amebiasis in Columbia, South America," Abstract No. 6, presented at the 18th Annual Meeting of the American Society of Tropical Medicine and Hygiene, Washington, D.C. (1969).

4. Cifarelli, P. "Acute Amoebic Colitis," *USARV Med. Bull.* **2**, 19-26 (1967).
5. Koifold, C. A., S. I. Kornhauser and J. T. Plate. "Intestinal Parasites in Overseas and Home Service Troops of the U.S. Army with Special Reference to Carriers of Amebiasis," *J. Amer. Med. Assoc.* **72**, 1721-1724 (1919).
6. World Health Organization. "Report of a WHO Expert Committee on Amebiasis," *World Health Org. Tech. Rep. Series* No. 421, Geneva, Switzerland (1969).
7. Chang, S. L. "Small, Free-Living Amoebas: Cultivation, Quantitation, Identification, Classification, Pathogenesis and Resistance," *Current Topics Comp. Pathobiol.* **1**, 201-254 (1971).
8. Fair, G. M., S. L. Chang and J. C. Morris. "Final Report on Disinfection of Water and Related Substances to the Committee on Medical Research of the Office of Scientific Research and Development," Washington, D. C. (1945).
9. Stringer, R. P. "Amoebic Cysticidal Properties of Halogens in Water," Doctor of Science Thesis, Department of Environmental Health, The Johns Hopkins University School of Hygiene and Public Health, Baltimore, Md. (1970).
10. Stringer, R. P. "New Bioassay for Evaluating Per Cent Survival of *Entamoeba histolytica* Cysts," *J. Parasitol.* **58**, 306-310 (1972).
11. Spector, B. K., J. R. Baylis and O. Gullans. "Effectiveness of Filtration in Removing from Water, and of Chlorine in Killing, the Causative Organism of Amoebic Dysentery," *U.S. Publ. Health Rep.* **49**, 786-800 (1934).
12. Bolduc, A. "Amoebic Dysentery," *Can. Publ. Health J.* **26**, 215 (1935).
13. Stone, W. S. "The Resistance of *Endamoeba histolytica* Cysts to Chlorine in Aqueous Solutions," *Amer. J. Trop. Med.* **17**, 539-551 (1937).
14. Chang, S. L. and G. M. Fair. "Viability and Destruction of Cysts of *Endamoeba histolytica*," *J. Amer. Water Works Assoc.* **33**, 1705-1815 (1941).
15. Gordon, E. "Purification of Sewage from Cysts of Intestinal Protozoa," *Med. Parasit. Parasitic Dis., Moscow* **10**, 236; *U.S. Publ. Health Eng. Abs.* **23**, 16 (1943).
16. Chang, S. L. and B. W. Kabler. "Detection of Cysts of *Endamoeba histolytica* in Tap Water by Use of a Membrane Filter," *Amer. J. Hyg.* **64**, 170-180 (1956).
17. Kessel, J. F., D. K. Allison, M. Kaime, M. Quiros, and A. Gloeckner. "The Cysticidal Effect of Chlorine and Ozone on Cysts of *Endamoeba histolytica* Together with a Comparative Study of Several Encystment Media," *Amer. J. Trop. Med.* **24**, 177 (1944).
18. Brady, F. J., M. F. Jones and W. L. Newton. "Effect of Chlorination of Water on the Viability of Cysts of *Endamoeba histolytica*," *War Med.* **3**, 409 (1943).
19. Chang, S. L. "Studies of *Endamoeba histolytica*. III. Destruction of Cysts of *Endamoeba histolytica* by a Hypochlorite Solution, Chloramines in Tap Water of Varying Degrees of Pollution," *War Med.* **5**, 46-55 (1944).

20. Becker, E. R., C. Burks and E. Keleita. "Cultivation of *Endamoeba histolytica* in Artificial Media from Cysts in Drinking Water Subjected to Chlorination," *Amer. J. Trop. Med.* **26**, 783 (1946).
21. Wenyon, C. M. and F. W. O'Connor. "The Carriage of *Endamoeba histolytica* and Other Protozoa and Eggs of Parasitic Worms by House-flies with Some Notes on the Resistance of Cysts to Disinfectants and Other Reagents," *J. Red Army Med. Corps.* **28**, 522-527 (1917).
22. Bercovitz, N. "Viability of Cysts of Human Intestinal Amoebas as Determined by Exposure to Various Substances and Subsequent Staining in Haematoxylin," *Univ. Calif. Publs. Zool.* **26**, 249-261 (1924).
23. Mills, R. G., C. L. Bartlett and J. E. Kessel. "The Penetration of Fruits and Vegetables by Bacteria and Other Particulate Matter, and the Resistance of Bacteria, Protozoan Cysts and Helminth Ova to Common Disinfection Methods," *Amer. J. Hyg.* **5**, 559-579 (1925).
24. Yorke, W. and A. R. D. Adams. "*Entamoeba histolytica.* The Longevity of Cysts *in Vitro*, and their Resistance to Heat and to Various Drugs and Chemicals," *Ann. Trop. Med. Parasit.* **20**, 317-326 (1926).
25. Liu, L. "The Comparative Lethal Effects of Certain Chemicals on Bacteria and Cysts of *Endamoeba histolytica* from Human Feces," *Chin. Med. J.* **42**, 568 (1928).
26. Garcia, E. Y. "Effects of Chlorinated Lime in Lethal Concentrations on *Entamoeba histolytica* Cysts," *Philipp. J. Sci.* **56**, 295 (1935).
27. Chang, S. L. and J. C. Morris. "Elemental Iodine as a Disinfectant for Drinking Water," *Ind. Eng. Chem.* **45**, 1009-1012 (1953).
28. Chang, S. L. and M. Baxter. "Studies on Destruction of Cysts of *Endamoeba histolytica.* I. Establishment of the Order of Reaction in Destruction of Cysts of *E. histolytica* be Elemental Iodine and Silver Nitrate," *Amer. J. Hyg.* **61**, 121-132 (1955).
29. Chang, S. L., M. Baxter and L. Eisner. "Studies on Destruction of Cysts of *Endamoeba histolytica.* II. Dynamics of Destruction of Cysts of *E. histolytica* in Water by Tri-iodide Ion," *Amer. J. Hyg.* **61**, 133-141 (1955).
30. Chang, S. L. "The Use of Active Iodine as a Water Disinfectant," *J. Amer. Pharm. Assoc.* **47**, 417-423 (1958).
31. Larson, T. and F. W. Sollo. "Determination of Free Bromine in Water," Annual Progress Report, U.S. Army Medical R & D Command, Contract No. DA-49-193-MD-2909 (1967).
32. Black, A. P. and G. P. Whittle. "New Colorimetric Determination of Halogen Residuals. I. Iodine. II. Free and Total Chlorine," *J. Amer. Water Works Assoc.* **59**, 471-490, 607-619 (1967).
33. Kinman, R. N., A. P. Black and W. C. Thomas. "Status of Water Disinfection with Iodine," *Proc. Natl. Specialty Conf. on Disinfection*, July 8-10, 1970, University of Massachusetts, Amherst (New York: American Society of Civil Engineers, 1970), pp. 11-35.

CHAPTER 11

DISINFECTION OF WATER AND WASTEWATER USING OZONE

E. W. J. Diaper

Crane Co.
Cochrane Environmental Systems
King of Prussia, Pa.

INTRODUCTION

Ozone has been used continuously in Europe for more than 70 years as a disinfectant in water treatment, principally for municipal supplies. In this country ozone is not common, chiefly because there is no residual in the distribution system. In the treatment of effluents, however, absence of a permanent ozone residual actually is an advantage because there are no toxic effects created in the receiving stream and no increase in the dissolved solids concentration. Other advantages of ozone include its high oxidizing potential, rapid reaction time and the high dissolved oxygen content in the effluent. This paper reviews the historical background and present-day use of ozone in water treatment, and present results of pilot tests on sewage effluents carried out in London, England, and Chicago, Illinois.

HISTORICAL BACKGROUND AND PRESENT-DAY USE

In 1785 Van Marum noticed an odor during an electric discharge. Schonbein (1840) showed that this odor was due to a particular gas, which he named ozone (Greek–*Ozein*, to smell). In 1857 Siemens constructed the first apparatus to produce ozone by electric discharge. Soret established the formula O_3 in 1867.

Sterilization of water with ozone was first carried out by De Meritens in 1886, and the first permanent installation was at Oudshoorn in Holland where Schneller, Van der Sleen and Tindal constructed a plant in 1893 for sterilizing Rhine River water after settlement and filtration.

In 1897 Dr. M. P. Otto of Paris published research results on ozone generation and laws governing its economic production from electricity. Four years later, in 1901, Siemens and Halske installed a plant in Wiesbaden, Germany, with 66,000 gph capacity.

In 1906 the city of Paris held a competition for sterilizing the River Marne water at St. Maur. Ozone was selected which led to rapid development, particularly in France. Since these early efforts ozone sterilization of water in France and other European countries has been continuous, interrupted only by two World Wars. There are now more than 1000 installations in 20 different countires including France, Germany, Switzerland, Norway, Holland, Russia, Canada and Mexico.

Some of the largest installations are used to supply water to the suburbs of Paris. Installations at Mery, Neuilly and Choisy treat water taken from the Oise, Marne and Seine Rivers, respectively. The combined capacity of these treatment works is 360 mgd and total output of ozone, which is produced from air, is approximately 10,000 lb/day. Because of the large quantity of ozone required to disinfect these municipal supplies, 20 of the world's largest ozone generating units were designed and installed, each measuring approximately 12 ft long by 6 ft in diameter and capable of 500 lb/day output on air and twice this amount on oxygen as the feed gas.

The most recent developments in water treatment using ozone have been the application of Microstraining™ (Crane Co.) and ozonation as complete treatment for colored waters in Scandinavia and Northern Europe. Because ozone will effectively bleach out organic color from water supplies, the conventional treatment process including chemical coagulation and settlement can be superceded by a relatively small dose of ozone, which can also ensure disinfection. A principal advantage of this system, besides economy, is the absence of chemical sludges.

Further development work is being carried out on polluted lakes in France and Switzerland. The city of Constance recently started up an installation comprising Microstraining and ozonation, followed by sand filtration. It is claimed that ozone not only removes color, disinfects and improves taste and odor, but also makes the organic colloids in the lake water more filterable.[1]

This development has been carried out on drinking water supplies, principally those to large cities. Very little work has been done on treatment of sewage effluents with ozone mainly because the need for disinfecting effluents has only recently arisen, primarily in the U.S.

There have been many investigations into the relative power of ozone and chlorine disinfection of bacteria and viruses. According to Bringman[2] destruction of bacteria in test conditions by ozone was between 600 and 3000 times more rapid than by chlorine. Ingols and Fetner[3] reported that the bactericidal action of chlorine is progressive, while that of ozone is sudden and total after a "threshold" is reached in dosage rate.

A report by Kessel & Associates showed that similar dilutions of poliovirus treated with chlorine to a residual of 0.5 to 1 ppm, and with ozone to a residual of 0.045 to 0.45 ppm, were rendered inert by chlorine in 1½ to 3 hours and by ozone in 2 minutes.[4]

A report presented at the 45th General Meeting of the Society of American Bacteriologists, indicated that bactericidal action of ozone was relatively unaffected by pH values whereas that of chlorine was considerably affected by changes in pH. Comparatively low concentrations of ozone produced rapid destruction of poliovirus and cysts of *Entamoeba*, whereas chlorine at higher concentrations was ineffective.[5] Ridenour and Ingols showed that ozone rendered a dilution of 1-500 virus inactive in four minutes.[6]

The comparison of ozone and chlorine is sometimes objected to because it is not possible to maintain an ozone residual in the water longer than a very short period. Chlorine, on the other hand, can give persistent residuals. Commenting on this Whitson observes that at Ashton, England, no deterioration in water quality has been noted in the sections supplied with ozonized water in 30 years and in any case the small amount of chlorine residual would have no real effect in the event of a major pollution accident within the distribution system.[7]

In practice an ozone residual is established and measured 5 minutes after injection at outlet from the contact column to ensure complete sterilization. This residual is maintained at 0.1-0.2 ppm to disinfect bacteria and 0.4 ppm to disinfect viruses. The applied dosage of ozone in drinking water would normally be between 0.5 and 1.5 ppm to obtain these residuals.

Although ozone is uncommon in the U.S. for the treatment of public water supplies, there are numerous installations for purifying bottled drinking water. The resulting additional palatability provided by ozone treatment is beneficial. Ozone is also used for food preservation in cold storage rooms, air conditioning and odor control from sewage treatment plants and industrial processes. In the chemical industry ozone is used to clean and bleach various organic materials such as fibers, flour, sugar and tobacco. Oxidation of furnace carbon black represents another important application for ozone. Some industrial wastes are treated with ozone, principally to remove phenols and cyanides.[8] Ozone is used in the purification and washing of shellfish, in the beverage industry for sterilizing containers and for the treatment of water used in zoos, medical baths and swimming pools.

PROPERTIES OF OZONE

Ozone is often referred to as an allotrope of oxygen. It is a pale blue gas with a distinct pungent odor that is apparent in concentrations well below 1 ppm in air. The color, however, is not apparent in concentrations normally produced and used. Although ozone is approximately 10 times more soluble than oxygen in water, the actual amount that can be dissolved under operating conditions is small. Ozone obeys Henry's Law and solubility is proportional to its partial pressure and the total pressure on the system. As ozone is normally produced in low concentrations in air or oxygen, its partial pressure is correspondingly low.

The molecular weight of ozone is 48; density at 0°C is 2.14 g/l. Its boiling point is -111.9°C and melting point is -251°C.

Because of the inherent instability of its molecular structure, ozone in liquid or gaseous phase can produce a series of virtually instantaneous reactions when in contact with oxidizable substances. The nature of these reactions is illustrated by the liberation of iodine from a solution of potassium iodide:

$$O_3 + 2KI + H_2O \rightarrow I_2 + O_2 + 2KOH$$

This reaction is used to determine the quantity of ozone in either air or water. If the determination is made on air, or oxygen, a measured quantity of the gas is drawn through a wash bottle containing potassium iodide solution. After lowering the pH with acid, titration is carried out with sodium thiosulfate, using starch solution as an indicator. A similar procedure may be followed to measure ozone dissolved in water although precautions must be observed for accuracy, since the amount of ozone being measured is normally very small.

Ozone residuals may also be measured electrically using silver and nickel electrodes placed in the flowing water. A small electric current is generated by the presence of ozone which can be measured on a milliammeter. The current reading is calibrated against chemical measurements, ozone residuals may be read off directly.

Although ozone is both toxic and corrosive it presents no safety or handling problems in properly designed operating systems. Unlike chlorine, which is normally stored under pressure, ozone is produced on site in low concentrations. Consequently any accidental leakage can easily be controlled. Its freedom from operating hazards is evident from its long background of trouble-free use in Europe.

The comparative oxidizing potentials of ozone, chlorine and chlorine dioxide are 2.07, 1.36 and 1.275 volts versus hydrogen, respectively. Ozone is the most powerful known oxidizing agent that can be used on a practical scale for water and waste treatment.

OZONE GENERATION AND INJECTION

Because of its instability, ozone cannot be bottled economically but must be generated on site. Electrical generation is the only practical method for large scale installations. An ozone generator incorporates a series of electrodes fitted with cooling arrangements and mounted in a gas-tight container. Metals used for the assemblies are stainless steel, aluminum or coated carbon steels.

The source gas, dry air or oxygen, is passed into the generator and through a narrow gap separating the high and low tension electrodes. Normally, high tension electrodes are provided with fuses while low tension electrodes are grounded. A glass insulating or dielectric plate prevents sparking and maintains a glow (corona) discharge in the air gap. Some oxygen is dissociated and reforms as O_3 following the reaction

$$3\ O_2 = 2\ O_3\ (-68.2K\ cal)$$

Commercial ozone generators are available with tube or plate electrodes and dielectrics.

Factors affecting ozone production include dryness of air, rate of air flow, air pressure within generator, power applied to electrodes, cooling water temperature, ozone concentration and frequency of applied voltage.

A typical performance curve for an ozone generator is shown in Figure 11-1. Increased ozone output can be obtained by using higher frequencies, or oxygen instead of air as the source gas.

When applied to water, special techniques must be used to introduce the ozonized air because of its low solubility. One system incorporates a Venturi throat through which the water to be treated passes downwards vertically into a contact chamber. The slight vacuum created sucks in the ozonized gas which forms an emulsion with the water created by the presence of very fine bubbles, across which the ozone activity takes place. This mixture of gas and water descends to the bottom of the vertical tube and rises through the contact chamber to the point of exit. This is the well-known "Otto" injection system. Another method incorporates porous diffusors which are laid at the bottom of the contact chamber. In some cases baffles are placed in the contact chamber to direct the flow of water, while the ozonized gas is introduced to the porous diffusors placed in the bottom of each compartment. In both systems the water depth is approximately 15 ft. The diffusor system has the advantage of small head loss, and changes in water flow rate can be accommodated more easily than with the emulsion column.

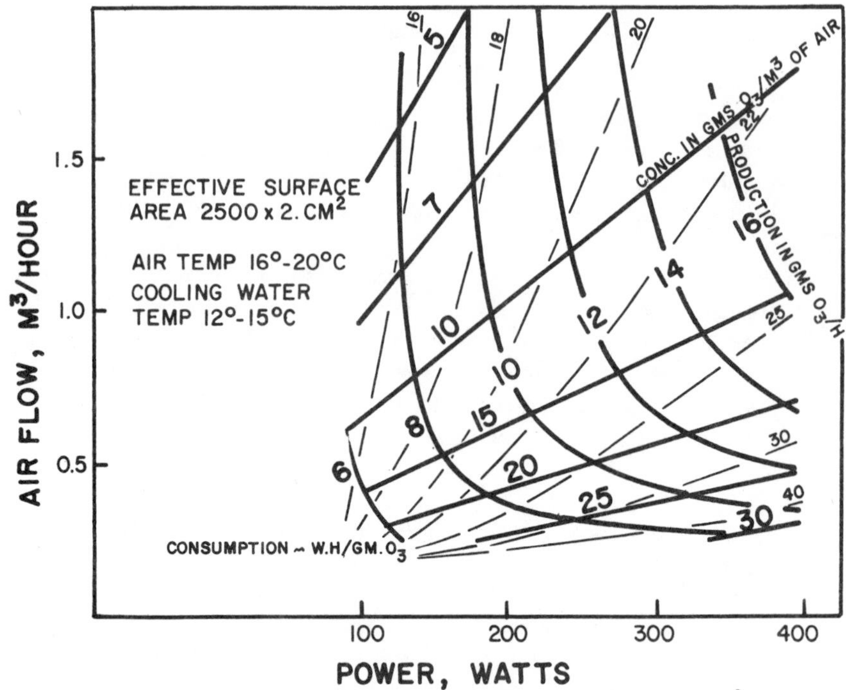

Figure 11-1. Typical performance curves for one ozonizer element.[9]

CONTROL OF DOSE IN DRINKING WATER

Addition of ozone to the inlet water can be controlled automatically by measuring the residual either at outlet from the contact chamber of at some intermediate point. Measurement of ozone residual is normally accomplished using electrodes placed in the flowing water; the signal obtained indicates the residual and controls the ozone generating equipment. This is normally done by regulating the primary side of the step-up transformer to control the voltage applied on the electrode plates.

OPERATING COSTS

Operating cost is directly related to electrical power costs and the concentration of ozone produced. Total power consumption to produce one pound of ozone from air at 1% weight concentration ranges from 10 to 12 kilowatt hours depending on efficiency of the ozone generator. Using oxygen the power consumption varies from 5 to 6 kilowatt hours per pound of ozone at 1.7% weight concentration. These figures do not include drying costs when air is used.

CAPITAL COSTS

Capital costs vary from approximately $1700 per pound of ozone produced from air for a skid mounted system under 12 lb/day capacity to $200-$300 per pound of ozone from air for systems with capacities over 1000 lb/day. If oxygen is used in place of air comparable costs are approximately 40% lower. Costs of producing dry air or oxygen must also be considered.

PILOT TESTS ON SEWAGE EFFLUENT

London, England

Description of Pilot Plant

Pilot plant tests were carried out at the Eastern Sewage Works, Redbridge, London, constructed in 1936 for a population of 36,800 and flow of 1.67 mgd. The present population is more than 50,000 and flow has risen to nearly 2.5 mgd. Raw sewage strength varies between 400 and 500 ppm suspended solids and 350 to 600 ppm BOD. Treatment comprises sedimentation, biological filtration at rates between 97 and 170 gal/cu yard per day followed by secondary settlement and Mark 1 Microstraining. Washwater from the microstrainer is recirculated. The works are overloaded by 40% volumetrically and by 60% in polluting load. British Royal Commission Standards of 30 ppm suspended solids and 20 ppm BOD are rarely obtained before Microstraining. After Microstraining these figures are better than the Royal Commission's Standards. Trade effluents comprise about 5% of the flow, some being pretreated. The outfall is to a small creek that flows into the estuary of the River Thames.

The pilot plant was installed at outlet from the sewage treatment works, taking effluent from the secondary sedimentation tanks upstream of the permanent Microstraining plant. Variable combinations of Mark 0 Microstraining, prechlorination, ozonation, coagulation and rapid sand filtration for a flow capacity of 2400 gal/hr were provided.

Equipment included a supply pump, 5 ft diameter x 1 ft wide Microstrainer, chlorine contact tank, circulating pump, total injection ozone contact column, inspection tank, second circulating pump and second ozone contact column, coagulation contact and inspection tank, rapid sand filter and final inspection tank. Two ozone contact columns in series were used because of the relatively high ozone dose (Figure 11-2).

Ozonation equipment included air refrigeration and dessication units, a 15-element CEO Otto Ozonator with high tension transformer, capable of producing a maximum ozone dose of 25 ppm at a concentration of 20 g of

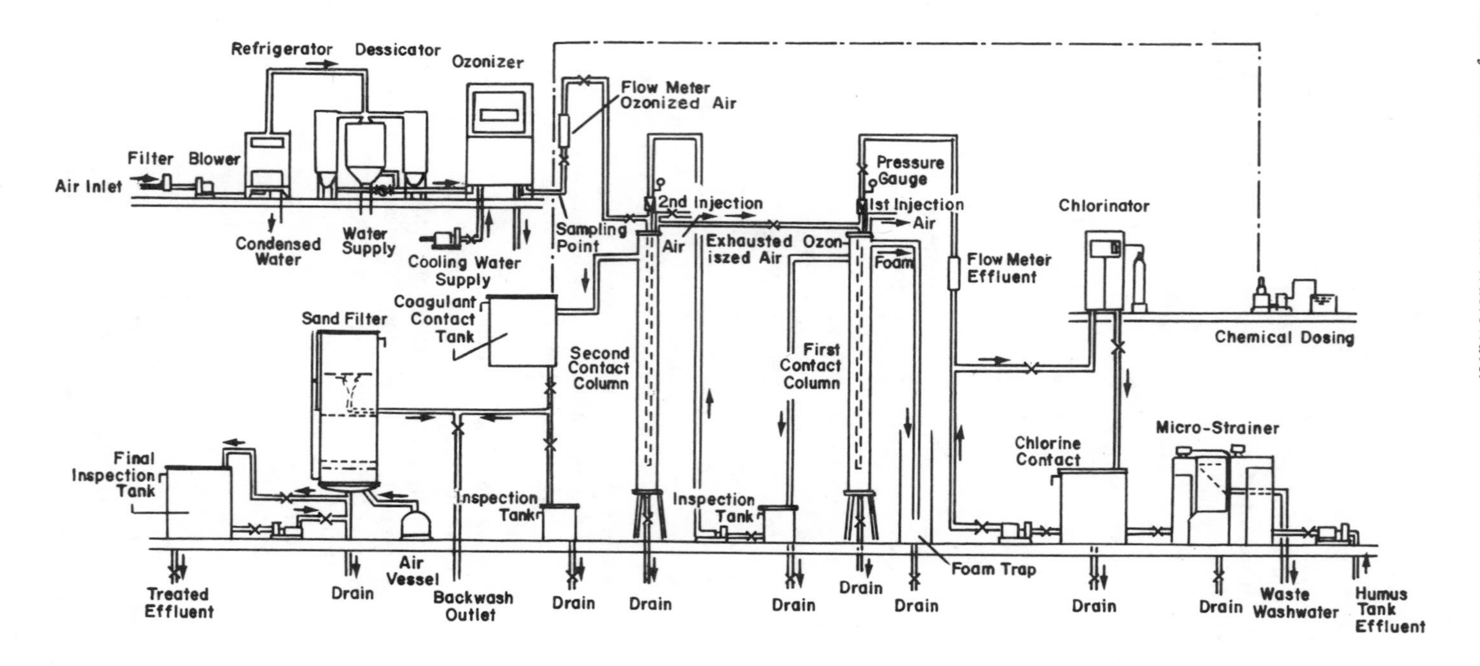

Figure 11-2. Pilot plant for investigation into water reclamation–Eastern Sewage Works, Redbridge, London.

ozone per cubic meter of air. The main ozone dose was applied in the second contact column, the residual being applied to the first contact column. The chlorinator could dose up to 20 ppm and the coagulation equipment up, to 60 ppm of aluminum sulfate. The sand filter was up to 4 gal/sq ft/min.

Experimental Results

Test work at the Redbridge Pilot Plant investigated the effects of Microstraining, ozonation and rapid sand filtration with and without prechlorination in modifying the physical, chemical and microbiological properties of a typical sewage effluent. Criteria for plant control was the color of the effluent leaving the plant and the residual ozone and/or chlorine at outlet from the ozone contact columns. Ozone dose was varied in accordance with the quality of the microstrained effluent to produce a treated water having color below 10° Hazen. To simplify operation and control the rate of water flow through the plant was maintained at 2400 gal/hr. Plant control data are summarized in Figure 11-3. When chlorination was employed a maximum dose of 20 ppm was applied with 5 minutes' reaction time before ozonation. For coagulation a dose of 2 ppm aluminum sulfate was applied. In early tests works effluents was used to backwash the sand filter but, since this procedure led to bacterial contamination of the ozonized effluent, the final treated water was used in later tests.

Table 11-1 shows the water composition at various stages of treatment using Microstraining, ozonation and sand filtration at 3 gal/sq ft/min. The humus tank effluent contained considerable quantities of suspended solids which the Microstrainer reduced by an average of 61%. The most obvious effect of ozonation was to remove the color of the effluent which, after sand filtration, was clear and sparkling, comparable in appearance with potable water. After ozonation the effluent was saturated with respect to dissolved oxygen.

Removal of detergent residue by ozone was appreciable. The anionic material was reduced from 1.1 to 0.2 ppm and the nonionic from 0.3 to 0.05 ppm. Some of this removal appeared to be due to foaming in the ozone column. The treated water still contained appreciable organic matter as indicated by COD and organic carbon figures.

Average results obtained during each of the four 2-week periods of operation are shown in Table 11-2. No improvement in physical and chemical quality resulted from chlorination, and its use did not reduce the dose of ozone required to produce a given degree of color reduction. The effect of the relatively small dose of coagulant employed in periods 3 and 4 was to reduce the suspended solids content and BOD to very low levels and to reduce the phosphate content. At the expense of shorter filter runs, better phosphate removal could have been obtained with larger doses of coagulant.

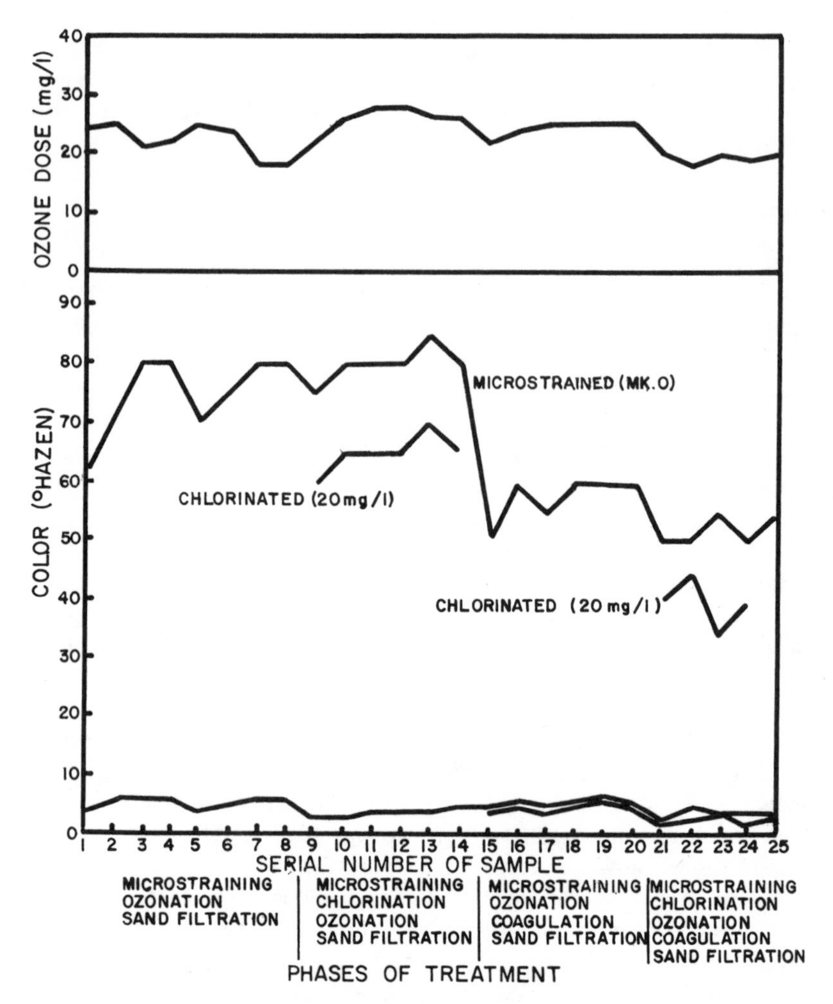

Figure 11-3. Summary of plant control measurements made between May and September, 1967.

The plant was operated during winter and summer conditions and performance was apparently unaffected by temperature.

The results of microbiological examinations are presented as median counts for all the samples collected at weekly intervals from January to September 1967 (Table 11-3). The results show that Microstraining alone had little effect on the counts of any group of organisms. Ozone killed the vast majority of organisms including all Salmonella and viruses. Chlorine produced lower counts than ozone and chlorine followed by ozone was

Table 11-1. Average Results Obtained at the Various Stages of Treatment[a] (Results mg/l, except where otherwise stated)

	Effluent From			
	Humus Tank	**Microstrainer**	**Ozonizer**	**Sand Filter**
Suspended solids	52	20	19	6
Total solids	927	–	–	896
BOD	16	10	10	8
COD	94	64	53	50
Permanganate value	17	11	8	8
Organic carbon	29	20	18	16
Surface-active matter:				
Anionic (as Manoxol OT)	1.1	1.1	0.2	0.2
Nonionic (as Lissapol NX)	0.3	–	0.05	0.05
Ammonia (as N)	4.6	4.8	5.3	5.2
Nitrite (as N)	0.4	0.4	0.03	0.01
Nitrate (as N)	25.7	25.7	26.2	26.7
Total phosphorus (as P)	9.2	–	–	7.6
Orthophosphate (as P)	7.6	–	–	7.3
Total hardness (as $CaCO_3$)	–	–	–	425
Chloride	122	–	–	121
Sulfate	192	–	–	194
Color (Hazen units)	41[b]	34[b]	7.5	7
Turbidity (ATU)	70	54	26	7
Total phenol	2.8	–	–	1.0
Temperature (°C)	15.2	14.9	15.6	15.7
Dissolved oxygen (% saturation)	54	57	99	96
Conductivity (μmho/cm^3)	1125	1118	1110	1150
Langelier Index	+ 0.10	–	–	+ 0.12
pH value	7.15	–	–	7.3
Pesticides (μg/l):				
α – BHC	–	Trace	Trace	–
γ – BHC	–	0.070	0.044	–
Dieldrin	–	0.042	0.033	–
pp–DDE	–	0.083	0.072	–
pp–TDE	–	<0.01	<0.01	–
pp–DDT	–	<0.01	<0.01	–

[a]Micellization-Demicellization pilot plant at the Eastern Sewage Works, Redbridge, London, 5-19 May 1967.

[b]Samples filtered through glass-fiber paper.

highly effective. All coliform and *E. coli* counts were zero. Results for the sand filter have been grouped together although some were after ozone only, some after chlorine only, some after ozone and chlorine and some after the use of a coagulant. The results show that filtration made little further improvement.

Table 11-2. Average Composition of Samples from Various Stages of Treatment During Operation of Water-Reclamation Plant
Flow through sand filter, 180 gal/ft^2 hr (175 m^3/m^2 day) (Results as mg/l except where stated otherwise)

Effluent From	Suspended Solids	BOD	Anionic Detergent (as Manoxol OT)	Nonionic Detergent (as Lissapol NX)	Phosphorus (as P)	Color (Hazen Units)	Turbidity (ATU)
Period 1 (May 5-19, 1967)							
Humus tank	52	16	1.1	0.3	9.2	41	70
Microstrainer	20	10	1.1	–	–	34	54
Ozonizer	19	10	0.2	0.05	–	7.5	26
Sand Filter	6	8	0.2	0.05	7.6	7	7
Period 2 (May 19-June 2, 1967)							
Humus tank	61	19	1.1	0.3	9.0	36	87
Microstrainer	27	12	1.1	–	–	37	61
Chlorinator	24	–	1.1	–	–	28	58
Ozonizer	19	–	0.3	0.003		3	37
Sand Filter	5	3	0.3	0.03	8.1	3	9
Period 3 (September 1-15, 1967)							
Humus tank	41	15	1.3	0.3	9.1	33	41
Microstrainer	13	8	1.3	–	–	34	24
Ozonizer	8	–	0.3	0.03	–	3.5	15
Coagulant and sand filter	2	2	0.3	0.03	6.5	4	4
Period 4 (September 15-29, 1967)							
Humus tank	43	14	1.2	0.2	8.1	38	58
Microstrainer	13	7	1.2	–	–	36	38
Chlorinator	12	–	1.2	–	–	28	35
Ozonizer	8	–	0.3	0.02	–	3	18
Coagulant and sand filter	1	1	0.3	0.01	5.7	2	1

Addition of Reagents (mg/l):

Period	Ozone	Chlorine	Coagulant (as $Al_2(SO_4)_3$)	Period	Ozone	Chlorine	Coagulant (as $Al_2(SO_4)_3$)
1	22	–	–	3	24	–	2
2	26	20	–	4	20	20	2

Table 11-3. Microbiological Examination Median Counts for All Samples

Treatment	Coliform Organisms per 100 ml	*E. coli* per 100 ml	37° Count per ml	22° Count per ml	Salmonella per liter	Virus PFU per liter
Works effluent	2,100,000	600,000	91,000	380,000	32	62
After microstrainers	1,700,000	700,000	92,000	440,000	32	58
Ozone only	90	32	10	162	0	0
Chlorine only	1	0	46	51	0	0
Chlorine + ozone	0	0	6	2	0	0
Sand filtrate	0	0	1	3	–	–

In presenting this work in England, Mr. G. A. Truesdale of the Water Pollution Research Laboratory summarized the results obtained by saying that ozone had produced an effluent which from a microbiological view point was far superior to river waters normally used as sources for domestic supplies and which would be suitable for a variety of industrial purposes without further treatment. In previous work his laboratory had demonstrated the feasibility of producing a clear sterile water from sewage effluents using the conventional process of coagulation, sand filtration and chlorination. Compared with this approach the MD process not only produced a product with better appearance and a lower salt content (because of the lower dose of coagulant) but had the advantage of using a smaller number of unit processes. In some cases treatment by Microstraining and ozone alone may be sufficient to permit water reuse.[10]

Chicago, Illinois

Description of Plant and Pilot

Pilot tests were performed in Chicago during 1970 on sewage effluent at the Hanover Park Plant operated by the Metropolitan Sanitary District of Greater Chicago. The Hanover water reclamation plant consists of conventional primary and secondary activated sludge processes followed by a tertiary facility to treat the entire plant flow. The input to the facility is basically domestic wastewater from a separate sewer system with an average flow of 1.5 mgd.

Effluent for the ozone tests was taken from downstream of a microstrainer, 10 ft diameter x 10 ft long, fitted with Mark 0 (23 micron aperture). At this point, the effluent contained between 1 and 7 ppm suspended solids and from 4 to 11 ppm BOD. Bacterial content was extremely variable ranging from 3000 to 1,500,000 cells per 100/ml.[11]

The purpose of the pilot tests was to evaluate ozone as a disinfectant and to determine whether its use had any other beneficial effects such as color reduction, BOD reduction, removal of suspended solids and other impurities. Operational characteristics of the ozone equipment were to be evaluated and the cost of installation and operation estimated.

The test equipment comprised a "Otto" plate-type ozonator (Type 4-63, 10 plates) operated at atmospheric pressure with a maximum output of approximately 200 g/hr at 4.5 kw. The ozone output could be varied by regulating the transformer to adjust voltage applied on the plates.

The air drying equipment included a filter, blower and refrigerator, followed by a two-cell dessicator. Regeneration of the dessicator was achieved by alternating the cells in service and blowing hot air through the exhausted media (silica). On the outlet from the dessicator a humidity cell was fitted to warn against moisture leakage. A recording watt meter was used to measure electrical energy supplied to the air blower, dessicator heater and blower, and circulating water pumps. Energy input to the ozone generator itself was obtained from readings on the ozonator instrument panel board.

Injection of ozone into the sewage effluent was achieved using the Otto "emulsion" system. Two contact columns were employed each measuring 30 inches inside diameter and 18 ft high containing a vertical pipe approximately 2 inches diameter fitted at the top with a Venturi throat. When sewage effluent was pumped through this throat, ozone was sucked from the generator. Two pumps were employed, one for each column and as these were constant speed pumps and somewhat overpowered for the duty, a bypass was fitted on each to allow flow regulation. Sewage effluent was pumped through the two contact columns in series and fresh ozone gas from the generator was introduced to the second contact column. Exhaust gas from this column was re-introduced into the first contact column. During operation, most of the ozone was completely used up in the second column and, therefore, little benefit was gained using two columns in series. Each column gave a contact time of five minutes at the flow rate used in the tests.

Tests were run each day, after a start-up period of approximately two hours to obtain steady state conditions. Water samples were taken at inlet to each contact column and the ozonized air was also sampled. Flow rates into the two columns were measured using rotometers and the flow of ozonized air was measured using an orifice meter. The optimum flow rate through the system was found to be 91 gpm.

Experimental Results

Figures 11-4 and 11-5 show that a dose of 6 ppm produced the Illinois state standards of 2000 cells per 100 ml fecal coliform and 5000 cells per 100 ml total coliform. At lower ozone doses, states standards were sometimes reached; a minimum of 4 ppm was required before any appreciable disinfection occurred, presumably because the initial ozone dose was used in reducing color and organics in the sewage effluent. At the ozone dosage required to meet the bacterial standards there was a 30% reduction observed in BOD content. (See Figure 11-6). Color in the sewage effluent before ozonation was 20°-30° Hazen. Ozonation at 6 ppm reduced this to 5° Hazen, approaching drinking water standards (Figure 11-7).

Other determinations were carried out on cyanide, phenol, ammonia, nitrite and suspended solids. Cyanide and phenol were almost completely removed but there was little effect on ammonia. Nitrite was oxidized to

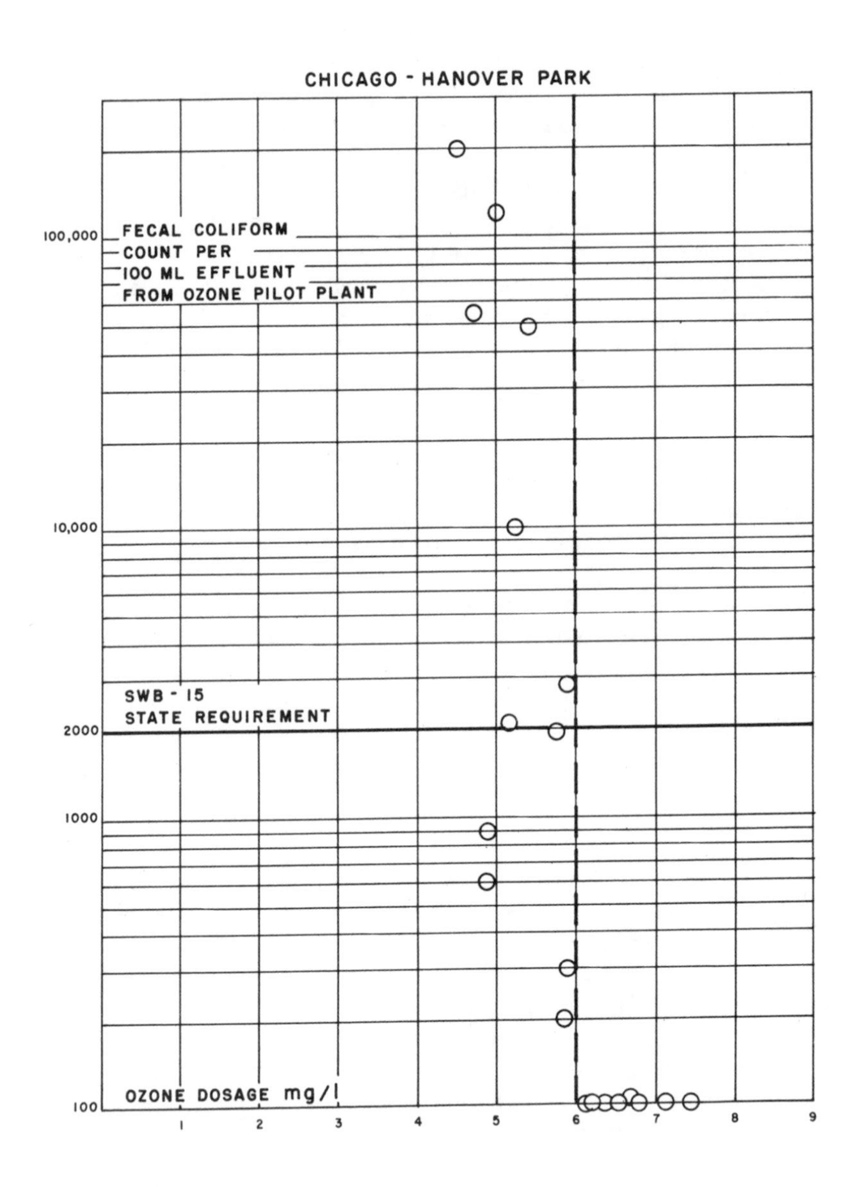

Figure 11-4. Plot of fecal coliform count leaving ozone pilot plant versus ozone dosage (for data where liquid flow rate is below 91 gpm)

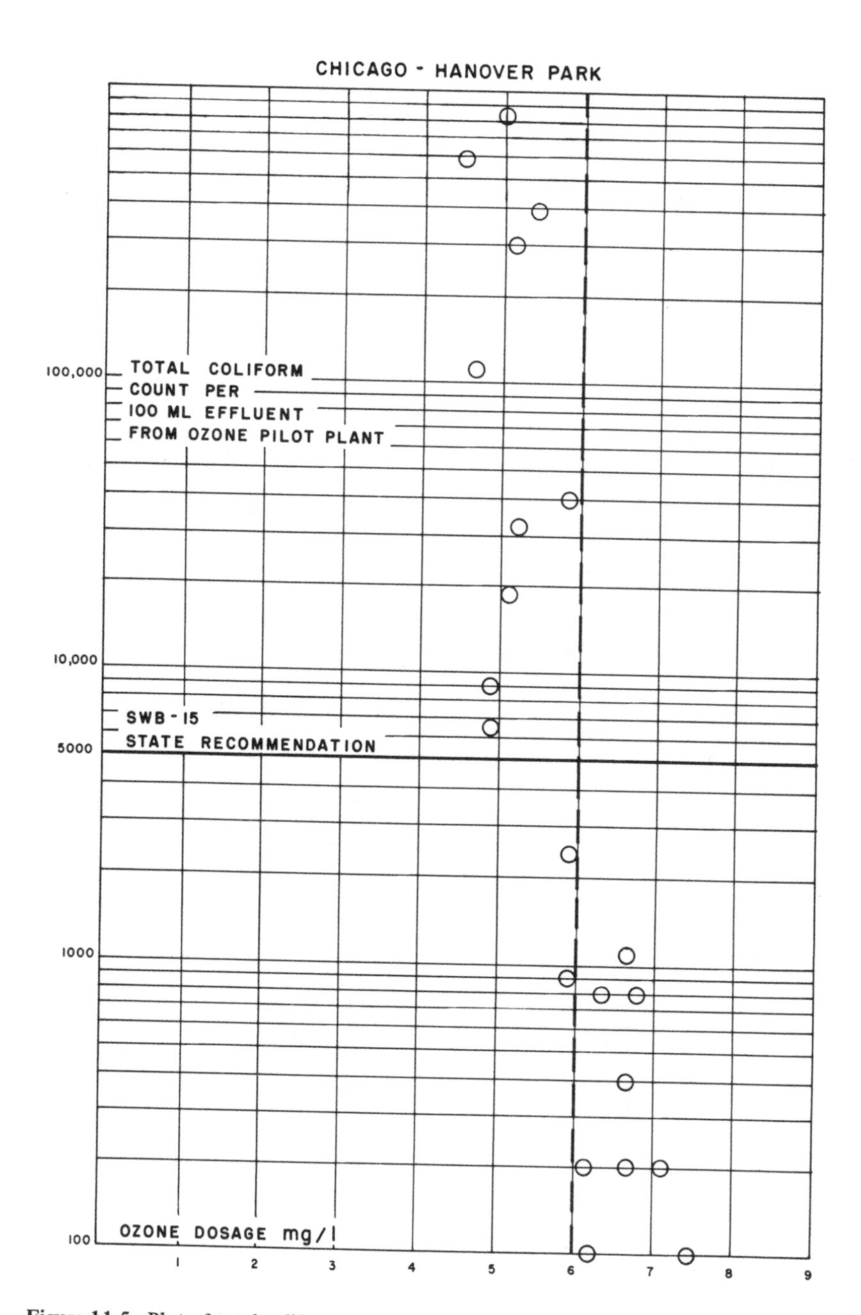

Figure 11-5. Plot of total coliform count leaving ozone pilot plant versus ozone dosage (effluent flow rate below 91 gpm)

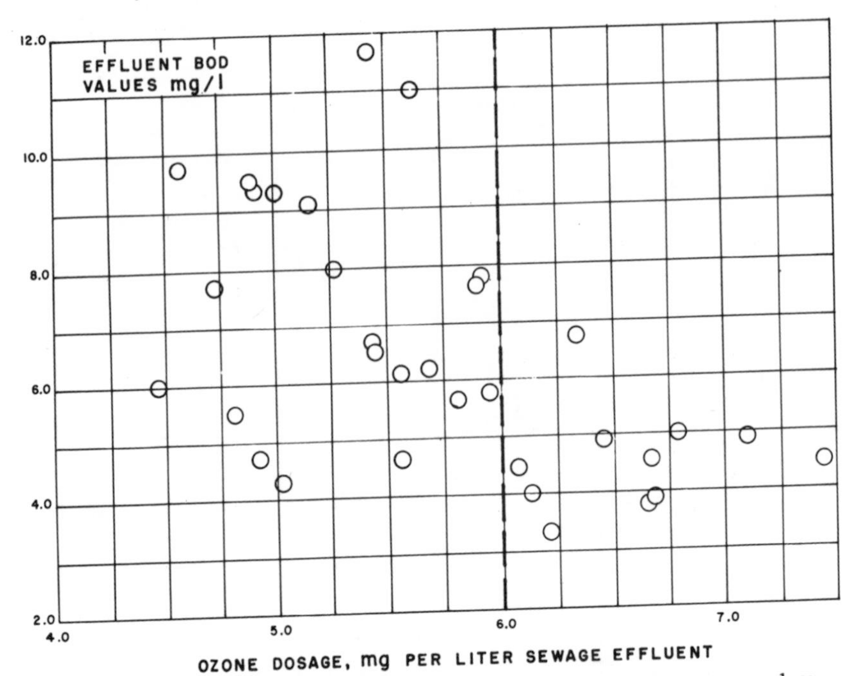

Figure 11-6. Graph of BOD values. Effluent from the pilot plant versus ozone dose. (Chicago–Hanover Park)

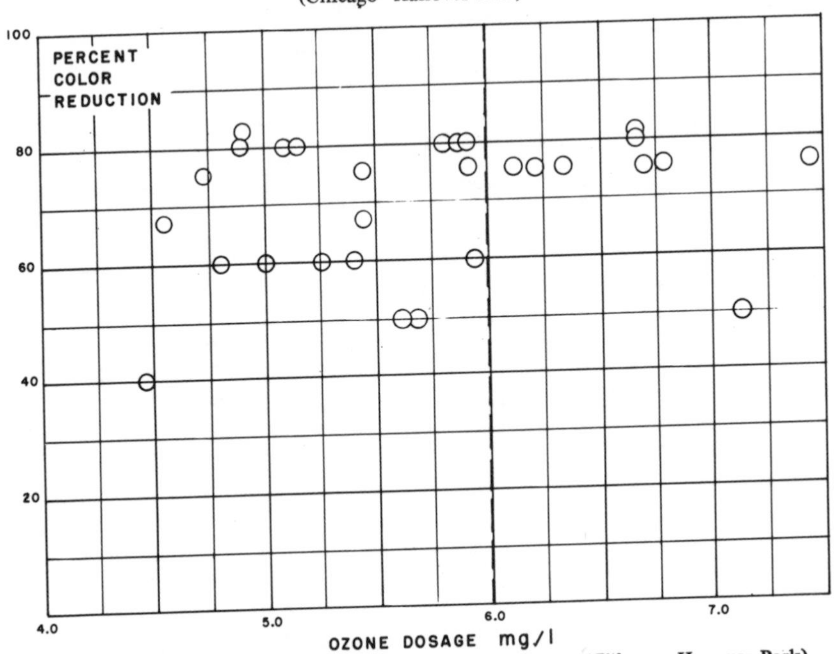

Figure 11-7. Per cent color reduction versus ozone dosage (Chicago–Hanover Park).

nitrate. Some removal of suspended solids was noted which may have been due to settlement, or frothing in the contact columns.

Operating Costs

Operating costs of the experimental installation were calculated using a charge of 0.8 cents per kilowatt hour. These were

Air Blower Cost	– 0.04 ¢/1000 gal
Dessicator Cost	– 0.35 ¢/1000 gal
Pumping Cost	– 0.80 ¢/1000 gal
Ozone Generation Cost	– 0.55 ¢/1000 gal

This figure is somewhat high since nearly half is for pumping costs which could be avoided in a full scale installation by using gravity flow.

Based on these and similar tests, it is possible to design an ozone installation to disinfect sewage effluent. This would include a contact tank of approximately 5 minutes' retention or longer if higher doses of ozone are to be employed. In this case the tank should be baffled and ozone admitted through porous diffusors into each compartment in series. The ozone generating equipment would include air drying processes and associated electrical controls. Oxygen, or air, could be used as the source gas. The same size ozone generator will produce twice as much ozone from oxygen as from air. Many large sewage treatment plants are considering oxygen in place of air for sewage treatment so ozonation for disinfection would then become more economical.

FUTURE DEVELOPMENTS

Further work on ozonation of sewage effluents is being carried out under Federal Research Contracts.

In Philadelphia, Microstraining and ozonation are being evaluated to treat combined sewers overflows. Initial results were very satisfactory, indicating that Microstraining will remove up to 90% of the suspended solids and a dose of 5 to 10 ppm of ozone will adequately disinfect.[12]

In Washington, D.C., ozone is being tested for chemical oxidation of refractory organics in activated sludge effluent and the effluent from a pilot physical/chemical treatment plant. Ozone is generated from oxygen and high doses (up to 50 ppm) are applied in a six-stage reactor to achieve color removal, BOD, COD reductions and disinfection.[13]

At the National Water Quality Research Laboratory in Duluth, Minnesota, tests are underway on fish life in streams receiving various effluents. It is found that chlorine residuals can be detrimental, even as low as 0.002 ppm,[14]

whereas ozone treatment, due to the higher dissolved oxygen content, actually improves living conditions for the fish.[15]

The first permanent installations for ozone disinfection of sewage effluents are presently in the final design stages. These will be used in conjunction with oxygen for activated sludge treatment.

SUMMARY AND CONCLUSIONS

The use of ozone, already well established in water treatment, is increasing for treatment of municipal and industrial effluents. Apart from rapid and effective disinfection, benefits include the removal of other undesirable impurities such as color, BOD, COD, phenols, cyanides and suspended solids (by frothing in the contact column). In addition effluents are usually saturated with dissolved oxygen, which can be beneficial to fish life in the receiving stream.

Dosage levels vary depending on the extent of contamination. For water supplies 0.5 ppm is usually sufficient for disinfection although color removal may require somewhat higher doses. In sewage treatment doses between 6 and 50 ppm have been employed to disinfect and oxidize organics, respectively. Ozone may be produced from air or oxygen, the latter being more economical in large projects, particularly if oxygen is also used as part of the sewage treatment.

REFERENCES

1. Diaper, E. W. J. "A New Method of Treatment for Surface Water Supplies," *Water Sewage Works* (November 1970).
2. Bringman, G. "Determination of the Lethal Activity of Chlorine and Ozone on *E. coli*," *Z. für Hyg* **139**, 130 (1954).
3. Ingols, R. S. and R. H. Fetner. "Some Studies of Ozone for Use in Water Treatment," *Proc. Soc. Water Treatment and Examination* **6**, 8 (1957).
4. Kessel, J. S., *et al.* "Comparison of Chlorine and Ozone as Viricidal Agents of Polio Virus," *Proc. Soc. Exp. Biol. Med.* **53**, 71 (1943).
5. Smith, W. W. and R. E. Bodkin. "The Influence of pH Concentration on the Bactericidal Action of Ozone and Chlorine," *J. Amer. Bacteriol.* **47**, 445 (1944).
6. Ridenour, G. M. and R. S. Ingols. "Inactivation of Polio Virus by Free Chlorine," *Amer. J. Pub. Health* **6**, 639 (1946).
7. Whitson, M. T. B. "The Use of Ozone in the Purification of Water," *J. Roy. Sanit. Inst.* **28**, 5 (1948).
8. Diaper, E. W. J. "Practical Aspects of Water and Wastewater Treatment by Ozone," *ACS 162nd National Meeting* (September, 1971).
9. Evans, F. L., III. *Ozone in Water and Wastewater Treatment.* (Ann Arbor, Mich.: Ann Arbor Science Publishers, Inc., 1972), p. 150.

10. Boucher, P. L., *et al.* "The Use of Ozone in the Reclamation of Water from Sewage Effluent," *J. Inst. Pub. Health Eng.* **4**, 11 (1968).
11. Zenz, D. "Ozonation Pilot Plant Studies at Chicago," *University of Wisconsin Seminar* (November 1971).
12. "Microstraining and Disinfection of Combined Sewer Overflows," Cochrane Division, Crane Co., *FWPCA Rep. 11023EVO* (June 1970).
13. Kirk, B. S., R. McNabney and C. S. Wynn. "Pilot Plant Studies of Tertiary Wastewater Treatment with Ozone," *ACS 162nd National Meeting* (September 1971).
14. Brungs, W. A. "Effects of Residual Chlorine on Aquatic Life," *J. Water Pollution Control Fed.* **45**, 2180-2193 (1973).
15. Arthur, T. W. and T. G. Eaton. "Chloramine Toxicity to the Amphipod and Fathead Minnow," *J. Fish. Res. Board Canada* (December 1971).

CHAPTER 12

ECONOMICAL WASTEWATER DISINFECTION WITH OZONE

Harvey M. Rosen, Frank E. Lowther, Richard G. Clark

Pollution Control Systems
W. R. Grace & Co.,
Columbia, Maryland

INTRODUCTION

Economical large scale disinfection of wastewater by ozone is rapidly approaching reality. Factors influencing the growing acceptance of ozone disinfection are related to the shortcomings of chlorine and the decreasing cost of ozonation. Chlorine has basic disadvantages as a disinfectant in wastewater, especially with respect to inactivation of viruses, toxicity of chlorinated effluents and safety. Ozone overcomes these technical problems and at the same time is becoming economically competitive. Currently, commercial techniques for ozone generation have demonstrated major improvements over older technology. However, there was an apparent need for more economical methods of application and this is the area addressed.

Wastewater disinfection using chlorine, as commonly practiced in the United States, has come under increasing scrutiny in the last several years as a result of greater environmental awareness. Increases in dosage, contact time and mixing efficiency, as well as stricter control of pH, sampling and monitoring necessary for chlorine to be an effective virucide, have been described.[1,2]

Chloramines and other chlorine residuals formed as a result of chlorine reactions in wastewater, have been reported to be toxic to aquatic life.[3-5] A similar report has been made regarding the toxicity of chlorinated hydrocarbons formed from the reaction of chlorine with organic material present in wastewater.[6] Improved chlorine disinfection practices will in many cases require the use of higher chlorine dosages. As a result, free chlorine and its reaction products may reach levels unacceptable for discharge to particular environments. In such cases, dechlorination with activated carbon, sulfur dioxide or thiosulfate salts will be required. Costs will rise significantly, and the use of SO_2 or thiosulfates may cause further pollution problems. Lastly, safety practices in the storage, application and transportation of chlorine are under review due to the many hazardous conditions created by its use.

The use of ozone for wastewater disinfection can overcome all the shortcomings of chlorine and provide additional benefits. The primary objection to ozone has been solely economic. Ozone, by essentially all reports[7-10] is a better virucide than chlorine and a much more efficient bacterial disinfectant.[11,12] Ozone use as a disinfectant results in other water quality improvements, such as reduction of oxygen-demanding wastes, color and turbidity, and in increases in effluent dissolved oxygen. Ozone produces no residuals to add to the pollution load nor does there appear to be any problem with effluent toxicity.[13]

Recent advances in generating ozone have eliminated, reliably and efficiently, one of the major obstacles to its economical use. The area that requires further investigation–to lower the cost of ozonation–is the reduction of dosage for specific applications.

OZONE DISINFECTION METHOD

Realizing the lack of understanding and techniques for the efficient application of ozone, a study was undertaken to define and optimize the parameters involved in ozone disinfection of wastewater. A conceptual model for ozone disinfection was postulated. Variables considered were: relative rates of reaction of ozone with components of wastewater; character, concentrations and distribution of these components; volume ratios of gas/liquid; ozone concentration and the disinfection mechanism.

The model assumes that ozone's rate of reaction with bacteria is very rapid and that ozone disinfection occurs by lysis. The rapid rate of disinfection is supported by essentially all previous work, and the lysis mechanism fits the observed kinetics. Certainly the difference in rate between halogen and ozone inactivation would indicate a different mechanism for each. It was further assumed that lysis could take place between an ozone bubble and a microorganism if the two could be brought into contact. For purposes of the model it is unimportant if there is mass transfer of ozone across an aqueous film surrounding the microorganism or if the attack is directly by ozone gas. Therefore, the model assumes that within reasonable limits disinfection is independent of ozone concentration and that the rate of disinfection is not limited by the bulk mass transfer rate of ozone into solution. For all practical purposes if a microorganism can be brought into contact with a bubble of ozone-containing gas, inactivation is instantaneous. This sets the prime design consideration for effluent ozone disinfection: all the water, or more specifically, all microorganisms, must be brought in contact with ozone as rapidly as possible.

A simplified physical picture of components in wastewater as a function of relative sizes of molecules, their character and distribution, was also assumed. From observations on effluent change after ozonation, it was predicted that the primary interferences as far as consuming ozone for side reactions would come from easily-oxidized dissolved organics (COD) including color bodies and colloidal matter (turbidity). Table 12-1 indicates the assumptions and calculations made in describing a regular distribution

Table 12-1. Distribution of Species in Wastewater[a]

	Turbidity	COD	Bacteria
Average linear dimension, L, cm	10^{-6}	10^{-7}	10^{-4}
Concentration, C	10 mg/l	10 mg/l	10^2-10^6 colonies/100 ml
Number of particles–molecules/l, N	10^{16}	10^{19}	10^3-10^7
Average separation, S, cm	0.5×10^{-4}	0.05×10^{-4}	1.0-0.05

[a]Assumes cubic species with a density of 1 g/cm^3.

$$N = \frac{C}{V x \rho} \qquad S = \frac{\sqrt{1000\ cm^3}}{\sqrt[3]{N}}$$

$$V = L^3, \rho = 1\ g/cm^3$$

of the species in wastewater. Figure 12-1 illustrates a physical picture of the distribution and the relative size of a bubble required to contact microorganisms in this oversimplified model. The shaded area represents COD and colloidal matter. Figure 12-2 illustrates the relative distribution and size of bubbles as the concentration of microorganisms is reduced from 10^6 to 10^2 colonies/100 ml, and illustrates the difficulties of limiting competition for ozone by oxidizable matter. According to the model, it is desirable to minimize bulk dissolved ozone. Thus, it was postulated that particular gas-to-liquid ratios at various gas and liquid pressures would be necessary to provide a size and distribution of bubbles that would selectively disinfect.

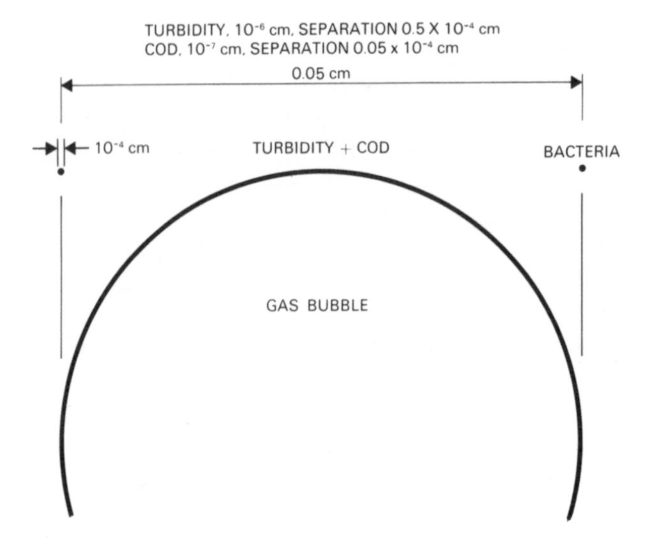

Figure 12-1. Schematic distribution of species in wastewater.

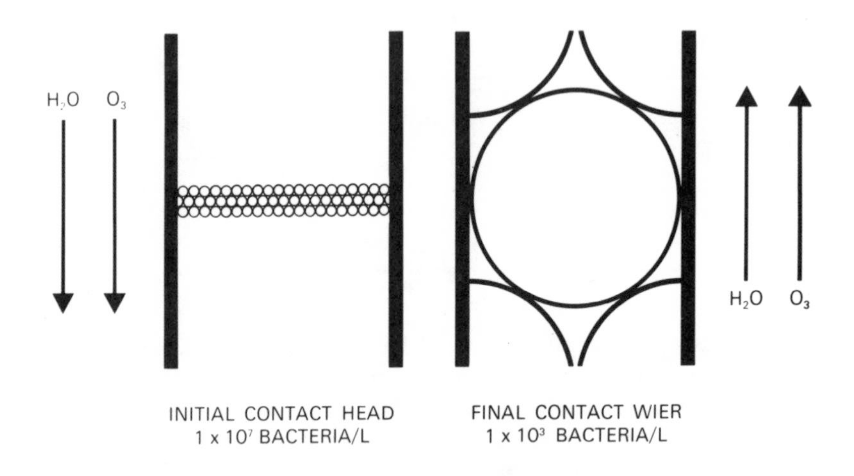

Figure 12-2. Gas bubble distribution.

CONTACTOR DESIGN

Once the model was defined a contact system was constructed based on the requirements of the model and the desirable engineering features (Table 12-2). Figure 12-3 represents the basic contactor (patent pending). It is constructed primarily of rigid polyvinylchloride for a nominal flow rate of 10 gpm and a nominal hydraulic range of 5 to 15 gpm. Ozone containing

Table 12-2. Engineering Variables in Design of Disinfection Contactor

1. Low cost
2. Simplicity
3. Ozone compatible materials of construction
4. No moving parts
5. Nothing to clog
6. Flexibility in hydraulics and disinfection range
7. Ease of control and automation
8. Ease of scale up
9. Minimization and/or use of waste ozone
10. Modular construction

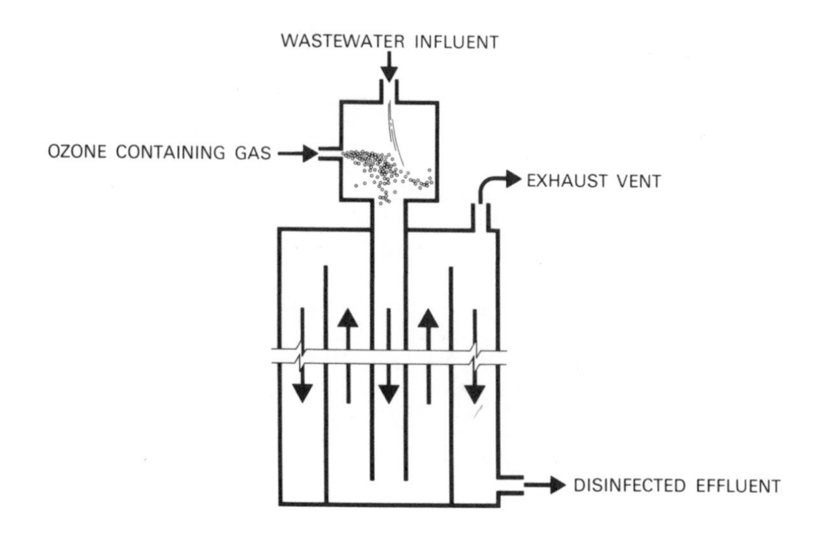

Figure 12-3. Disinfection contactor.

gas and water mix in a specially-designed positive pressure injector and flow cocurrently through the system before the gas and liquid are separated. Figure 12-4 shows the operation of a 100-gpm scale up of this original design in which the outer receiver has been removed to show the mixing action obtained.

Figure 12-4. Mixing action in a 100-gpm contactor.

DISINFECTION STUDIES

Initial disinfection studies were conducted at the Westgate Wastewater Treatment Plant in Fairfax County, Virginia. This site was chosen because the oxygen activated sludge system consistently produced a high quality effluent. Effluent for disinfection tests was taken directly from the secondary clarifier. A Grace Model LG-2-L2 ozonator fed by compressed oxygen provided the ozone feed, and applied ozone dosage was determined by calibrating the ozonator against a standard KI titration procedure.[14] Ozone dosage was varied between 3 and 10 mg/l while liquid and gas pressures and volume ratios were also varied. Adequate disinfection was obtained at gas and liquid pressures of 2-3 psig and 4-5 mg/l ozone dosage. All data were taken from the Westgate plant records[15] using procedures specified in *Standard Methods.*[16] Figures 12-5, 12-6 and 12-7 show results of over three weeks' continuous ozone disinfection. The data and comparisons with other published data[17,18] on pilot scale ozone disinfection studies are summarized in Table 12-3.

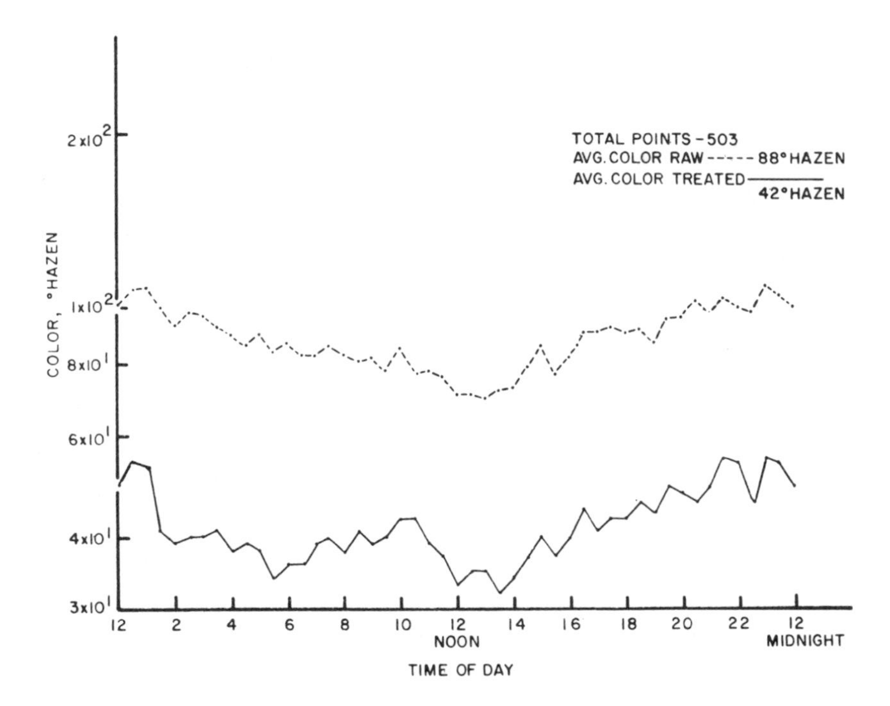

Figure 12-5. Ozone dose 5 mg/l–Ozonation of oxygen activated sludge effluent from secondary clarifier.

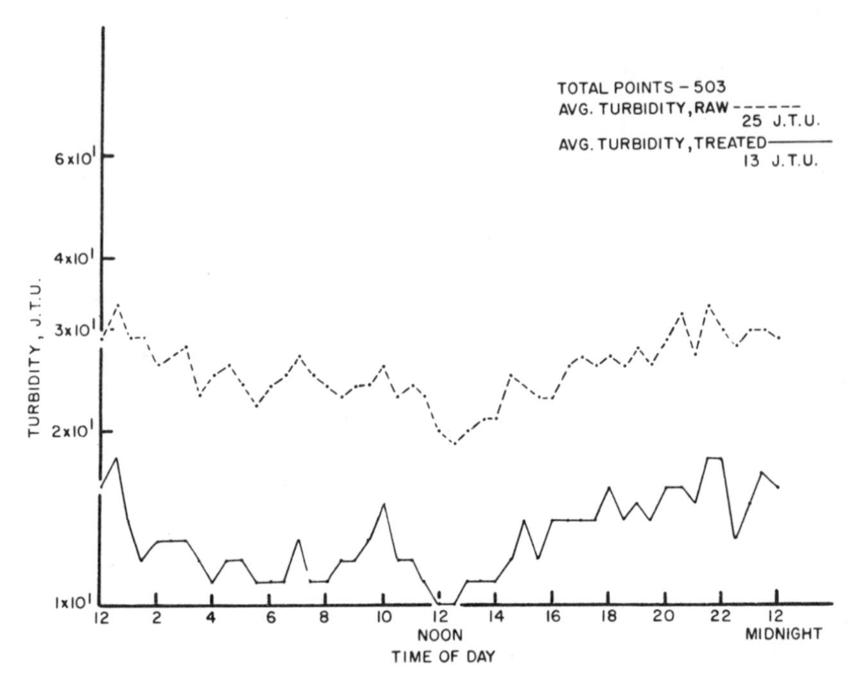

Figure 12-6. Ozone dose 5 mg/l–Ozonation of oxygen activated sludge effluent from secondary clarifier.

OTHER PILOT DATA

Dallas, Texas

Another pilot plant at 25 gpm and a constant 5 mg/l dosage is being run intermittently at the Dallas Water Reclamation Research Center by the city of Dallas and Texas A&M University.[19,20] Wastewater ozonation at Dallas is generally being conducted in the same manner as was Fairfax. However, this work includes inactivation of total coliforms, studies on total bacteria and, in some runs, f_2 phage bacterial virus. Also three sources were studied, including physical/chemical treated using high alum in a densator, filtered activated sludge, and alum/densator-activated carbon column effluents. Disinfection data are averages of eight points for each run, except raw counts which are averages of initial, final and composite samples. Each point in a run was taken under different conditions of gas/liquid ratio to study disinfection efficiency as a function of contact efficiency (which was expected to vary with changing gas flow rate) while keeping dosage

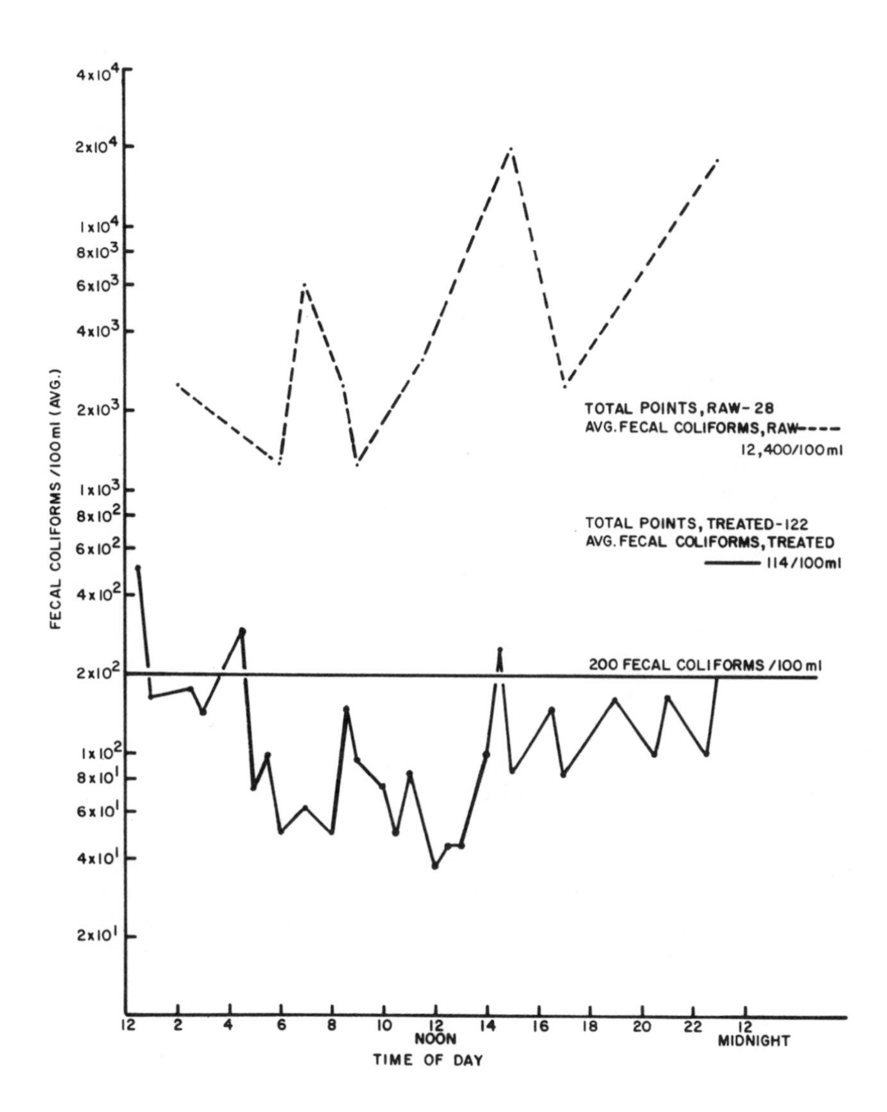

Figure 12-7. Ozone dose 5 mg/l–Ozonation of oxygen activated sludge effluent from secondary clarifier.

Table 12-3. Ozone Disinfection Comparative Data[a]

Pilot Plant[b]	A	B	Grace
Influent Water Quality	**Microstrained Sec/Tert**	**Packaged Plant Sec**	**Oxygen Activated Sludge, Sec**
BOD (mg/l)	7	7 (sol)	10 (sol)
COD (mg/l)	91	28 (sol)	47 (sol)
Color (°H)	22	61 (APHA)	88
Turbidity (J.T.U.)	4	29	26
Suspended Solids (mg/l)	4	20	12
Ozone Application	**Single Stage Cocurrent Aspiration**	**Multi-Stage Cocurrent/Counter-current Sparging**	**Single-Stage Cocurrent Injection**
Dosage (mg/l)	6.6	15	5
Contact time (min)	8	5-22	<1
Effluent Water Quality			
Total Coliform (counts/100 ml)	480	774	735 (extrap)
Fecal Coliform (counts/100 ml)	100	153	114
COD Reduction (%)	5	18	17
Turbidity Reduction (%)	13	69	48
Color Reduction (%)	75	72	52

[a]Average figures for duration of optimized Fairfax pilot plant studies
[b]Data from studies using other methods of contact.[17,18]

and liquid flow constant. However, no effect of this parameter was observed. Table 12-4 summarizes the bacterial disinfection data and Table 12-5, the bacterial virus runs.

In general these data show that the higher quality of the effluent prior to ozonation, the higher the degree of sterilization achieved. The most obvious comparison is between Columns B and C (Table 12-4) and A and B (Table 12-5). As the effluent quality is improved by carbon adsorption, all measurements of microorganisms are zero. It is also evident that ozone inactivates f_2 phage extremely rapidly and at relatively low dosages. Another observation is the relative resistance to ozone inactivation of gram-positive rods compared with coliform and phage. The data presented here are preliminary, and studies are continuing at Dallas. Data will be extended to include animal viruses and variations in ozone dosage. It is also hoped that resistance to ozone disinfection between specific species may be developed and an indicator organism for viral inactivation identified.

Table 12-4. Dallas Bacterial Disinfection Data (5 mg/l, 25 gpm)

Feed	A Filtered Activated Sludge	B Physical-Chemical (P-C) High Alum, Densator	C (P-C)/Activated Carbon
No. of Tests	3	8	1
Avg Influent Quality			
BOD mg/l	<10	<10	<10
COD mg/l	20	24	9
SS mg/l	4	9	1
NH_3 - N mg/l	2.4	1.4	1.9
Organic - N mg/l	3.1	2.4	1.8
pH	7.3	7.9	7.4
Total Bacteria/100 ml	2.8×10^6	2.6×10^6	0.1×10^6
Total Coliform/100 ml	7.2×10^3	—	60×10^3
Disinfected Effluent			
Total Bacteria/100 ml[a]	37×10^2	80×10^2	0
Total Coliform/100 ml	5	–	0

[a]Primarily gram positive rods.

Table 12-5. Dallas Bacterial Virus Disinfection Data (5 mg/l, 25 gpm)

Feed	A Physical-Chemical (P-C) High Alum, Densator	B (P-C)/Activated Carbon
No. of Tests	2	1
Avg Effluent Quality		
BOD mg/l	<10	<10
COD mg/l	21	9
SS mg/l	7	1
NH_3 - N mg/l	3.6	1.9
Organic - N mg/l	2.1	1.8
pH	7.4	7.4
f_2 phage - Pfu/ml, seeded	1700	52,500
Disinfected Effluent		
f_2 phage - Pfu/ml, seeded No. samples - Pfu/ml sample	13/0, 3/2	6/0

St. Paul, Minnesota

A 10-gpm pilot study was conducted under conditions similar to the Fairfax County study. This pilot was run in preparation for a 1-mgd pilot plant scheduled for early 1974. Effluent was taken directly from a channel between the secondary clarifier of the Seneca Sewage Treatment Plant (STP) and the chlorination facilities. The Seneca STP[21] is designed to operate as a complete mix activated sludge plant and was operating significantly under its 24-mgd average design flow during the pilot runs. Table 12-6 summarizes the St. Paul data.

Table 12-6. St. Paul Disinfection Data

Effluent Quality	Average		Range	
BOD mg/l	22		18-25	
BOD mg/l	90		65-106	
SS mg/l	20		9-39	
Fecal coliform/100 ml	5,000		2,000-7,000	
Disinfected Effluent				
No. of runs	4	9	11	1
Avg O_3 dose/mg/l	5.4	4.8	4.2	2.5
Avg fecal coliform/100 ml	0	1.3	5.7	26

In this pilot significant fecal coliform counts did not appear until ozone dosage was 2.5 mg/l. However, effluent quality in terms of COD and suspended solids was low compared with the Fairfax and Dallas studies. No post-ozonation COD analyses were done so it is not known how much demand the COD had for ozone. It would appear that demand was small based on the disinfection efficiency. Initially it was expected that certain components of COD would have a high ozone demand and that high solids would tend to shield microorganisms from the disinfectant. Neither assumption appears true based on the St. Paul data, but the Dallas data tends to support these assumptions. It must be concluded that there are not enough data to determine the effect of various parameters of wastewater quality on disinfection by ozone with the Grace contacting system. Further studies of the effect of effluent quality on disinfection efficiency are in progress.

ECONOMICS

The question of economics is a function of the individual installation. However, an economic analysis for a 10-mgd average flow plant requiring 5 mg/l ozone for disinfection to a level of 200 fecal coliform/100 ml is presented in Tables 12-7 and 12-8. This analysis is based on a specific situation and would vary considerably for other installations. A major cost savings *could* be realized such as on a plant installed in conjunction with oxygen-activated sludge treatment where dry, low cost, on-site oxygen was available as feed gas to the ozonators. Capital and operating costs for gas preparation equipment would be eliminated and ozonator capital and operating costs reduced significantly.

Table 12-7. Costs of Ozone Disinfection (Assumptions)

1. 10 mgd flow, 23 mgd peak, <14 mgd flow 90% of time.
2. Flow distribution based on a plant in the Northeast U.S. for last three years.
3. Effluent quality to allow 5 ppm dosage to achieve 200 fecal coliform or less/100 ml.
4. Operating characteristics of Grace Ozone Generators and Contactors.
5. Power costs @1.5¢/kwh.
6. Designed for air feed to 14 mgd with liquid oxygen addition to ozonator feed for peak shaving to 23 mgd.
7. Two-day supply of liquid oxygen for maximum peak usage.
8. Three psig gravity feed for wastewater to contactor.

Data for breakpoint chlorination, activated carbon dechlorination and reaeration (Table 12-9) taken from a plant designed for ammonia removal must be considered when comparing these economics with those for ozonation. It must be noted also that the ozonation cost is based on a 10-mgd plant while chlorination data are based on 6 mgd.

Chlorination/dechlorination figures are probably a maximum for the cost of disinfection assuming breakpoint chlorination was used to produce free chlorine for elimination of bacteria and viruses. Current practice, which in many cases does not achieve the proposed EPA disinfection specification of 200 fecal coliform/100 ml, is not effective on viruses and leaves a relatively high total chlorine residual. EPA and state authorities are now considering that dechlorination to fractional mg/l levels of total chlorine residual be required as a result of chloramine toxicity.

Table 12-8. Costs of Ozone Disinfection

Capital Costs	$ (000)
Ozone generators (585 lb/d, air feed)	150
Compressor/dryer @ 27% of ozone generator	40
Oxygen storage tank (15T)	20
Contactors (installed)	160
Building @ $25/sq ft	38
Controls	25
Installation (piping)	50
Engineering & contingencies	32
Total	515
Direct Operating Costs	**¢/1000 gal**
Ozone generators @ ~8.0 kwh/lb ozone	0.51
Compressor/dryers @ 3.5 kwh/lb ozone	0.21
Oxygen, 135T/yr @ $30/T (including usage and evaporation loss)	0.11
Operation and maintenance	0.24
Total	1.07

Table 12-9. Cost of Chlorination-Dechlorination and Reaeration[22]
(Basis: 6 mgd plant, Cl_2:NH_3 feed rate = 9:1, 12 mg/l NH_3)

Capital Costs	$(000)	
Chlorination-Dechlorination		
Structure	376	
Equipment	674	1,050
Reaeration		
Structure	17	
Equipment	21	38
Total		$1,088
Operating Costs	**¢/1000 gal**	
Chlorination-Dechlorination		
Chemicals	3.32	
Supplies and Utilities	0.30	
O & M Labor	0.64	4.26
Reaeration		
Supplies and Utilities	0.07	
O & M Labor	0.64	0.71
Total		4.97

Generally operating costs for current chlorination practices run between 0.5¢ and 1.0¢/1000 gal. Chemical dechlorination will add significantly to these costs bringing ozonation into the range of competitive-to-more-economical without considering the value of ozone as a virucide, reduction of other pollutants and a high DO in the effluent.

ACKNOWLEDGMENT

Some of the information presented here was previously published in *Water and Wastes Engineering* **11** (July 1974). The authors are grateful to Donnelly Publishing Corp. for permission to present the information in this volume.

REFERENCES

1. Liu, O. C. "Effect of Chlorination on Human Exteric Viruses in Partially Treated Water from Potomac Estuary," A Progress Report under contract from the Army Corps of Engineers by the Environmental Protection Agency (November 1970).
2. Kruse, C. W., *et al.* "Improvement of Terminal Disinfection of Sewage Effluents," *Water Sewage Works* **120**, 57 (June 1973).
3. "Chlorinated Municipal Waste Toxicities to Rainbow Trout and Fathead Minnows," Environmental Protection Agency, Report 18050GZZ (October 1971).
4. Zillich, J. A. "Toxicity of Combined Chlorine Residuals to Freshwater Fish," *J. Water Pollution Control Fed.* **44**, 212 (1972).
5. Esvelt, L. A., *et al.* "Toxicity Assessment of Treated Municipal Wastewaters," *J. Water Pollution Control Fed.* **45**, 1558 (1973).
6. "The Effect of Chlorination on Selected Organic Chemicals," Environmental Protection Agency, Report 12020EXG (March 1972).
7. Pavoni, S. L., *et al.* "Virus Removal from Wastewater Using Ozone," *Water Sewage Works* **59** (December 1972).
8. Evison, L. "Inactivation of Viruses in Water with Ozone," *British Water Supply* **14** (September 1972).
9. Schaffernoth, T. J. "High Level Inactivation of Poliovirus in a Biologically Treated Wastewater by Ozonation," M.Sc. Thesis, University of Maine, Orono (June 1970).
10. Gomella, C. "Ozone Practices in France," *J. Amer. Water Works Assoc.* **64**, 39 (1972).
11. Fetner, R. H. and R. S. Ingols. "A Comparison of the Bactericidal Activity of Ozone and Chlorine Against *Escherichia coli* at 1°," *J. Gen. Microbiol.* **15**, 381 (1956).
12. Scott, D. B. M. and E. C. Lesher. "Effect of Ozone on Survival and Permeability of *Escherichia coli*," *J. Gen. Microbiol.* **85**, 567 (1963).
13. Arthur, J. W. "Toxic Responses of Aquatic Life to an Ozonated, Chlorinated and Dechlorinated Municipal Effluent," University of Wisconsin short course, Ozonation in Sewage Treatment, Milwaukee, Wisconsin (November 1971).

14. Birdsall, C. M., *et al.* "Iodometric Determination of Ozone," *Anal. Chem.* **24**, 662 (April 1952).
15. Morrison, R. Manager, Westgate W. W. T. Plant, Fairfax Co., Virginia, personal communication.
16. American Public Health Association. *Standard Methods for Examination of Water and Wastewater*, 13th ed. (New York: American Public Health Association, 1972).
17. Nebel, C. W. "Ozone Treatment of Secondary Effluent," University of Wisconsin short course, Ozonation in Sewage Treatment, Milwaukee, Wisconsin (November 1971).
18. Zenz, D. "Ozonation Pilot Plant Studies at Chicago," University of Wisconsin short course, Ozonation in Sewage Treatment, Milwaukee, Wisconsin (November 1971).
19. Esmond, S. and A. Petrasek, Jr. "Removal of Heavy Metals by Wastewater Treatment Plants," The WWEMA Industrial Water and Pollution Conference, Chicago, Illinois (March 1973).
20. Petrasek, A., Jr., *et al.* "Municipal Wastewater Qualities and Industrial Requirements," AICHE National Conference on Complete Water Reuse, Washington, D.C. (April 1973).
21. Robinson, J. H. "New Regional Plant Built for the Future," *Water Wastes Eng.* **10**, 35 (August 1973).
22. Atkins, P. F., Jr., *et al.* "Ammonia Removal in a Physical-Chemical Wastewater Treatment Plant," 27th Purdue Industrial Waste Conf., Purdue University (May 1972).

CHAPTER 13

CHLORINE AND OXYCHLORINE SPECIES REACTIVITY WITH ORGANIC SUBSTANCES

David H. Rosenblatt

U.S. Army Medical Bioengineering
Research and Development Laboratory
Ft. Detrick, Md.

INTRODUCTION

Sanitary engineers, microbiologists, environmental toxicologists and water chemists are interested in reactions between chlorine-containing oxidants and organic compounds because if these oxidants are to be used most efficiently as water supply disinfectants, the chemical processes of disinfection must be understood. Further, organic compounds in water and wastewater compete with pathogens for disinfectants, and this competition could be decreased beneficially if quantitative rate predictions could be made for all reactions of interest over a suitable range of conditions. Finally, toxicity, and taste and odor of both the primary organic pollutants and the products of their reactions with disinfectants must be considered. We are only beginning to understand the nature of the primary organic pollutants, and little information is available on the compounds actually formed by chlorine reacting with these organics in real water samples.

INTERRELATIONS AMONG CHLORINE SPECIES

Chlorine is one of three elements (bromine and iodine are the others) that uniquely react reversibly with water. In these rapid equilibria the oxidizing power of the parent element, Cl_2, is preserved in hypochlorous acid, hypochlorite ion, chlorine monoxide, trichloride ion, and probably hypochlorous acidium ion (H_2OCl^+) (Figure 13-1 and Table 13-1), but not

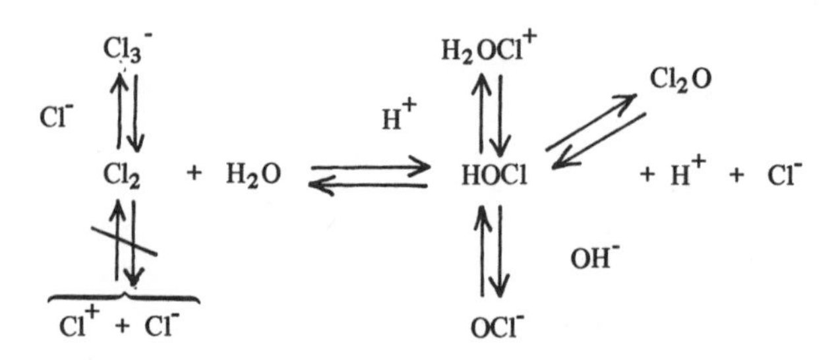

Figure 13-1. Fundamental equilibria for aqueous chlorine.

Table 13-1. Equilibrium Data for Chlorine Species in Water at 25°C

Reaction	Equilibrium Constant	Reference
1. $Cl_2 + H_2O \rightleftharpoons HOCl + H^+ + Cl^-$*	$K = [HOCl][H^+][Cl^-]/[Cl_2] = 3.944 \times 10^{-4}\ M^2$	1
2. $HOCl \rightleftharpoons H^+ + OCl^-$	$K = [H^+][OCl^-]/[HOCl] = 2.88 \times 10^{-8}\ M$	2
3. $2\ HOCl \rightleftharpoons Cl_2O$	$K = [Cl_2O]/[HOCl]^2 = 0.0112\ M^{-1}$	3
4. $Cl_2 + Cl^- \rightleftharpoons Cl_3^-$	$K = [Cl_3^-]/[Cl_2][Cl^-] = 0.169\ M^{-1}$	4
5. $H_2OCl^+ \rightleftharpoons H^+ + HOCl$	$K = [H^+][HOCl]/[H_2OCl^+] \sim 10^3\text{-}10^4\ M$ or $4 \times 10^{11}\ M$	5,6
6. $HClO_2 \rightleftharpoons H^+ + ClO_2^-$	$K = [H^+][ClO_2^-]/[HClO_2] = 1.1 \times 10^{-2}\ M$	7
7. $ClO_2 + ClO_2^- \rightleftharpoons Cl_2O_4^-$	$K = [Cl_2O_4^-]/[ClO_2][ClO_2^-] \cong 1.6\ M$	8
8. $HClO_3 \rightleftharpoons H^+ + ClO_3^-$	$K = [H^+][ClO_3^-]/[HClO_3] > 1\ M$	9
9. $HClO_4 \rightleftharpoons H^+ + ClO_4^-$	$K = [H^+][ClO_4^-]/[HClO_4] = 38\ M$	10

*The kinetics and mechanism of this reaction at 20°C have been elucidated by Eigen and Kustin.[11]

chlorinium ion,[12] (Cl^+) which probably does not exist. Under appropriate conditions in aqueous solution, irreversible reactions occur whereby additional oxychlorine species may be formed. Thus, especially near neutral pH, $2\ HOCl + OCl^- \overset{K_{eq}}{\rightleftharpoons} [H_2Cl_3O_3^-] \xrightarrow{k_1} 2\ H^+ + 2\ Cl^- + ClO_3^-$. Here $K_{eq} \cong 83\ M^{-2}$ and $k_1 \cong 1.7\ hr^{-1}$ at 25°C. Chlorate ion, which is produced in this normally slow reaction, is very slowly oxidized by hypochlorous acid to perchlorate.[13] At low pH chlorate ion interacts with hydrochloric acid to form chlorine and chlorine dioxide. The latter is a relatively stable and water-soluble free radical:

$$4\ H^+ + 2\ Cl^- + 2\ ClO_3^- \rightleftharpoons Cl_2 + 2\ ClO_2 + 2\ H_2O$$

This reaction is quite complex, and the yield of chlorine dioxide is usually less than that indicated by the equation.[7] In waterworks, it should be added, chlorine dioxide is generated from sodium chlorite (but this chlorite salt is always manufactured by the reduction of chlorine dioxide originally generated from a chlorate salt):

$$HOCl + 2\ ClO_2^- \rightarrow 2\ ClO_2 + Cl^- + OH^-$$

A side-reaction producing chlorate ion always occurs when chlorine dioxide is generated in this manner:

$$HOCl + ClO_2^- \rightarrow H^+ + ClO_3^- + Cl^-$$

At a low pH chlorite ion is quite unstable and decomposes by a complex chain mechanism, with the approximate stoichiometry:

$$4\ HClO_2 \rightarrow Cl^- + 2ClO_2 + ClO_3^- + 2\ H^+ + H_2O$$

Chlorine dioxide is not particularly stable and disproportionates, especially at strongly alkaline pH, to give chlorite and chlorate ions in equal amounts:[7]

$$2\ ClO_2 + 2\ OH^- \rightarrow ClO_2^- + ClO_3^- + H_2O$$

Neglecting the effects of basic ions, such as carbonate, the rate law for this reaction[1] can be expressed as:

$$-d[ClO_2]/dt = (5 \times 10^5\ e^{-7350/T} + 1.67 \times 10^8\ e^{-7350/T}[OH^-])[ClO_2]$$

$$+ (2.67 \times 10^9\ e^{-7350/T} + 7.37 \times 10^6\ e^{-3875/T}[OH^-])[ClO_2]^2$$

where t is in seconds and T is the temperature (°K). This expression predicts a half-life for $10^{-4} M$ ClO_2 of 3¼ hours at pH 12 and 25°C. Half-life predictions for decomposition at neutral pH seem inordinately low, in view of the author's laboratory experience.

In addition to the above-mentioned reactive species, two related to chlorine dioxide deserve notice. One of these, Cl_2O_2, must arise as an unstable intermediate in various chlorine dioxide formation reactions;[7]

$$HOCl + ClO_2^- \rightarrow \left[HO\text{-}Cl\text{-}Cl\genfrac{}{}{0pt}{}{\nearrow O}{\searrow O}\right]^- \rightarrow \left[Cl\text{-}Cl\genfrac{}{}{0pt}{}{\nearrow O}{\searrow O}\right] + OH^-$$

After this,

$$2\ Cl_2O_2 \rightarrow ClO_2 + Cl_2,$$

as well as

$$Cl_2O_2 + H_2O \rightarrow 2\ H^+ + Cl^- + ClO_3^-$$

The other strange species, $Cl_2O_4^-$, is readily evident by the brown color that appears when chlorine dioxide is added to a concentrated chlorite solution (see Table 13-1).

THE USE OF OXIDATION POTENTIALS AND OTHER THERMODYNAMIC EXPRESSIONS TO PREDICT REACTION RATES

Basic Concepts

Oxidation potentials, specifically those of some of the chlorine species discussed above, are useful for predicting chemical reactivity. Their actual utility, however, is extremely limited. To understand the limitations, it is necessary to review some basic concepts.

The standard half-cell potential of an oxidant is properly combined with the standard half-cell potential of a reductant to give a standard potential value proportional to the logarithm of an equilibrium constant, namely, $\log K_{eq} = 5040\ E^\circ/T$ (where T is in degrees Kelvin); when E° is zero, $K_{eq} = 1$. Thus, the potential predicts the reactant ratio of a system truly at equilibrium and is a *thermodynamic* constant. The calculation of the acid dissociation constant of hypochlorous acid is an example of the use of standard potentials:[14]

	E° (volts)	nE° (volts)
$2\ OH^- + Cl^- = OCl^- + H_2O + 2\ e^-$	-0.89	-1.78
$HOCl + H^+ + 2\ e^- = Cl^- + H_2O$	1.49	2.98
$2\ H_2O + 2\ e^- = 2\ OH^- + H_2$	-0.83	-1.66
$H_2 = 2\ H^+ + 2\ e^-$	0.0	0.0
Net: $HOCl \rightleftharpoons H^+ + OCl^-$		-0.46

$\log K_a = 16.9 \times (-0.46) = -7.77$ (reasonably close to the accepted -7.54).

A second example is the estimation of the standard potential for monochloramine. From Granstrom's data, reported by Morris,[15] the hydrolysis equilibrium constant, $K_{hyd} = 4.1 \times 10^{-12} M$ and

$$2\,E^\circ = \log K_{hyd}/16.9 = -11.39/16.9 = -0.67 \text{ volts}$$

	E° (volts)	nE° (volts)
$NH_2Cl + H_2O = NH_3 + HOCl$	-0.33	-0.67
$2\,e^- + HOCl + H^+ = H_2O + Cl^-$	1.49	2.98
Net: $NH_2Cl + H^+ + 2\,e^- = NH_3 + Cl^-$	1.16	2.31

It might be concluded that just as the oxidation potential (1.16 volts) is less for monochloramine than for hypochlorous acid (1.49 volts), the reactivity of monochloramine with organic compounds and its effectiveness as a disinfectant should be lower than for hypochlorous acid. Although this generally seems to be true, such reasoning is not infallible, as will be shown later in this chapter.

Very few organic oxidation-reduction (redox) reactions exhibit true equilibria; for this reason, the redox potentials of only a handful of organic compounds have been measured, notably those of quinone-hydroquinone systems.

Calculations of half-cell potentials at other than unit activities must consider the concentrations (or more correctly the activities) of the reacting species. For the oxidant hypochlorous acid, consider the half-cell reaction at 25°C, for which $E^\circ = 1.08$ volts:

$$HOCl + 2\,e^- = Cl^- + OH^-$$

The half-cell potential at pH 4, $[Cl^-] = 10^{-1} M$, and $[HOCl] = 10^{-2} M$ would be:

$$E = 1.08 - (0.059/2) \log ([Cl^-][OH^-]/[HOCl]) = 1.34 \text{ volts}$$

If the pH of the solution were raised to 9, where most of the hypochlorous acid would be dissociated to hypochlorite ion, it would be necessary to use the dissociation constant,

$$K_a = 2.9 \times 10^{-8} M$$

as follows:

$$E = 1.08 - (0.059/2) \log \frac{[Cl^-][OH^-]([H^+] + K_a)}{[H^+][HOCl]_{total}} = 1.15 \text{ volts}$$

The potential for this solution could also be arrived at by using the hypochlorite ion half-cell:

$$OCl^- + H_2O + 2e^- = Cl^- + 2\ OH^-$$

where $E^\circ = 0.89$ volts. The half-cell potential would then be

$$E = 0.89 - (0.059/2) \log ([Cl^-][OH^-]^2/[OCl^-]$$
$$= 0.89 - (0.059/2) \log (10^{-1} \times 10^{-10}/10^{-2})$$
$$= 1.16 \text{ volts}$$

It is especially noteworthy that even with other concentrations equal, half-cell potentials for hypochlorous acid (or hypochlorite ion) oxidations are affected significantly by pH.

Oxidation potentials are generally valid only for electron-transfer equilibria, or where establishment of an equilibrium is the governing (rate-controlling) quantitative effect. A reaction such as the chlorination of a phenol, which certainly involves changes in the oxidation states both of a phenolic ring carbon and of the chlorinating reagent, should be considered an oxidation; but it involves atomic rather than electron transfer and is affected by factors not normally associated with electron transfer.

An equilibrium expression describes the distribution of reactants in a system. This distribution results from the exact balancing of forward and reverse reactions. If the kinetic rate constants for the forward and reverse reactions are k_f and k_r, then $K_{eq} = k_f/k_r$. (The rate of disappearance of reactant A, in the reaction $aA + bB + cC \rightarrow$ products, might be expressed as $-d[A]/dt = k_f[A]^a\ [B]^b\ [C]^c$, provided the reverse reaction could be ignored.) Two comparable reactions could have the same value for K_{eq}, but the values of k_f for the forward reaction could vary by orders of magnitude, with appropriate compensation by the values of k_r. Sometimes it is possible to obtain meaningful thermodynamic relationships using information that is mathematically related to half-cell potentials, such as polarographic peak potentials and ionization potentials.[16,17] In summary the equilibrium constant, but not the two rate constants, for a particular reaction at equilibrium would be governed solely by the redox potential. This distinction between *thermodynamics* and *kinetics* is crucial.

A hypothetical reaction may not actually take place, even though it has a positive E° value (derived from other E° values determined by reversible electrode processes or calculated from heats of combustion or decomposition), which predicts a favorable equilibrium constant. Starting with standard oxidation potentials, let us consider two hypothetical reactions of ferrous ion with perchlorate ion at 25°C. In the first of these

$$2\ Fe^{+2} + ClO_4^- + 2\ H^+ \rightarrow 2\ Fe^{+3} + ClO_3^- + H_2O$$

	E° (volts)	nE° (volts)
$2\ Fe^{+2} = 2\ Fe^{+3} + 2\ e^-$	-0.771	-1.542
$ClO_4^- + 2\ e^- + 2\ H^+ = ClO_3^- + H_2O$	+1.19	+2.38
$\text{Log } K_{eq} = 16.907 \times 0.838 = 14.168$		+0.838

The equilibrium relationship would be:

$$K_{eq} = [Fe^{+3}]^2[ClO_3^-]/[Fe^{+2}]^2[ClO_4^-][H^+]^2 = 1.47 \times 10^{14}$$

$$\text{or } [ClO_3^-]/[ClO_4^-] = ([Fe^{+2}][H^+]/[Fe^{+3}])^2 \times 1.47 \times 10^{14}$$

Assuming that enough extra Fe^{+2} has been added to equal the Fe^{+3} formed, and that the pH is 2 (*i.e.*, $[H^+] = 10^{-2}$), the ratio of $[ClO_3^-]$ to $[ClO_4^-]$ should be $\sim 10^{10}$.

In the second reaction

$$8\ Fe^{+2} + ClO_4^- + 8\ H^+ \rightarrow 8\ Fe^{+3} + 4\ H_2O + Cl^-.$$

	E° (volts)	nE° (volts)
$8\ Fe^{+2} \rightarrow 8\ Fe^{+3} + 8\ e^-$	-0.771	-6.168
$ClO_4^- + 7\ e^- + 8\ H^+ \rightarrow \frac{1}{2}Cl_2 + 4\ H_2O$	1.34	9.38
$\frac{1}{2}Cl_2 + e^- \rightarrow Cl^-$	1.36	1.36
		4.57

$$\text{Log } K_{eq} = 16.907 \times 4.57 = 77.26$$

$$K_{eq} = [Fe^{+3}]^8[Cl^-]/[Fe^{+2}]^8[H^+]^8[ClO_4^-] = 2 \times 10^{77}$$

$$[Cl^-]/[ClO_4^-] = ([Fe^{+2}][H^+]/[Fe^{+3}])^8 \times 2 \times 10^{77}$$

Assuming enough extra Fe^{+2} has been added to equal the Fe^{+3} formed, and pH is 2, the ratio of $[Cl^-]$ to $[ClO_4^-]$ should be $\sim 10^{61}$. There is no evidence that ferrous salts react with perchlorate ion at pH 2 to form either chlorate or chloride ions.

Thus, thermodynamic constants represent the necessary, but not the sufficient, criteria for the chemical reactivity of a system. Sometimes a catalyst is required to permit a thermodynamically favored reaction to proceed.

Despite the important distinction between thermodynamic and kinetic factors, there are numerous examples of excellent correlations between data of the two types. These are usually referred to as "linear free energy relationship."[18] Unfortunately, only a relatively few linear-free energy relationships involving oxidations have been studied, of which a sampling will be discussed below.[15-17,19] Most linear free energy relationships correlate kinetic rate constants with equilibria (or equilibria vs equilibria or rate constants vs rate constants).

Correlation of Thermodynamics with Reaction Rates

Oxidation potentials are generally available only for inorganic species. In a simple experiment one may use such potentials to demonstrate linear free energy relationships for the reactions of several such inorganic reagents with a single organic substrate. Evidently this only applies when the reagents are very similar in behavior, for instance when all are one-electron oxidants.[17] Here we are dealing with triethylamine (Figure 13-2). The one oxidant that deviated widely from the plot was ferricyanide ion, and this anomaly was attributed to the experimental difficulty caused by the presence of ferrocyanide ion, which strongly retards the reaction. Choice of oxidants for this investigation (and therefore of data points) was quite limited; few one-electron oxidants of the appropriate potentials could be found.

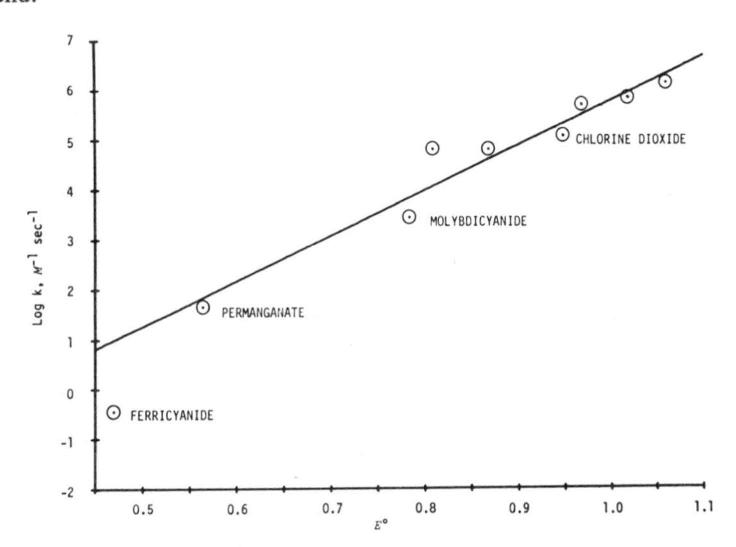

Figure 13-2. Correlation of oxidant E° values with logarithms of rate constants for one-electron oxidations of triethylamine. Unmarked points correspond to various substituted ferric phenanthrolines.

A larger number of data points for a linear-free energy plot can generally be produced if the kinetic experiments involve reaction of a single inorganic reagent with a suitable group of organic compounds. The organic compounds should differ from one another only in the presence of different substituent groups, whose function is to increase or decrease the electron density at the common reactive site. Care must be taken to avoid introduction of effects not primarily electronic in origin, especially of steric (bulk) effects. Unlike the example given in Figure 13-2, oxidation potentials cannot be used (since they are not available for the organic reactants). The plots are made against such thermodynamically related parameters as pK_b,[15,19] Hammett substituent constants,[16,20] and polarographic half-wave peak potentials.[16] A good example of this is the Brønsted plot of log k_2 vs amine pK_a for the oxidation of substituted benzyldimethylamines by chlorine dioxide[19] (Figure 13-3). It is indeed possible to devise a more

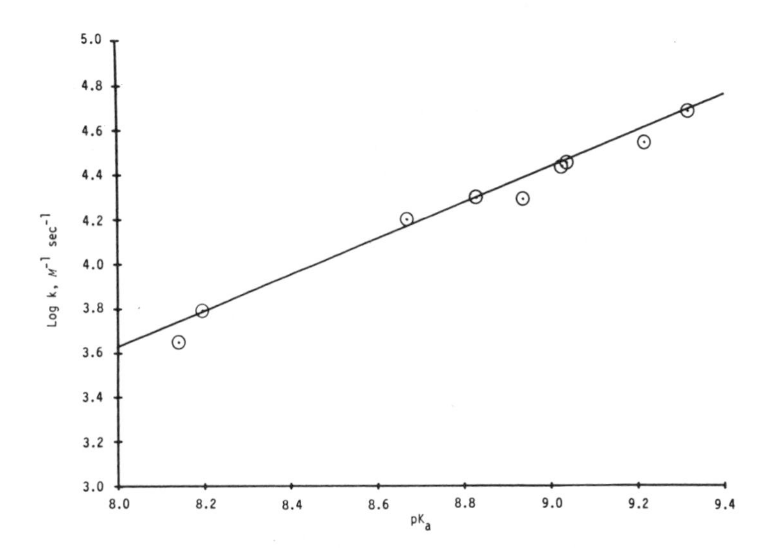

Figure 13-3. Brønsted relationship for reaction of chlorine dioxide with substituted benzyldimethylamines at 27°C.

general correlation in which the reactivities of the inorganic oxidants of a given class and the organic reductants of a given class are correlated in a single equation.[17] Thus, it proved feasible to use E° values for the oxidants and either the photoionization potentials (IP) or the Taft $\Sigma\sigma^*$ values for the reductants (tertiary amines), according to the equations:

$$\log k = -7.84\,E^{o} - 5.43\,\mathrm{IP} + 3.85$$

and

$$\log k = -7.64\, E^{o} - 4.78\Sigma\sigma^{*} - 3.47$$

Careful distinction should be drawn between reversible and irreversible reactions. In the former, one can sometimes correlate the forward rate constant (k_f) with a thermodynamic property of the reactant group. Thus, Morris[15] reported A. G. Friend's studies on the rates of N-chlorination of amines and amino acids with hypochlorous acid. The rate constants for these reactions can be expressed by the linear equation

$$k_2 = 10.19 + 0.52 \log K_B$$

Irreversible reactions can present problems. Valid comparison is generally achieved only if the rate-determining factors are constant. This was evidently true for the substituted benzyldimethylamines mentioned above.[19] The essential features of the kinetics could be considered in terms of:

$$X\text{-}C_6H_4\text{-}CH_2\text{-}\ddot{N}(CH_3)_2 + ClO_2 \underset{\text{slow}}{\rightleftharpoons} X\text{-}C_6H_4\text{-}CH_2\text{-}N^{\cdot +}(CH_3)_2 + ClO_2^- \xrightarrow{\text{fast}} \text{Intermediates} \xrightarrow{ClO_2} \text{Products}$$

With excess amine, and in the absence of a build-up of chlorite ion, the disappearance of the chlorine dioxide was strictly first-order. The slow step was rate-determining and it had changed from the first to the second stage of reaction as the substituent X was varied, it is likely that the plot of Figure 13-3 would not have been linear. Such mishaps are more likely in such multistage reactions than in the much simpler overall-reversible reactions.

The frequent futility of trying to predict chemical reactivity solely on the basis of redox potentials, even as between inorganic reactants, has been demonstrated above. The same point can be made regarding the reactivity of halogen oxidants with organic species. Thus it is well-known that elemental chlorine reacts rapidly with most alkenes, whereas the hypochlorite ion, OCl^-, does not. Suitable substituents on an alkene, however, completely reverse this picture, so that o-chlorobenzylidenemalononitrile (which could be called 1,1-dicyano-2-*o*-chlorophenylethylene) is oxidized rapidly by hypochlorite ion though it does not react with chlorine:[21]

$$o\text{-}Cl\text{-}C_6H_4\text{-}CH{=}C(CN)_2 \xrightarrow{OCl^-} o\text{-}Cl\text{-}C_6H_4\text{-}\overset{O}{\widehat{C\text{—}C}}(CN)_2$$

Experiments to determine the reactive species require adjusting conditions to favor the presence of one species or another. Thus, a solution of chlorine gas in water with much added hydrochloric acid would favor the presence of elemental chlorine; a relatively dilute chlorine water solution at pH 4 would contain mostly hypochlorous acid, and the same solution adjusted to pH 9 would have hypochlorite ion as the dominant species. Chlorine, of course, has a much higher standard oxidation potential, 1.36 V, than either hypochlorite ion (at 0.89 V) or hypochlorous acid (at 1.08 V for $HOCl + 2e^- = OH^- + Cl^-$).

As might be expected on the basis of oxidation potentials, chlorine dioxide, with $E^\circ = 0.95$ V, oxidizes triethylamine much more rapidly than does permanganate ion, with $E^\circ = 0.564$ V; the k_2 values are $1.2 \times 10^5\ M^{-1}$ sec^{-1} and $46\ M^{-1}\ sec^{-1}$, respectively.[16] The roles are sharply reversed, however, towards olefins, though exact rate information is lacking; here permanganate is a vigorous oxidant,[22,23] with $k_2 = 10^2 - 10^3\ M^{-1}\ sec^{-1}$ at 25°C, whereas chlorine dioxide appears to react quite slowly.[24,25] These results are all obtainable at pH 8.0. In practice, the pH is adjusted somewhat, and suitable calculations are made to convert the data to standard conditions. The values of E° are considered constant throughout the pH range in question, and are independent of pH since there are no hydronium or hydroxide terms in the half-cell equations:

$$ClO_2 + e^- = ClO_2^-, \text{ and } MnO_4^- + e^- = MnO_4^{-2}$$

Even acid-base reactions can show apparent anomalies that make kinetic predictions difficult, if not futile. Displacement of positive chlorine from the N-chloroquinuclidinium ion is a case in point.[26] Although hydroxide ion is 10^5 times more basic than dimethylamine towards protons, it is 100 times less reactive towards the positive chlorine:

$$\text{(H-C(CH}_2\text{CH}_2)_3\text{N}^+\text{-Cl)} + OH^- \xrightarrow{k = 1.82 \times 10^5\ M^{-1}\ sec^{-1}} \text{(H-C(CH}_2\text{CH}_2)_3\text{N:)} + HOCl$$

$$\text{(H-C(CH}_2\text{CH}_2)_3\text{N}^+\text{-Cl)} + (CH_3)_2NH \xrightarrow{k = 2.02 \times 10^7\ M^{-1}\ sec^{-1}} \text{(H-C(CH}_2\text{CH}_2)_3\text{N:)} + (CH_3)_2^+NHCl$$

Thermodynamics and Disinfection Rates

Considering these precautions one is somewhat skeptical about attempting to correlate the efficacy of disinfectants with oxidation potentials. The initial impetus in this direction was apparently provided by Hallum and Youngner,[27] who demonstrated loss of infectivity of five types of viruses at the surface of a platinum electrode. The rates of inactivation were clearly related to the applied potentials. These authors pointed out that the rates of the reactions between oxidants and viruses are strongly affected by catalysts, especially metal ions. For this reason the controlled potential electrolysis study was undertaken followed by a series of papers by Ebba Lund and co-workers on the oxidative inactivation of viruses.[28-38] In the first paper[28] several solutions of different measured oxidation potential ranges were prepared, and the time determined for 50% decrease in poliomyelitus virus activity at 37°C. The reciprocal of the 50% virus inactivation time rose significantly with rising voltage in a very regular fashion over the oxidation potential interval (250-500 mv). The oxidants for this interval were potassium permanganate (four concentrations to give the desired potentials) and cystine (one concentration). Later papers showed fairly good adherence to the initial linear plot by such diverse oxidants as chloramine-T, bromate ion, silver ion, mercuric ion and iodine. By contrast to the amine oxidations cited above,[17] the oxidants were mixed with the corresponding reduced species where possible to produce poised systems of hopefully constant potentials, *e.g.,*[28] MnO_4^- with Mn^{+2}. Some good data points were obtained for hypochlorous acid in the absence of high concentrations of organic reductants. Two of the oxidants tested behaved anomalously. The first of these, hydrogen peroxide, acted towards the potentiometer in a poliomyelitis virus suspension as if it were a reductant; yet it was a moderately effective disinfectant.[31] The author concluded that "the process could not be an oxidation of the same type as found for purely oxidizing compounds." Solutions of sodium periodate gave very high measured potentials, which were, however, far less effective than predicted from the behavior of most other oxidants.[33] The addition of a low concentration of permanganate ion (too low to destroy the virus rapidly by its own action) accelerated the virus kill considerably. The author provided an extremely dubious explanation for this effect:[33] "If no mutual potential exists for two redox systems . . . then no appreciable interaction can take place, but if a mediator system is added which has a normal potential between the two first systems so that the curves have mutual potentials with both the original systems, then the over-all reaction between these two systems may be brought about through stepwise reactions between the systems." Notwithstanding these exceptions, the relation of the speed of action of oxidative disinfectants–including hypochlorous acid,[29,34,36]

N-chloroamides[29,30,32,34,36] and monochloramine[37]–to the observed redox potentials may be of some use. Furthermore, although most of the work was done on viruses, there was some indication of its validity for washed suspensions of bacteria.[37] It may be concluded that the same problems exist in correlating redox potentials for disinfection as exist in the linear free energy correlation for the reactions of organic compounds.

VARIETY IN THE REACTIONS OF HALOGEN SPECIES WITH ORGANIC COMPOUNDS

Formation of a given product or set of isomeric products by the action of a single oxidizing reagent (which may be a mixture of species in equilibrium–for example, chlorine water at some particular pH) with a single organic compound can be quite complicated. Consider the following examples.

Reagents Containing Species in Equilibrium with Hypochlorous Acid

When phenol is chlorinated with aqueous Cl_2 at low pH, it forms a complex that proceeds to chlorophenols by three distinct pathways:[39]

$$C_6H_5OH + Cl_2 \xrightarrow[pH<2]{} C_6H_5OH \cdot Cl_2 \left\{ \begin{array}{l} \xrightarrow{H^+} \\ \longrightarrow \\ \xrightarrow[H^+,\ Cl^-]{} \end{array} \right\} \text{Monochlorophenols}$$

As the pH is raised, the two acid-catalyzed pathways diminish in importance, and the concentration of elemental chlorine also decreases; thus all these reactions are decelerated. At the same time the dissociation of phenol to phenoxide permits reaction to take place by the following alternative mechanism:[40]

$$C_6H_5O^- + HOCl \xrightarrow[\text{pH 7 to 12}]{} \text{Monochlorophenols}$$

Regardless of which mechanisms are favored, the monochlorophenols can react further to form polychlorophenols or other oxidation products. The final product assortment depends on several factors, including the initial reactant concentrations (and their ratios), pH, temperature, competing organic substances and possibly the way in which the chlorine is mixed in.

When anisole is chlorinated in aqueous solution,[12] the kinetic evidence suggests participation of three species, involving four mechanisms:

2 HOCl → Cl_2O ← 2 HOCl + H^+

Cl_2O + anisole (C_6H_5–OCH_3) → *o*- and *p*-Chloroanisoles

HOCl + H^+ ⇌ H_2OCl^+ (fast; K_{eq} not known)

H_2OCl^+ + anisole (C_6H_5–OCH_3) → *o*- and *p*-Chloroanisoles

Cl_2 + anisole (C_6H_5–OCH_3) → *o*- and *p*-Chloroanisoles

As pH decreases, the content of *p*-chloroanisole in the product decreases from 67.6% at 5 x 10^{-3} *M* H_3O^+ to 57.9% at 1 *M* H_3O^+. This is probably due to the low positional selectivity of H_2OCl^+ as a chlorinating agent, compared with Cl_2 or Cl_2. The extent to which the reagent species determine the proportions of isomers produced is further illustrated by a comparison of the chlorination of toluene by Cl_2 in acetic acid (a polar, and hence water-like solvent) to its chlorination in water by H_2OCl^+.[41] The product ratio, *o*-chlorotoluene: *p*-chlorotoluene, os 0.75 for Cl_2 and 1.6 for $HOCl^+$.

The chlorine-free radical is probably the most selective chlorinating species of all:[12]

Cl · (from SO_2Cl_2 and benzoyl peroxide) + anisole (C_6H_5–OCH_3) → 80% *p*-Chloroanisole

The chlorine water oxidation of O,O,S-triethyl thiophosphate in aqueous solution appears to involve at least three mechanistic pathways.[42] These are (1) acid- as well as chloride-dependent Cl_2 formation, (2) chloride-dependent Cl_2 formation, and (3) hypochlorite-catalyzed hydrolysis. A simplified representation shows the roles of the reactants:

HOCl ⇌ Cl_2 (with Cl^-; and with Cl^-, H^+)

Cl_2 + $(C_2H_5O)_2P(O)SC_2H_5$ → Final products (multistep, k = 1.5 x 10^8 M^{-1} min^{-1})

HOCl → OCl^-

OCl^- + $(C_2H_5O)_2P(O)SC_2H_5$ → $(C_2H_5O)_2P(O)O^-$ + C_2H_5SH

$(C_2H_5O)_2P(O)O^-$ + C_2H_5SH → Final products (OCl^-, fast)

A fourth possible pathway would involve an unproved termolecular reaction of triethyl thiophosphate with HOCl and OCl^-; it is needed to explain the chloride ion-independent reaction at pH 7.[42]

Note that the hypochlorite-catalyzed hydrolysis of the third pathway is not actually an oxidation, but exemplifies the role of hypochlorite ion as a fairly potent nucleophile–in this case a nucleophilic catalyst for hydrolysis. Further examples of this behavior are the catalyzed hydrolyses of the nerve gas sarin[43] (isopropyl methylphosphonofluoridate), of *p*-nitrophenyl acetate,[44] and of activated nitriles.[21,45] The nucleophilicity of hypochlorite ion is further manifest in the mechanism proposed for the previously-cited oxidation of *o*-chlorobenzylidene-malononitrile:[21]

$$\underset{H}{\overset{(C_6H_4Cl)}{>}}C{=}C\underset{CN}{\overset{CN}{<}} \xrightarrow[\text{slow, rate-determining}]{OCl^-} \left[\underset{H}{\overset{(C_6H_4Cl)}{>}}\underset{OCl}{\underset{|}{C}}-\ddot{C}\underset{CN}{\overset{CN}{<}}\right]^- \xrightarrow{\text{rapid}} \text{epoxide} + Cl^-$$

In its nucleophilic reactions hypochlorite obviously attacks with the oxygen. Hypochlorous acid, which is an electrophile, attacks with the chlorine end, for example in the chlorination of amino or amide nitrogen and carbanions.

It is interesting to compare the nucleophilicities of the hypochlorite and chlorite ions. Experiments in the author's laboratory have indicated that chlorite is generally a poor nucleophile, especially in catalyzed hydrolyses, which would be consistent with its low basicity. Nevertheless, chlorite is a uniquely reactive oxidant towards aldehydes, where it is believed to attack initially as a nucleophile:[46]

$$R-C\underset{O}{\overset{H}{\lessgtr}} + H^+ \rightarrow R-C\underset{\overset{+}{OH}}{\overset{H}{\lessgtr}} \xrightarrow{ClO_2^-} \left[R-\overset{H}{C}\underset{O\cdots Cl}{\overset{O-H\cdots O}{\lessgtr}}\right]$$

$$\longrightarrow R-C\underset{O}{\overset{O-H}{\lessgtr}} + H^+ + OCl^-$$

One might postulate a similar reaction for hypochlorous acid, but such a reaction, if it occurs, must be very slow:

$$R-C\underset{\overset{+}{OH}}{\overset{H}{\lessgtr}} \xrightarrow{OCl^-} \left[R-\overset{H}{C}\underset{O\cdots Cl}{\overset{O-H}{\lessgtr}}\right] \longrightarrow R-C\underset{O}{\overset{O-H}{\lessgtr}} + H^+ + Cl^-$$

Chlorine Dioxide as an Oxidant

The oxidation of secondary and tertiary amines by aqueous chlorine dioxide is an excellent way to illustrate the diversity of reactions between a single reagent and a class of compounds whose members might be considered quite similar. Chlorine dioxide attacks benzyl-*t*-butylamine, for example, by two different mechanisms.[47] The first (which applies also to the reaction of trialkylamines) begins with removal of an electron from the nitrogen, followed by loss of a proton:

$$ClO_2 + C_6H_5\text{-}CH_2\text{-}\ddot{N}H\text{-}C(CH_3)_3 \underset{}{\overset{\text{slow}}{\rightleftharpoons}} ClO_2^- + C_6H_5\text{-}CH_2\text{-}\dot{N}^+H\text{-}C(CH_3)_3$$

$$\xrightarrow{\text{fast } (-H^+)} C_6H_5\text{-}\dot{C}H\text{-}NH\text{-}C(CH_3)_3$$

$$C_6H_5\text{-}\dot{C}H\text{-}NH\text{-}C(CH_3)_3 \xrightarrow[\text{fast}]{ClO_2} C_6H_5\text{-}CH{=}\overset{+}{N}H\text{-}C(CH_3)_3 + ClO_2^-$$

$$C_6H_5\text{-}CH{=}\overset{+}{N}H\text{-}C(CH_3)_3 \xrightarrow{H_2O} C_6H_5\text{-}CHO + (CH_3)_3C\text{-}NH_3^+$$

Overall stoichiometry is

$$C_6H_5\text{-}CH_2NC(CH_3)_3 + 2\ ClO_2 + H_2O \longrightarrow C_6H_5\text{-}CHO + (CH_3)_3CNH_2 + 2\ ClO_2^- + 2\ H^+$$

The second reaction avoids the first intermediate, proceeding to the second directly, by removal of a hydrogen atom:

$$ClO_2 + C_6H_5\text{-}CH_2\text{-}\ddot{N}H\text{-}C(CH_3)_3 \longrightarrow HClO_2 + C_6H_5\text{-}\dot{C}H\text{-}NH\text{-}C(CH_3)_3$$

(captured hydrogen atom: one benzylic H)

This second reaction appears to be irreversible (unlike the first) because the chlorous acid formed can immediately dissociate to chlorite ion and hydronium ion. Interestingly, both reaction pathways start and end with exactly the same set of species. They can be distinguished from each other because only the first is subject to retardation by chlorite ion (which tends to reverse the initial equilibrium) and because the second is subject to a much larger kinetic isotope effect *i.e.*, a decrease in rate, when the benzylic hydrogens are replaced by deuterium atoms. First-step hydrogen atom removal is not generally seen with tertiary amines.

When a tertiary amine cannot assume a planar configuration for bonds around the nitrogen, specifically in the case of quinuclidine,[48] reaction kinetics become considerably slower, and the kinetics and mechanism are exceedingly complex. Both chlorate and chloride ions (rather than chlorite) are formed as the dominant inorganic products. Instead of the C–N cleavage previously encountered, oxidation produces quinuclidine N-oxide:

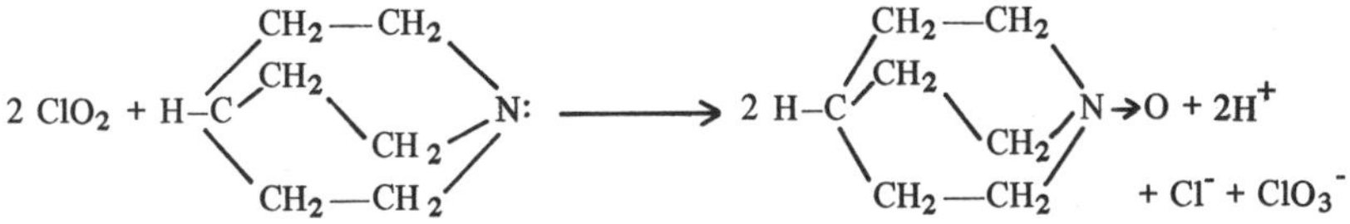

$$2\,ClO_2 + \text{H–C}(CH_2\text{–}CH_2)_3\text{N:} \longrightarrow 2\,\text{H–C}(CH_2\text{–}CH_2)_3\text{N}\rightarrow\text{O} + 2H^+ + Cl^- + ClO_3^-$$

It is not too difficult to explain why this slow reaction (which may also participate very slightly in the oxidation of other tertiary amines) is seen in this instance. The usual oxidation route is blocked because it would require a double bond between the "bridgehead" nitrogen and an adjacent carbon. The formation of a double bond at a bridgehead is considered "forbidden" (*i.e.*, energetically and probabilistically very difficult) according to Bredt's rule;[49] it would require a distortion of bonds, somewhat like this:

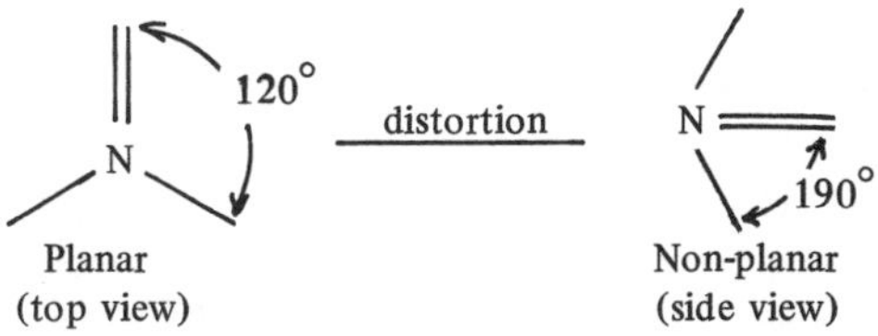

120° — distortion — 190°

Planar (top view) — Non-planar (side view)

During early experiments, the supply of quinuclidine, not commercially available then, was expended. The attempt to substitute triethylenediamine for quinuclidine led to the discovery of yet a fourth general oxidative mechanism, exemplified by the reaction between triethylenediamine and chlorine dioxide,[50-52] but characteristic of other 1,2-diamines and other oxidants:

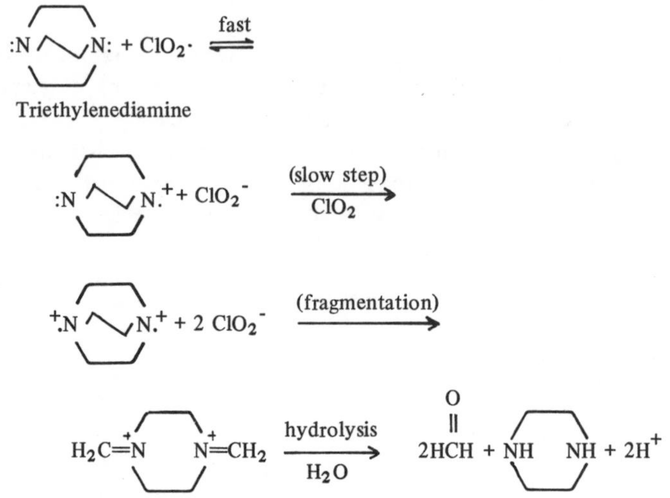

$$:N(C_2H_4)_3N: + ClO_2\cdot \overset{\text{fast}}{\rightleftharpoons}$$

Triethylenediamine

$$:N(C_2H_4)_3N\cdot^{+} + ClO_2^{-} \xrightarrow[ClO_2]{\text{(slow step)}}$$

$$^{+}\cdot N(C_2H_4)_3N\cdot^{+} + 2\ ClO_2^{-} \xrightarrow{\text{(fragmentation)}}$$

$$H_2C{=}\overset{+}{N}(C_2H_4)_2\overset{+}{N}{=}CH_2 \xrightarrow[H_2O]{\text{hydrolysis}} 2HCH(=O) + NH(C_2H_4)_2NH + 2H^{+}$$

Variety in Amine Oxidations

Still other special cases of tertiary amine oxidation by chlorine dioxide or hypochlorous acid have been identified,[53] though not studied in the detail of those previously discussed:

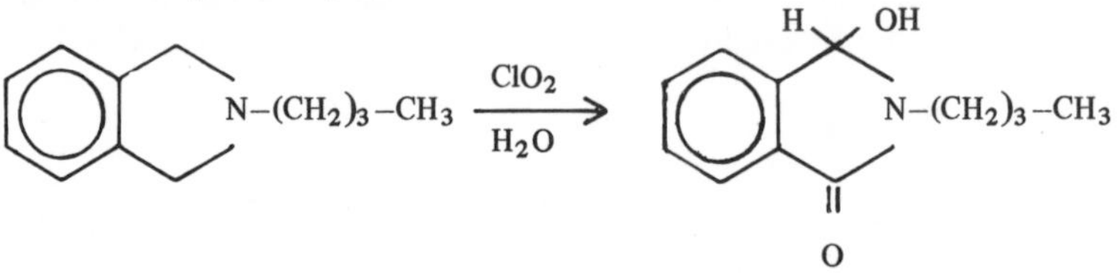

$$C_6H_4(CH_2)_2N{-}(CH_2)_3{-}CH_3 \xrightarrow[H_2O]{ClO_2} C_6H_4(CH(OH))(C{=}O)N{-}(CH_2)_3{-}CH_3$$

The starting material for the above reaction, N-butylisoindoline, is notable for its susceptibility to air oxidation.[54] With a 6-membered, rather than the above 5-membered used ring, oxidations proceed somewhat differently,[53]

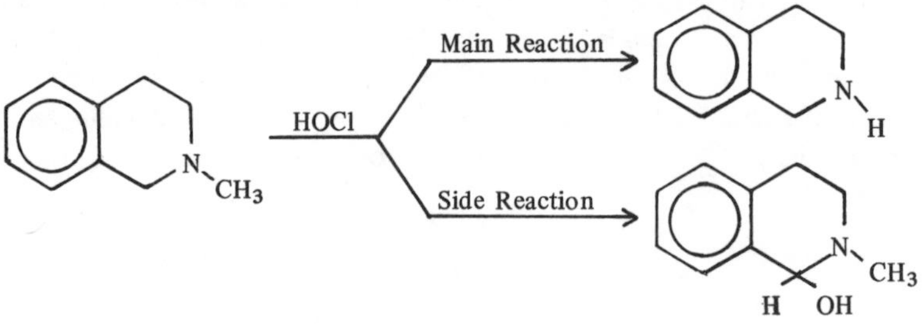

N-methyltetrahydroisoquinoline (N–CH_3) + HOCl:

- Main Reaction → tetrahydroisoquinoline (N–H)
- Side Reaction → 1-hydroxy-N-methyltetrahydroisoquinoline (H, OH; N–CH_3)

Essentially the same reaction products as were produced by hypochlorous acid would be expected with chlorine dioxide.

Finally, there is the reaction that might be called "oxidative hydrolysis":[53]

$$(C_6H_5)_2CH-C(=O)-O-CH_2-CH_2-CH_2-N\langle\text{piperidine}\rangle \xrightarrow[pH = 5]{HOCl} (C_6H_5)_2CH-CO_2H + CH_2{=}CH-CHO + HN\langle\text{piperidine}\rangle$$

Related esters with the three-carbon chain between carboxyl and nitrogen reacted in like manner with chlorine dioxide, but only the carboxylic acids were identified.[53]

SEQUENTIAL REACTIONS OF RELATED SPECIES WITH ORGANIC COMPOUNDS

As an organic substrate undergoes attack by a reagent such as aqueous chlorine dioxide or chlorine, different molecular or ionic species may become key factors in the overall process. For example, yellow chlorine dioxide vapors mixed with air were passed into an unbuffered aqueous solution of triethylamine.[55] The color of the chlorine dioxide vanished, pH of the amine solution gradually dropped, and acetaldehyde could be detected by odor or chemical test in the air emerging from the absorbing solution. The indicated stoichiometric reaction (similar to that shown above for benzyl-*t*-butylamine) was:

$$(C_2H_5)_3N + 2\ ClO_2 + H_2O \rightarrow (C_2H_5)_2NH + CH_3CHO + 2\ ClO_2^- + 2\ H^+$$

Unexpectedly, however, the pH continued to drop and the color of chlorine dioxide appeared in the solution. A secondary reaction–the acid-catalyzed reaction of acetaldehyde with chlorite ion had begun:

$$H^+ + CH_3CHO + 3\ ClO_2^- \rightarrow CH_3CO_2^- + Cl^- + 2\ ClO_2 + H_2O$$

Although this reaction uses up acid, its ultimate effect is to produce more acid, since the overall stoichiometry is:

$$H_2O + 2\ ClO_2 + 3\ (C_2H_5)_3N \rightarrow 3\ (C_2H_5)_2NH + CH_3CHO + 2\ CH_3CO_2H + 2\ HCl$$

In addition, diethylamine $(C_2H_5)_2NH$ itself reacts with chlorine dioxide, but at a much reduced rate. A further complication of this system is the incompletely understood manner in which chlorine dioxide arises from the interaction of chlorite ion with acetaldehyde. If, as discussed previously, hypochlorite is the initial inorganic reaction product,[46] then it must combine with chlorite in a reaction (see above) in which the transient intermediate Cl_2O_2 plays a part. In the oxidation of triethylamine, this intermediate does not seem to appear under ordinary circumstances. Yet there is reason to believe it might appear in a well-chosen experiment. Thus, the presence of Cl_2O_2 might have to be postulated to explain why a high benzyl-to-alkyl cleavage ratio for p-chlorobenzyldimethylamine was observed when chlorine dioxide was generated *in situ* from formaldehyde and chlorite ion at pH 6.6, instead of being introduced into the amine solution as preformed ClO_2.[19]

Similarly, Grimley and Gordon[56] found it necessary to postulate a role for $Cl_2O_4^-$ in the oxidation of phenol by chlorine dioxide to account for the accelerated disappearance of chlorine dioxide due to added chlorite. This can be compared with the deceleration of the disappearance of chlorine dioxide by chlorite ion in amine oxidations.[47]

A final example is the sequence of reactions whereby malononitrile is converted to dichloracetamide,[45] which probably involves nine major steps:

$$(CN)_2CH_2 \xrightarrow[\text{Step 1}]{-H^+} (CN)_2CH^- \xrightarrow[\text{Step 2}]{HOCl} (CN)_2CHCl \xrightarrow[\text{Step 3}]{-H^+}$$

$$(CN)_2CCl^- \xrightarrow[\text{Step 4}]{HOCl} Cl_2C(CN)_2 \xrightarrow[\text{Step 5}]{OCl^-,\ H_2O} Cl_2C(CN)(CONH_2) \xrightarrow[\text{Step 6}]{OCl^-,\ H_2O}$$

$$Cl_2C(CONH_2)_2 \xrightarrow[\text{Step 7}]{HOCl,\ OH^-} [Cl_2C(CONH_2)(CONHCl)] \xrightarrow[\text{Step 8}]{OCl^-,\ H_2O}$$

$$[Cl_2C(CONH_2)(CO_2H)] \xrightarrow[\text{Step 9}]{} Cl_2CH{-}CONH_2$$

Steps 1 and 3 typify the relatively slow dissociation of pseudoacids. *Steps 2 and 4* are chlorinations of carbanions by hypochlorous acid. *Steps 5 and 6* provide examples of the hypochlorite-catalyzed hydrolysis of activated nitrile groups. In *Step 7* one sees the N-chlorination of an amide, probably involving HOCl. In *Step 8*, hypochlorite ion hydrolyzes the N-chloroamide to a carboxylic acid group, which spontaneously decarboxylates in *Step 9*. Thus, six of the nine steps appear to involve chlorine species.

OCCURRENCE OF CHLORINATED ORGANIC COMPOUNDS IN WASTEWATER

Within the present decade we have become aware of the potential of chlorination not only to improve water quality, but also to degrade it. This process can lead to derivatives more toxic to aquatic organisms and more refractory to biodegradation than the organic constituents originally present in the wastewater. Ironically, the formation of such derivatives usually leads to a reduction in BOD (biochemical oxygen demand); normally, reduction in BOD would be considered beneficial, but not necessarily through chlorination.

In one pioneering study,[57] 14 industrially important compounds known to appear in wastewater were exposed to aqueous chlorine under realistic treatment conditions of 8-20 mg/l of organic compound and 5-20 mg/l of chlorine for periods of up to 24 hours. Five compounds showed definite reactivity, and partially successful attempts were made to characterize the products (Table 13-2); but product identifications were not definitive. Static bioassays with fathead minnows (*Pimphales promelas*) indicated the following TL_m ranges (in mg/l) for three of the probable reaction products: 2,4,6-Trichlorophenol, 0.1-1.0; 2,4,6-trichloroaniline, 1.0-10.0; 4-chloro-3-methylphenol, 0.01-0.1. Thus, the chlorine-containing derivatives were significantly toxic to fish.

A group from the University of Minnesota[58] has begun a very fundamental and wide-ranging approach to the presence of chlorination products in water. Three types of reactions that form such products were considered—electrophilic substitutions on aromatic systems, chlorohydrination of olefins, and free-radical reactions (especially benzylic halogenation). An EPA study by L. Keith, leading to the identification of fish-killing chlorinated guaiacols in Kraft paper mill wastes, provided an example of a toxic mixture formed by electrophilic aromatic substitution. Over long enough periods or under sufficiently drastic conditions, the same class of reactions converted biphenyl to polychlorinated biphenyls (PCB). The chlorohydrination reaction was demonstrated at wastewater treatment concentrations with oleic acid and with α-terpineol. Chlorinations carried out with resin acids (abietic and dehydroabietic) at pH 2 and 10-100 mg/l of Cl_2 were used to illustrate free-radical mechanisms. Based on studies by Hansch and co-workers,[59,60] the Minnesota group developed excellent structure–activity relationships for the lethality of substituted phenols to *Daphnia magna*; the most important physico-chemical parameter in these correlations is the partition coefficient between water and a suitable lipophilic solvent, such as l-octanol. Thus, serious consideration is being given to the prediction of toxic effects on the basis of calculable physico-chemical properties.

Table 13-2. Fourteen Industrially Significant Organic Waste Components and Some Probable Reaction Products with Chlorine Under Wastewater Treatment Conditions[57]

Chemicals Selected for Study	Probable Reaction Products
Alcohols	
Methanol	None
Isopropyl alcohol	None
tert-Butyl alcohol	None
Ketones	
Acetone	None
Benzene and Derivatives	
Benzene	None
Toluene	None
Ethylbenzene	None
Benzoic acid	None
Phenol and Phenolics	
Phenol	*o*- and *p*-Chlorophenols; 2,4- and 2,6-dichlorophenols; 2,4,6-trichlorophenol; non-aromatic oxidation products
m-Cresol	Three monochlorocresols; three dichlorocresols; 2,4,6-trichloro-3-methylphenol; non-aromatic oxidation products
Hydroquinone	*p*-Benzoquinone; non-aromatic oxidation products
Organic Nitrogen Compounds	
Aniline	*o*- and *p*-Chloroaniline; 2,4- and 2,6-dichloroaniline; 2,4,6-trichloroaniline; non-aromatic oxidation products
Dimethylamine	N-Chlorodimethylamine; oxidation products
Nitrobenzene	None

One noteworthy aspect of the foregoing study is the inclusion of free-radical reactions. Although very little attention has been paid to the chemistry of chlorine atoms (which are free radicals) in aqueous solution, they may be more common than generally supposed. Thus, they are believed to arise in the decomposition of triethylenediamine chlorammonium[52] and triethylchlorammonium[61] cations:

$$R_3N^+{-}Cl \rightarrow R_3N\cdot^+ + Cl\cdot$$

Undoubtedly they play a very important role in ultraviolet-catalyzed chlorine oxidations, such as those of cornstarch[62] and of surfactants.[63] This suggests that experimental studies whose purpose is to predict the fate of organic compounds in chlorinated wastewater, should always include real or simulated irradiation with sunlight.

R. L. Jolley,[64] in a recently published doctoral dissertation, listed organic wastewater constituents from a variety of sources. He claimed to identify 16 or 17 chlorine-bearing compounds in treated domestic sewage (Table 13-3), apparently resulting from chlorination. It is especially significant that the list includes the chlorination products of a pyrimidine and a purine–uracil and guanine—found in ribonucleic acids (RNA).

Table 13-3. Tentative Identifications and Concentrations of Chlorine-Containing Constituents in Chlorinated Effluents[64]

Organic Compound	Concentration (μg/l)
5-Chlorouracil	4.3
5-Chlorouridine	1.7
8-Chlorocaffeine	1.7
6-Chloroguanine	0.9
8-Chloroxanthine	1.5
2-Chlorobenzoic Acid	0.26
5-Chlorosalicylic Acid	0.24
4-Chloromandelic Acid	1.1
2-Chlorophenol	1.7
4-Chlorophenylacetic Acid	0.38
4-Chlorobenzoic Acid	1.1
4-Chlorophenol	0.69
3-Chlorobenzoic Acid[a]	0.62
3-Chlorophenol[a]	0.51
4-Chlororesorcinol	1.2
3-Chloro-4-hydroxybenzoic Acid	1.3
4-Chloro-3-methylphenol	1.5

[a]One or both of these compounds

Surprisingly little is known about the reactivity of purines and pyrimidines with aqueous chlorine, despite the strong possibility that the mechanisms of disinfection of viruses and other microbes might involve the modification or destruction of DNA or RNA (which contain them). The monochlorination of cytosine appears to proceed through rapid initial and reversible formation of 4-N-chlorocytosine, followed by conversion to 5-chlorocytosine during product isolation.[65,66] It is not yet clear that these reactions are fast enough to explain observed rates of disinfection. Much remains to be done, therefore, to establish a plausible mechanism for disinfection, as well as to define the conditions for forming such products as 5-chlorouracil and 6-chloroguanine in chlorinated domestic waste effluents.

HALOFORM FORMATION IN DRINKING WATER

Chloroform and bromoform and their intermediates have been identified after chlorination of Rhine River water for the water supply of Rotterdam[67] and Mississippi River water for the water supply of New Orleans.[68] Rook has suggested these compounds are formed as the end products of the oxidation of humic substances.[67] To support this hypothesis, he found CH_3Cl was formed from 5 mg/l solutions of pyrogallol, phloroglucinol and several polyphenols including two natural products.

SUMMARY

Chlorine and the several oxychlorine species capable of existing in water form an interesting family of reagents. Each has its own pattern of reactivity with organic compounds. The energetic feasibility for reactions between these species and organic compounds can be expressed in thermodynamic terms (and at least partly in terms of redox potential). But thermodynamic feasibility is only a necessary and not a sufficient criterion for reactivity. For the reactions to occur, appropriate pathways must also exist; their relative importances are expressed through kinetic rate equations. Examples have been given to show how organic compounds may react in diverse ways with chlorine-containing oxidizing species—or with groups of such species in equilibrium with one another. These reactions proceed in parallel with, and also in succession to one another. The literature is still far too sparse to accurately predict the fate of many reactive organic compounds when they are put in water containing chlorine and oxychlorine species. However, an appreciation of the rich variety of chemistry in this area is essential to further progress.

REFERENCES

1. Granstrom, M. L. and G. F. Lee. "The Disproportionation of Chlorine Dioxide," Department of Sanitary Engineering, School of Public Health, University of North Carolina, Chapel Hill, N.C. (November 15 1957).
2. Morris, J. C. "The Acid Ionization Constant of HOCl from 5° to 35°," *J. Phys. Chem.* **70**, 3798-3805 (1966).
3. Roth, W. A. "On the Thermochemistry of Chlorine and Hypochlorous Acid," *Z. Physik. Chem.* **A 145**, 289-297 (1929).
4. Sherrill, M. S. and E. F. Izard. "The Solubility of Chlorine in Aqueous Solutions of Chlorides and the Free Energy of Trichloride Ion," *J. Amer. Chem. Soc.* **53**, 1667-1674 (1931).
5. Arotsky, J. and M. C. R. Symons. "Halogen Cations," *Quart. Rev.* **16**, 282-297 (1962).

6. Hine, J. *Physical Organic Chemistry*, 2nd ed. (New York: McGraw-Hill Book Co., 1962), pp. 362-363.
7. Gordon, G., R. G. Kieffer and D. H. Rosenblatt. "The Chemistry of Chlorine Dioxide," In *Progress in Inorganic Chemistry,* S. J. Lippard, Ed., Vol. 15 (New York: John Wiley & Sons, Inc., 1972), pp. 202-286.
8. Gordon, G. and F. Emmenegger. "Complex Ion Formation Between ClO_2 and ClO_2^-," *Inorg. Nucl. Chem. Lett.* 2, 395-398 (1966).
9. Downs, A. J. and C. J. Adams. "Chlorine, Bromine, Iodine and Astatine," In *Comprehensive Inorganic Chemistry*, Vol. 2, J. C. Bailor, Jr., H. J. Emeleus, R. Nyholm and A. F. Trotman-Dickenson, Eds. (Oxford: Pergamon Press, Ltd., 1973), p. 1421.
10. Redlich, O. and G. C. Hood. "Ionic Interaction, Dissociation and Molecular Structure," *Disc. Faraday Soc.* **24**, 87-93 (1957).
11. Eigen, M. and K. Kustin. "The Kinetics of Halogen Hydrolysis," *J. Amer. Chem. Soc.* **84**, 1355-1361 (1962).
12. Swain, C. G. and D. R. Crist. "Mechanisms of Chlorination by Hypochlorous Acid. The Last of Chlorinium Ion, Cl^{+1}," *J. Amer. Chem. Soc.* **94**, 3195-3200 (1972).
13. D'Ans, J. and H. E. Freund. "Kinetic Investigations. I. On Chlorate Formation from Hypochlorite," *Z. Elektrochem.* **61**, 10-18 (1957).
14. "Potentials of the Elements and their Compounds at 25° C," In *Lange's Handbook of Chemistry*, 11th ed., J. A. Dean, Ed. (New York: McGraw-Hill Book Co., 1973), pp. 6-6 - 6-16.
15. Morris, J. C. "Kinetics of Reactions Between Aqueous Chlorine and Nitrogen Compounds," In *Principles and Applications of Water Chemistry*, S. D. Faust and J. V. Hunter, Eds. (New York: John Wiley & Sons, Inc., 1967), pp. 23-53.
16. Hull, L. A., G. T. Davis, D. H. Rosenblatt and C. K. Mann. "Oxidations of Amines. VII. Chemical and Electrochemical Correlations," *J. Phys. Chem.* **73**, 2142-2146 (1969).
17. Hull, L. A., G. T. Davis and D. H. Rosenblatt. "Oxidations of Amines. IX. Correlation of Rate Constants for Reversible One-Electron Transfer in Amine Oxidation with Reactant Potentials," *J. Amer. Chem. Soc.* **91**, 6247-6250 (1969).
18. Laidler, K. J. *Chemical Kinetics*, 2nd ed. (New York: McGraw-Hill Book Co., 1965), pp. 246-253.
19. Rosenblatt, D. H., L. A. Hull, D. C. DeLuca, G. T. Davis, R. C. Weglein and H. K. R. Williams. "Oxidations of Amines. II. Substituent Effects in Chlorine Dioxide Oxidations," *J. Amer. Chem. Soc.* **89**, 1158-1163 (1967).
20. Taft $\Sigma\sigma^*$ values are conveniently listed in H. K. Hall, Jr. "Correlation of the Base Strengths of Amines," *J. Amer. Chem. Soc.* **79**, 5441-5444 (1957).
21. Rosenblatt, D. H. and G. H. Broome. "Direct Epoxidation of o-Chlorobenzylidenemalononitrile with Hypochlorite Ion," *J. Org. Chem.* **28**, 1290-1293 (1963).

22. Lee, D. G. and J. R. Brownridge. "The Oxidation of Cinnamic Acid by Permanganate Ion. Spectrophotometric Detection of an Intermediate," *J. Amer. Chem. Soc.* **95**, 3033-3034 (1973).
23. Wiberg, K. B., C. J. Deutsch and J. Rocek. "Permanganate Oxidation of Crotonic Acid. Spectrometric Detection of an Intermediate," *J. Amer. Chem. Soc.* **95**, 3034-3035 (1973).
24. Lindgren, B. O., C. M. Svahn and G. Widmark. "Chlorine Dioxide Oxidation of Cyclohexene," *Acta Chem. Scand.* **19**, 7-13 (1965).
25. Lindgren, B. O. and C. M. Svahn. "Reactions of Chlorine Dioxide with Unsaturated Compounds, II. Methyl Oleate," *Acta Chem. Scand.* **20**, 211-218 (1966)
26. Hussain, A., T. Higuchi, A. Hurwitz and I. Pitman. "Rates of Hydrolysis of *N*-Chlorinated Molecules," *J. Pharm. Sci.* **61**, 371-374 (1972).
27. Hallum and J. S. Youngner. "The Effect of Oxidative Controlled Potential Electrolysis on Viruses," *Virology* **12**, 283-308 (1960).
28. Lund, E. and E. Lycke. "The Effect of Oxidation and Reduction on the Infectivity of Poliomyelitus Virus," *Arch. Ges. Virusforsch.* **11**, 100-110 (1961).
29. Lund, E. "Inactivation of Poliomyelitis Virus by Chlorination at Different Oxidation Potentials," *Arch. Ges. Virusforsch.* **11**, 330-342 (1961).
30. Lund, E. "Effect of pH on the Oxidative Inactivation of Poliovirus," *Arch. Ges. Virusforsch.* **12**, 632-647 (1963).
31. Lund, E. "The Significance of Oxidation in Chemical Inactivation of Poliovirus," *Arch. Ges. Virusforsch.* **12**, 648-660 (1963).
32. Lund, E. "Oxidative Inactivation of Poliovirus at Different Temperatures," *Arch. Ges. Virusforsch.* **13**, 375-386 (1963).
33. Lund, E. "Permanganate as a Mediator for the Inactivation of Poliovirus by Means of Periodate," *Arch. Ges. Virusforsch.* **13**, 387-394 (1963).
34. Lund, E. "The Rate of Oxidative Inactivation of Poliovirus and its Dependence on the Concentration of the Reactants," *Arch. Ges. Virusforsch.* **13**, 395-412 (1963).
35. Lund, E. "Oxidative Inactivation of Different Types of Enterovirus," *Amer. J. Hyg.* **80**, 1-10 (1964).
36. Lund, E. "The Oxidation Potential Concept of Inactivation of Poliovirus in Sewage," *Am. J. Epidemiol.* **81**, 141-145 (1965).
37. Kjellander, J. and E. Lund. "Sensitivity of *Esch. Coli* and Poliovirus to Different Forms of Combined Chlorine," *J. Amer. Water Works Assoc.* **57**, 893-900 (1965).
38. Lund, E. "Oxidative Inactivation of Adenovirus," *Arch. Ges. Virusforsch.* **19**, 32-37 (1966).
39. Grimley, E. and G. Gordon. "Kinetics and Mechanism of the Reaction Between Chlorine and Phenol in Acidic Aqueous Solution," *J. Phys. Chem.* **77**, 973-978 (1973).
40. Lee, G. F. "Kinetics of Reactions Between Chlorine and Phenolic Compounds," In *Principles and Applications of Water Chemistry*, S. D. Faust and J. V. Hunter, Eds. (New York: John Wiley & Sons, Inc., 1967), pp. 54-74.

41. Berliner, E. "The Current State of Positive Halogenating Agents," *J. Chem. Educ.* **43**, 124-133 (1966).
42. Lordi, N. G. and J. Epstein. "Kinetics and Mechanism of Chlorination of Triethylphosphorothiolate in Dilute Aqueous Media at 25°," *J. Amer. Chem. Soc.* **80**, 509-515 (1958).
43. Epstein, J., V. E. Bauer, M. Saxe and M. M. Demek. "The Chlorine-Catalyzed Hydrolysis of Isopropyl Methylphosphonofluoridate (Sarin) in Aqueous Solution," *J. Amer. Chem. Soc.* **78**, 4068-4071 (1956).
44. Jencks, W. P. and J. Carriuolo. "Reactivity of Nucleophilic Reagents Toward Esters," *J. Amer. Chem. Soc.* **82**, 1778-1786 (1960).
45. Rosenblatt, D. H. and G. H. Broome. 'Reaction of Malononitrile with Chlorine Near Neutral pH," *J. Org. Chem.* **26**, 2116-2177 (1961).
46. Isbell, H. S. and L. T. Sniegoski. "Tritium-Labeled Compounds. XI. Mechanism for the Oxidation of Aldehydes and Aldoses *-l-t* With Sodium Chlorite," *J. Res. Nat. Bur. Stand.* **68A**, 301-304 (1964).
47. Hull, L. A., G. T. Davis, D. H. Rosenblatt, H. K. R. Williams and R. C. Weglein. "Oxidation of Amines. III. Duality of Mechanism in the Reaction of Amines with Chlorine Dioxide," *J. Amer. Chem. Soc.* **89**, 1163-1170 (1967).
48. Rosenblatt, D. H., M. M. Demek, W. H. Dennis, Jr. and G. T. Davis. "Reaction of Quinuclidine with Chlorine Dioxide," In preparation.
49. Bredt, J. "On Steric Hindrance in Bridge Rings (Bredt's Rule) and on the *Meso-Trans-* Position in Condensed Ring Systems of the Hexamethylenes," *Justus Liebigs Ann. Chem.* **437**, 1-13 (1924).
50. Dennis, W. H., Jr., L. A. Hull and D. H. Rosenblatt. "Oxidations of Amines. IV. Oxidative Fragmentation," *J. Org. Chem.* **32**, 3783-3787 (1967).
51. Hull, L. A., W. P. Giordano, D. H. Rosenblatt, G. T. Davis, C. K. Mann and S. B. Milliken. "Oxidations of Amines. VIII. Role of the Cation Radical in the Oxidation of Triethylenediamine by Chlorine Dioxide and Hypochlorous Acid," *J. Phys. Chem.* **73**, 2147-2152 (1969).
52. Rosenblatt, D. H., M. M. Demek and G. T. Davis. "Oxidations of Amines. XI. Kinetics of Fragmentation of Triethylenediamine Chlorammonium Cation in Aqueous Solution," *J. Org. Chem.* **37**, 4148-4151 (1972).
53. Dennis, W. H., Jr. and D. H. Rosenblatt. Unpublished results.
54. Kochi, J. K. and E. A. Singleton. "Autoxidation of N-Alkylisoindolines. Solvent Effects and Mechanisms," *Tetrahedron* **24**, 4649-4665 (1968).
55. Rosenblatt, D. H., A. J. Hayes, Jr., B. L. Harrison, R. A. Streaty and R. A. Moore, *J. Org. Chem.* **28**, 2790-2794 (1963).
56. Grimley, E. and G. Gordon. "The Kinetics and Mechanism of the Reaction Between Chlorine Dioxide and Phenol in Acidic Aqueous Solution," *J. Inorg. Nucl. Chem.* **35**, 2383-2392 (1973).
57. Barnhart, E. L. and G. L. Campbell. "The Effect of Chlorination on Selected Organic Chemicals," Water Pollution Control Research Series 12020 EXG 03/72, Office of Research and Monitoring, U.S. Environmental Protection Agency, Washington, D.C. (March 1972).
58. Carlson, R. M. "Organic Compounds Produced During Wastewater Chlorination," presented at the Symposium on the Identification and Transformation of Aquatic Pollutants, Athens, Georgia (April 8, 1974).

59. Hansch, C. "A Quantitative Approach to Biochemical Structure-Activity Relationships," *Accounts Chem. Res.* **2**, 232-239 (1969).
60. Leo, A., C. Hansch and D. Elkins. "Partition Coefficients and Their Uses," *Chem. Rev.* **71**(6), 525-616 (1971).
61. Horner, L. and G. Podschus. "Reply on the Above Inquiry," *Angew. Chem.* **63**, 531-532 (1951).
62. Meiners, A. F. and F. V. Morriss. "The Light-Catalyzed Oxidation of Starch with Aqueous Chlorine," *J. Org. Chem.* **294**, 449-452 (1964).
63. Conroe, K. E., I. Kaplan, J. S. Roscoe and S. I. Trotz. "Light-Promoted Reactions of Hypochlorite with Organics," *Water Sewage Works* **113**, 237-240 (1966).
64. Jolley, R. L. "Chlorination Effects on Organic Constituents in Effluents from Domestic Sanitary Sewage Treatment Plants," Ph.D. Thesis, University of Tennessee, 1973, 342 pp. (Graduate Program in Ecology, Oak Ridge National Laboratory, Oak Ridge, Tennessee, ORNL-TM-4290).
65. Patton, W., V. Bacon, A. M. Duffield, B. Halpern, Y. Hoyano, W. Pereira and J. Lederberg. "Chlorination Studies. I. The Reaction of Aqueous Hypochlorous Acid with Cytosine," *Biochem. Biophys. Res. Comm.* **48**(4), 880-884 (1972).
66. Hayatsu, H., S.-K. Pan and T. Ukita. "Reaction of Sodium Hypochlorite with Nucleic Acids and their Constituents," *Chem. Bull. Japan* **19** (10), 2189-2192 (1971).
67. Rook, J. J. "Formation of Haloforms During Chlorination of Natural Waters," *Water Treat. Exam.* **23**, Part 2, 234-243 (1974).
68. Harris, R. H. "The Implications of Cancer Causing Substances in Mississippi River Waters," Environmental Defense Fund, Washington, D.C. (1974).

CHAPTER 14

THE CHEMISTRY OF AQUEOUS NITROGEN TRICHLORIDE

Jose Luis S. Saguinsin and J. Carrell Morris

Division of Engineering and Applied Physics
Harvard University
Cambridge, Massachusetts

INTRODUCTION

Nitrogen trichloride formation during water and wastewater chlorination is particularly interesting because of its nuisance properties.[1] At very low concentration, it contributes undesirable taste and odor to drinking water and irritates the mucous membranes when contacted in swimming and bathing waters. Knowledge of conditions for the formation of NCl_3 and of its reactions should enable operators of water treatment plants to avoid its formation or at least to minimize its adverse effects.

The effectiveness of NCl_3, or the lack of it, as a disinfectant, is unknown. Although it retains full capacity to respond to reagents for determination of oxidizing chlorine, its ability to kill or inactivate microorganisms such as bacteria, viruses and cysts has never been evaluated. Data on the disinfecting efficiencies of HOCl, OCl^-, NH_2Cl and $NHCl_2$ against a variety of microorganisms have been compiled,[2-5] but so far no tabulation has included NCl_3. Like the other chloramines it is probably a much less potent disinfectant than free chlorine. NCl_3 formation is thus a waste of chlorine above the dosage necessary to maintain a necessary level of free chlorine for efficient disinfection. Or, in another sense, its formation increases the level of residual chlorine required for insuring a safe, uncontaminated water fit for drinking and other important purposes.

More information on the dynamics of reactions involving NCl_3 in water and wastewater is needed because it is generated to a greater or lesser extent during breakpoint chlorination,[6-8] the ideal way to achieve persistent levels of effective free residual chlorine.

In addition to disinfection, breakpoint chlorination has also been employed to reduce the level of ammonia-nitrogen in water and to minimize its nutrient effects.[9] Since a complex sequence of reactions is involved in breakpoint chlorination, added knowledge of NCl_3 reactions should contribute to a better understanding of this important process. Because of interrelationships among chloramination reactions, additional information on reactions involving NCl_3 will augment generally the knowledge of water chlorination and chloramination chemistry.

Previous Work on Chloramines

Three chloramines are formed successively from the reaction of chlorine and ammonia in dilute aqueous solutions as shown by the following equations:

$$NH_3 + HOCl \rightarrow NH_2Cl + H_2O \quad (1)$$

$$NH_2Cl + HOCl \rightarrow NHCl_2 + H_2O \quad (2)$$

$$NHCl_2 + HOCl \rightarrow NCl_3 + H_2O \quad (3)$$

The ammonia that reacts with HOCl may come from an actual dosing of water with NH_3 as in the old practice of chloramination or from nitrogenous substances in raw or waste waters. Enzymatic degradation of urea, proteins and other organic nitrogenous matter normally yields ammonia as the nitrogenous product. Ammonia itself may be directly introduced in rainfall or by leaching of fertilizer. A classic and common source of ammonia is urine. A typical analysis of fresh urine yields a concentration of about 555 mg/l of NH_3 and 23,800 mg/l of urea, making it a rich source of actual and potential ammonia,[10] not only in municipal wastewater, but also in bathing waters.

Successive chlorination of NH_3 to the total substitution of all the attached hydrogen atoms by chlorine (reactions 1-3) have been of longstanding interest to researchers in water chlorination. Kinetic studies on the formation and interconversion of monochloramine and dichloramine have been undertaken by Morris and co-workers.[11-14] Specifically, formation and disproportionation reactions of NH_2Cl and $NHCl_2$ and the hydrolysis of NH_2Cl were investigated, but very little information on the formation and decomposition reactions of NCl_3 has been gathered.

The literature is replete with data on the properties and preparation of NCl_3,[15-17] but there is no mention of the dynamics of its reactions.

Moreover, many data are not very relevant to water chlorination practice since concentrated reactants were used rather than a few parts per million of aqueous chlorine. Observations on the conditions favorable to formation and decomposition of NCl_3 have been reported by D. B. Williams[1,18-20] in connection with attempts to solve taste and odor problems encountered during water treatment, but the reactions were never quantitatively studied on a kinetic basis. Samples attempted to study the decomposition of NCl_3 using solutions prepared from liquid NCl_3,[21] but terminated his investigations after residues of liquid NCl_3 exploded. Information has been obtained in this investigation on the dynamics of formation and decomposition of NCl_3. The results provide data useful in minimizing the formation and undesirable effects of NCl_3 during water treatment. They also serve to elucidate other aspects in water chlorination chemistry.

EXPERIMENTAL

Preparation of Water and Reagent Solutions

Water free of chlorine demand was obtained by dosing distilled water with 10 ppm of HOCl, allowing it to stand overnight and subsequently dechlorinating it by exposure to intense radiation from two 275-watt G.E. sunlamps and a 30-watt germicidal lamp for several days. Complete dechlorination was assured by testing for the presence of free and combined available chlorine by the DPD-FAS method.[22]

The water was then redistilled in a borosilicate glass distilling apparatus equipped with a double vapor bulb. Concentrated stock solutions of reagents such as H_3PO_4, KH_2PO_4 and Na_2HPO_4 used to buffer reactant solutions were prepared and rendered chlorine demand-free by a treatment similar to that for distilling water. Because only relatively small portions of these stock solutions were added to buffer the reactant solutions, their contributions of chloride ion, which amounted to not more than $10^{-7}M$ in the reactant solution, were considered negligible and were disregarded. Stock solutions of HOCl were prepared by neutralizing portions of bleach solution with KH_2PO_4 to pH near 7 then distilling the mixtures under reduced pressure at 60°C in a Buchler rotary evaporator, catching the distillate in a glass vessel immersed in ice water.

Stock HOCl solutions were about 0.19M and virtually chloride-free when tested with silver nitrate. The solutions were then stored in a dark bottle and refrigerated.

Formation of NCl_3

Initially, it was hoped that the rate of NCl_3 formation could be studied by mixing stoichiometric proportions of NH_4^+ and HOCl in dilute solutions buffered at low pHs, and by following the course of the reaction by UV absorption spectrophotometry. However, NH_2Cl formation is slower than that of $NHCl_2$ or NCl_3 and therefore rate limiting at experimentally suitable pH values. Accordingly, it was necessary to start with preformed solutions of dichloramine to measure the direct rate of NCl_3 formation. This procedure presents the difficulty that dichloramine is unstable in the presence of HOCl at pH values where it is most readily formed from ammonia and HOCl. Reasonably stable $NHCl_2$ can be obtained from the disproportionate of NH_2Cl near pH 5,[14] but then there is excess of NH_4^+ in solution as shown by the equation:

$$2NH_2Cl + H^+ \rightleftharpoons NHCl_2 + NH_4^+ \quad (4)$$

The remaining alternative is to preform $NHCl_2$ just prior to its reaction with HOCl to form NCl_3. Preforming $NHCl_2$ was accomplished by mixing buffered solutions (KH_2PO_4-Na_2HPO_4) at pH 4.91 of $2 \times 10^{-4} M$ NH_4^+ and $4.1 \times 10^{-4} M$ HOCl. The NH_4^+ solution was prepared by diluting a stock 0.1*M* $(NH_4)_2SO_4$ solution with the buffer to the final desired concentration. Similarly the HOCl solution was prepared by diluting 0.19*M* stock HOCl with buffer. The final HOCl solutions were checked for HOCl concentration using the DPD-FAS method; no adjustments in concentration were considered necessary.

After equal volumes of the reactant solutions were mixed for 20 seconds, a portion of the reacting mixture was transferred into a 5-cm quartz spectrophotometer cell and the progress of $NHCl_2$ formation was followed with a Beckman DK-2 spectrophotometer by measurements of the changing absorption at 220 nm, where both $NHCl_2$ and NCl_3 absorb strongly. At this wavelength NCl_3, $NHCl_2$ and HOCl have molar absorptivities of 8200, 1200 and 67.5 respectively.

Reaction was allowed to proceed until absorbance reached a maximum or steady condition, and the lapsed time required was noted for each temperature at which the experiments were conducted. At maximum absorbance NH_2Cl has presumably reacted completely to form $NHCl_2$ at nearly the maximum practicable concentration. Then the reaction mixtures were analyzed quickly for HOCl by the DPD-FAS method and for $NHCl_2$ and NCl_3 by absorbance readings at 220 nm and 295 nm. Prior to computation of concentrations of $NHCl_2$ and NCl_3 absorbance readings were corrected for the contribution of HOCl.

Table 14-1 lists typical concentrations found for the three chlorine species after the indicated reaction times at 20°, 15°, 10° and 5°C. It can be observed that at lower temperatures more time is required to reach the desired degree of reaction. The problem is to wait until NH_2Cl has reacted fully, without having formed excessive NCl_3. Some NCl_3 formation is inescapable.

Table 14-1. Typical Concentrations of Chlorine Species in Preformed Dichloramine Solutions at pH 4.9[a]
$(NH_4^+)_0 = 1 \times 10^{-4}\,M$; $(HOCl)_0 = 2.05 \times 10^{-4}\,M$

Temperature, °C	20	15	10	5
Time of reaction, min	9.5	15	24	33.5
HOCl, $M \times 10^5$	1.77	2.31	1.92	1.69
$NHCl_2$, $M \times 10^5$	5.57	3.76	4.42	4.22
NCl_3, $M \times 10^5$	1.24	1.69	1.77	1.91

[a]The initial or added concentrations are shown with zero subscripts.

Based on the analytical data (Table 14-1), HOCl solutions were prepared with concentrations equal to the difference in concentrations between $NHCl_2$ and HOCl in the prereacted solution. These HOCl solutions also contained necessary buffer capacities so mixing with an equal volume of the preformed $NHCl_2$ solution at pH 5 gave pH values of 2.3, 3.2, 3.97 and 4.5. Buffer containing H_3PO_4 and KH_2PO_4 was used with the HOCl solutions. Constant temperatures were maintained in the reaction vessel by means of a water-bath equipped with heating and cooling coils and in the spectrophotometer cell compartment by circulating water from the water-bath through the copper-tubing loop in the cell holder.

Reaction began by mixing equal volumes of preformed $NHCl_2$ and HOCl solutions for 20 seconds. A portion of the reaction mixture was then transferred quickly into a 5-cm quartz cell. Formation of NCl_3 was measured by following the absorbance at 220 nm as a function of time. Complete absorption spectra from 346 to 218 nm were measured periodically to follow both the increase in NCl_3 and decrease in $NHCl_2$ concentration.

The ideal situation would have been to start with a dichloramine solution at pH 5, free of either NCl_3 or HOCl. The removal of NCl_3 from the preformed $NHCl_2$ solution (by extraction with CCl_4) was therefore attempted, but in the process the $NHCl_2$ solution became saturated with CCl_4 which strongly contributed to the total absorbance at 220 nm. This made measurements of reaction rate impracticable. Moreover, the partition coefficient of NCl_3 between CCl_4 and water is only about 32,[23] and a number of time-consuming extractions had to be undertaken before nearly complete NCl_3 removal was achieved. NCl_3 was thus allowed to remain in the preformed $NHCl_2$ solution because, as will be shown later, the presence of NCl_3 at the outset of the reaction did not appear to affect reaction rates.

Formation studies were conducted at four different temperatures, as indicated previously, four different pH values, several different initial concentrations of reactants and for different initial ratios of reactants. Calculation of the specific rate parameter for each run was based on the initial rate of reaction and initial concentrations of $NHCl_2$ and HOCl derived from the analyses of preformed $NHCl_2$ solutions.

The effect of the presence of chloride ions on the reaction rate was also observed. As in the other studies of the formation of NCl_3 from $NHCl_2$ the same steps were taken and, additionally, appropriate quantities of stock NaCl solution (0.1*M*) were added to the $NHCl_2$ solution just prior to mixing it with an equal volume of the HOCl solution to give final chloride concentrations in the reaction mixture from 2.5×10^{-5} to $5 \times 10^{-4}M$. Addition of the portions of stock NaCl solution just prior to the onset of the reaction was intended to minimize loss of HOCl through conversion to aqueous Cl_2 and volatilization. These studies permitted investigation of the relative rates of reaction of $NHCl_2$ with HOCl and with Cl_2, since the effect of the chloride at pH 3 is to convert much of the HOCl to Cl_2.

Decomposition of NCl_3

Decomposition of NCl_3 was studied by first preparing stock NCl_3 solutions by reacting $4.5 \times 10^{-5}M$ NH_4^+ and $1.35 \times 10^{-4}M$ HOCl (1:3 molar ratio) at pH 2.3. The reacting mixture was $2 \times 10^{-3}M$ in H_2SO_4 to give the low pH. The mixtures were stored in glass-stoppered bottles in the dark for a week prior to further use. They were then analyzed for concentrations of NCl_3 and HOCl. Other chloramines were not present. NCl_3 concentrations formed, about $1.2 \times 10^{-5}M$ were less than stoichiometric, probably because of some decomposition of $NHCl_2$ during the initial stages of the formation. The excess concentration of HOCl, about $1.9 \times 10^{-5}M$, present with the NCl_3 can also be explained in this way, for the redox decomposition reactions of $NHCl_2$ have a reaction ratio of Cl:N equal to about 2 instead of the initial reactant ratio of 3.

Decomposition of NCl_3 was observed by mixing portions of the above stock solution of NCl_3 with equal volumes of Na_2HPO_4 or Na_2HPO_4--$Na_2B_4O_7 \cdot 10H_2O$ buffers to give final pH values near 7, 8.3 or 9. The decline of NCl_3 concentration was then observed by means of changes of absorbance with time at 220 nm. As in the studies of the formation of NCl_3 the complete spectrum from 346 nm to 218 nm was measured occasionally to show products formed as a result of the progressive decrease in NCl_3 concentration.

Additionally, the reaction was followed at three different initial concentrations of NCl_3. The initial rates of decomposition were measured to determine the order of the decomposition process.

The effect of ammonia on the rate of NCl_3 decomposition was studied by adding appropriate concentrations of a stock NH_4^+ solution (0.1*M*) to the Na_2HPO_4 or Na_2HPO_4-$Na_2B_4O_7 \cdot 10H_2O$ buffer solutions to give final NH_3 concentrations in reaction mixtures from 2×10^{-5} to 5×10^{-4}*M*. The effects of ammonia were expected to help define the role of hydrolysis of NCl_3 in the decomposition process.

RESULTS AND DISCUSSION

Formation of NCl_3

Reaction orders for the formation of NCl_3 from $NHCl_2$ and HOCl were measured by the van't Hoff initial rate method.[24] Initial concentrations of each reactant were varied for fixed initial concentration of the other and the changes in initial rate were determined. These experiments gave the reaction order with respect to each reactant. Initial rates were also obtained as a function of total initial concentrations to give the overall reaction order. The results showed that the formation of NCl_3 is overall second-order, and first-order with respect to each of the reactants, $NHCl_2$ and HOCl. The kinetic equation may then be written

$$d[NCl_3]/dt = k[HOCl][NHCl_2] \quad (5)$$

Values of the specific rate parameter, k, indicate that the formation reaction of NCl_3 is intrinsically slow, much slower than the formation of NH_2Cl or $NHCl_2$. Values of the specific rate are shown in Figure 14-1 as a function of pH for each temperature. Figures 14-2 and 14-3 show typical changes in absorption spectra during NCl_3 formation for experiments at 20° and pH 2.3 and 3.2, respectively. The progressive increase in absorbance at 220 nm due to the accumulation of NCl_3 and the decrease at 295 nm due to the depletion of $NHCl_2$ can be seen readily.

Figure 14-1 shows that the formation reaction is not strongly pH dependent in the acid range until pH is less than 3.2. Earlier assumptions of strong pH dependence as found in the formation of NH_2Cl or $NHCl_2$ are therefore not valid.[8]

The instability of $NHCl_2$ in the presence of HOCl prevented study of NCl_3 formation rates at pH 5 and greater. There seems to be a slight

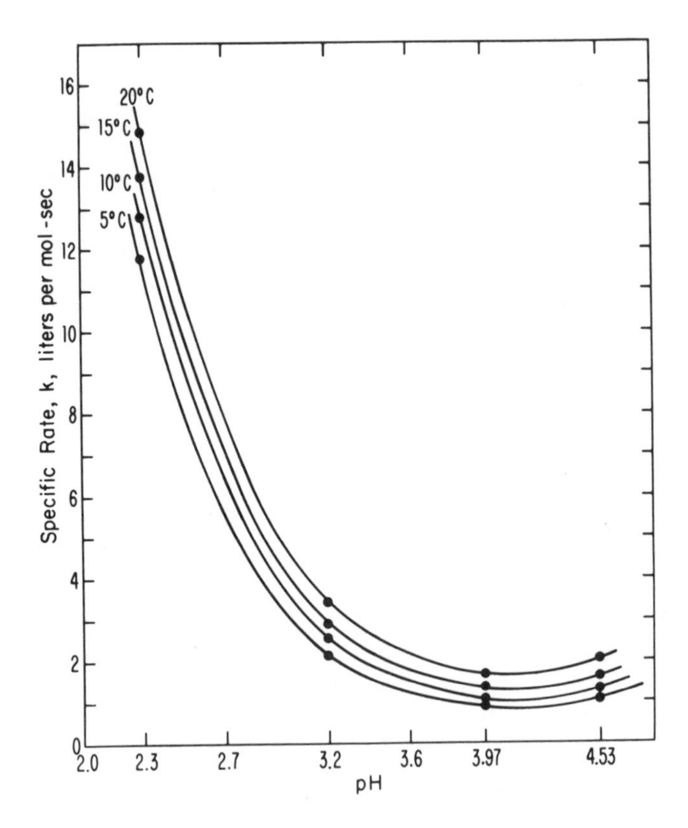

Figure 14-1. Specific rates of NCl_3 formation as a function of pH and temperature.

consistent increase in k from pH 4 to 4.5; this may indicate some base catalysis of the reaction, but the magnitude of the effect is uncertain.

In Table 14-2 values obtained for the specific rates of formation of NCl_3 are compared with the Weil and Morris values for monochloramine and dichloramine formation. At 20°C the specific rate of formation of NH_2Cl is about 10^4 times that for the formation of $NHCl_2$ and about 10^6 times that for NCl_3 formation based on direct reaction between HOCl and the uncharged amine.

Morris[25] has postulated that the specific rates of N-chlorination with a given chlorine donor such as HOCl should be directly related to the nucleophilicity of the nitrogen atom being chlorinated. Nucleophilicity toward Cl^+ should parallel basicity toward H^+ and thus a direct relation between basicity and rate of N-chlorination can be anticipated. Slowness of NCl_3

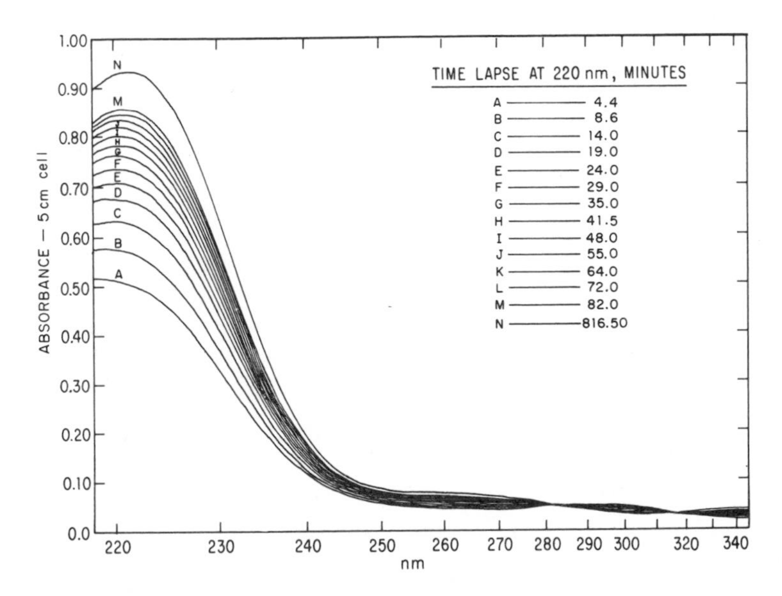

Figure 14-2. Absorption spectra during the formation of NCl_3 from $NHCl_2$ and HOCl at 20°C and pH 2.3; $(NHCl_2)_o = (HOCl)_o = 2.78 \times 10^{-5} M$; cell length = 5 cm.

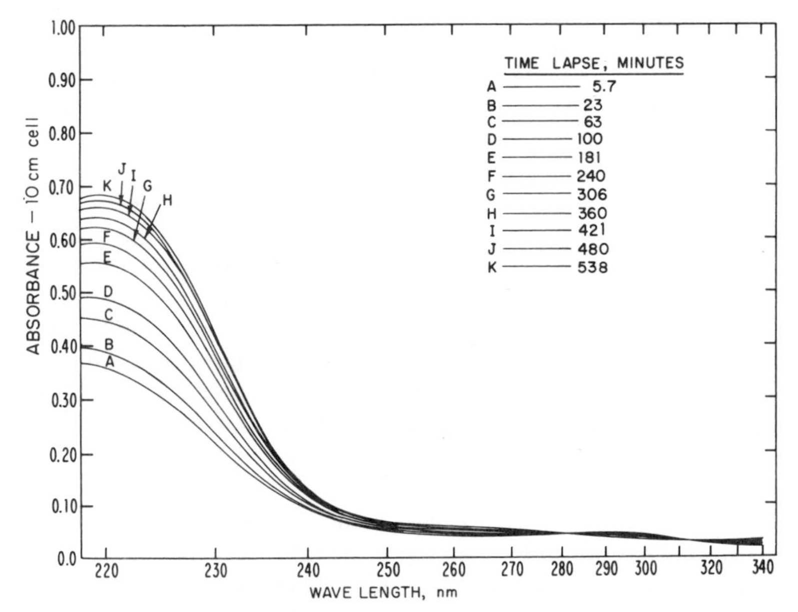

Figure 14-3. Absorption spectra during the formation of NCl_3 from $NHCl_2$ and HOCl at 20°C and pH 3.2; $(NHCl_2)_o = (HOCl)_o = 2.78 \times 10^{-5} M$; cell length = 10 cm.

Table 14-2. Comparison of Specific Rates for Chloramine Formation at 20°C

		k, M^{-1} sec^{-1}	General Expression for k, liter/mole/sec
NH_2Cl		5.6×10^6 [a]	$k = 9.7 \times 10^8 \exp(-3000/RT)$
	(pH 3.2)	4.9 [b]	
	(pH 2.3)	0.6 [b]	
$NHCl_2$		2.7×10^2 [c]	$k = 7.6 \times 10^7 \exp(-7300/RT)$
NCl_3	(pH 2.3)	14.9	$k = 1.15 \times 10^3 \exp(-2500/RT)$
	(pH 3.2)	3.4	$k = 2.62 \times 10^4 \exp(-5200/RT)$
	(pH 3.97)	1.6	$k = 7.83 \times 10^4 \exp(-6270/RT)$
	(pH 4.53)	2.0	$k = 3.43 \times 10^5 \exp(-7000/RT)$

[a]Calculated assuming reaction between neutral molecules
[b]Extrapolated k_{obs}
[c]Rate for non-acid-catalyzed reaction

formation is thus consistent with the fact that $NHCl_2$ is not demonstratably basic with an acidity constant much like that of organic imides with a Ka of about 10^{-8}. It is also possible that multiple substitution for the hydrogen atoms of ammonia leads to steric effects that reduce the reactivity of $NHCl_2$ toward Cl^+ as compared with reactivity toward H^+.

In Figure 14-4 logarithms of the specific rates at each pH value are shown plotted against reciprocals of the Kelvin temperature to obtain a value for the activation energy at each of the four pH levels. The values, computed from the slopes of the lines through the points, are listed in Table 14-2. Decrease in activation energy is observed for the acid-catalyzed reaction at pH 2.3 as might be expected; the values for pH 4.5 and 4.0 are the same within experimental error, but at pH 3.2 seem to show a bit of the acid effect.

Experiments to test the effect of chloride on the rate of NCl_3 formation were conducted at a number of chloride ion concentrations at 20°C and pH 2.3, 3.2, 3.97 and 4.5 with specific rates evaluated as in the previous sections. The specific rates found are plotted as a function of the concentration of chloride at Figure 14-5 and are shown in Table 14-3 for the reactions at pH 3.2.

Table 14-3 shows the acceleration with chloride ion to be expected if Cl_2 is a more active chlorinating agent toward $NHCl_2$ than HOCl. The overall initial rate $(d[NCl_3]/dt)_o$ in these circumstances may be expressed by the equation

$$(d[NCl_3]/dt)_o = (k_1[HOCl] + k_2[Cl_2])[NHCl_2]_o \tag{6}$$

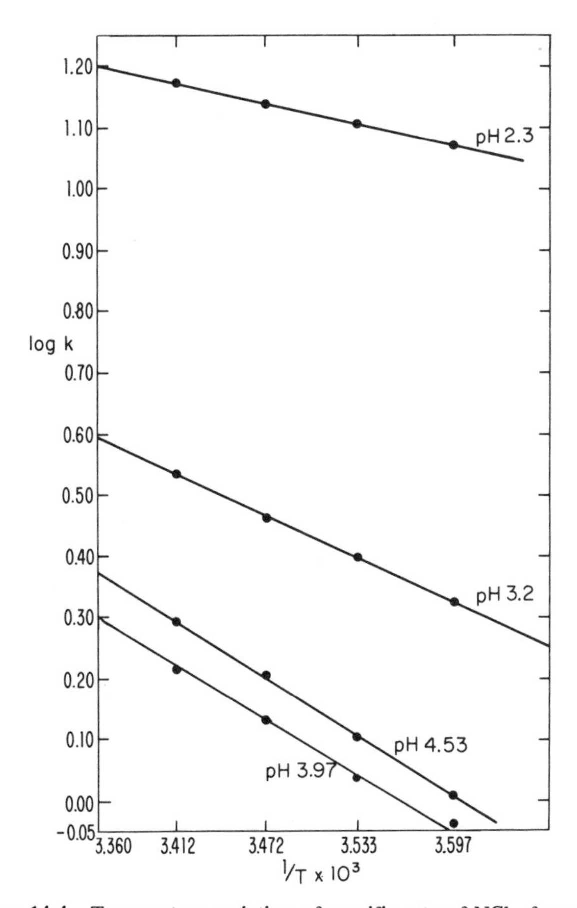

Figure 14-4. Temperature variation of specific rate of NCl_3 formation.

where the k_1 refers to the specific rate in the absence of chloride and k_2 is the specific rate for the reaction

$$NHCl_2 + Cl_2 \rightarrow NCl_3 + H^+ + Cl^- \tag{7}$$

In the presence of chloride the distribution between Cl_2 and HOCl is governed by the equilibrium process

$$HOCl + Cl^- + H^+ = Cl_2 + H_2O \tag{8}$$

with K_H, the hydrolysis constant, being given by

$$K_H = \frac{[HOCl](H^+)[Cl^-]}{[Cl_2]} \tag{9}$$

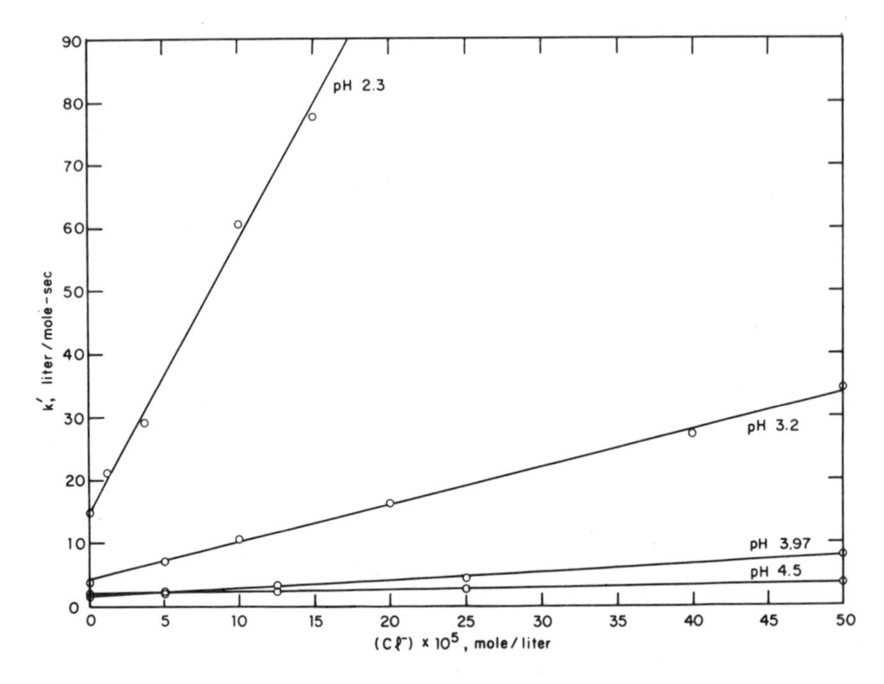

Figure 14-5. Dependence of specific rate of NCl_3 formation on chloride concentration at 20°C.

Table 14-3. Effect of Chloride on the Specific Rate of Formation of NCl_3 (pH 3.2; 20°C; $[HOCl]_o = [NHCl_2]_o = 2.8 \times 10^{-5}$ *M*)

$[Cl^-] \times 10^5$, mole/l	k', l/mole-sec
0.0	3.40
5.0	6.93
10.0	10.51
20.0	16.45
40.0	26.90
50.0	34.80

Substitution into equation (6) for $[Cl_2]$ then gives

$$(d[NCl_3]/dt)_o = \left(k_1 + \frac{k_2(H^+)[Cl^-]}{K_H}\right) [HOCl]\,[NHCl_2]_o \qquad (10)$$

The measured free chlorine, T, is the sum of $[Cl_2]$ and [HOCl], there being no measurable ionization of HOCl in these acid solutions. Then, with

$$[Cl_2] = T - [HOCl]$$

Equation (9) gives

$$[HOCl] = T\left(1 + \frac{(H^+)[Cl^-]^{-1}}{K_H}\right) \qquad (11)$$

But, at pH 3.2 $(H^+) = 6.3 \times 10^{-4}$ and with $K_H = 3.8 \times 10^{-4}$ at 20° for the ionic strength of the buffered solutions,[26] the second term in parentheses is less than 0.01 so long as $[Cl^-]$ does not exceed 5×10^{-4} M, the maximum value for the present experiments. Accordingly,

$$[HOCl] \sim T \qquad (11a)$$

and

$$(d[NCl_3]/dt)_o = \left(k_1 + \frac{k_2(H^+)[Cl^-]}{K_H}\right) T\,[NHCl_2]_o \qquad (10a)$$

The term in parentheses is the observed specific rate, k′.

$$k' = k_1 + \frac{k_2(H^+)}{K_H}[Cl^-] \qquad (12)$$

The observed specific rate, k′, therefore varies linearly with the chloride ion concentration.

The linear plots in Figure 14-5 show equation (12) fits the chloride catalysis of NCl_3 formation. The slope of the plot in Figure 14-5 for the reaction at pH 3.2, 5.94×10^4 (liters per mol)^2 sec^{-1}, should be equal to $k_2(H^+)/K_H$ or

$$k_2 = \frac{5.94 \times 10^4\, K_H}{(H^+)} \qquad (13)$$

Then, with $K_H = 3.8 \times 10^{-4}$ and with $(H^+) = 6.3 \times 10^{-4}$,

$$k_2 = 3.58 \times 10^4 \text{ liter/mole/sec} \qquad (14)$$

Since, at this pH value and temperature, $k_1 = 3.4$, the reactivity of Cl_2 toward $NHCl_2$ is about 10^4 times that of HOCl.

Decomposition of NCl_3

Measurements on the decomposition of NCl_3 in neutral and mildly alkaline solutions showed it to be first-order as a function of time as shown by linear semi-logarithmic plots of $[NCl_3]$ against time; an example is given in Figure 14-6. Absorbance measurements were made at three different initial

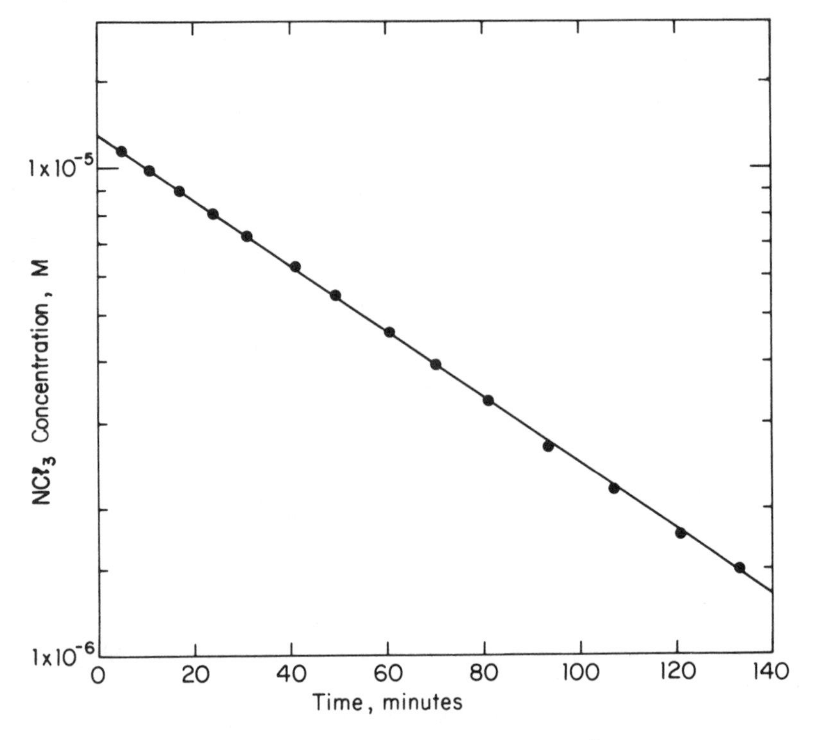

Figure 14-6. Rate of decomposition of NCl_3 at 20°C, pH 9; $(NCl_3)_o = 1.15 \times 10^{-5} M$.

concentrations of NCl_3 (Figure 14-7). The initial slopes or rates of reaction were directly proportional to initial concentration of NCl_3 within experimental error. This confirms the first-order dependence on $[NCl_3]$.

The rate of decomposition increases greatly as the pH increases from 7 to 9, as shown by the plots of absorbance at 220 nm against time given in

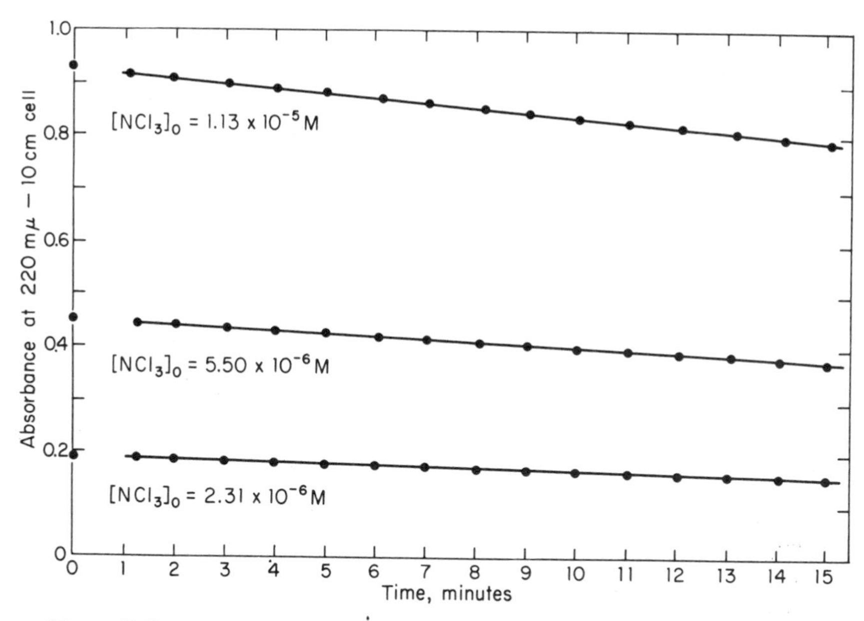

Figure 14-7. Rates of decomposition of NCl_3 as a function of initial concentration at 20°C, pH 9.

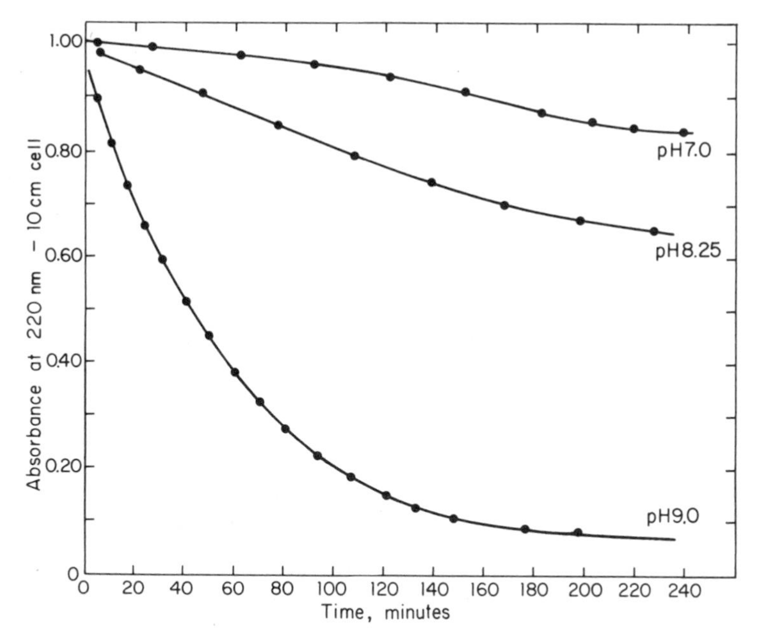

Figure 14-8. Effect of pH on rate of decomposition of NCl_3 at 20°C, $(NCl_3)_o = 1.20 \times 10^{-5} M$.

Figure 14-8. At 20°C the first-order specific rates for pH 7.0, 8.25, and 9.0 had values of 3.4×10^{-5}, 6.3×10^{-5}, and 2.2×10^{-4} sec^{-1}, respectively. Decomposition depended directly on OH^- concentration as shown by the linearity of the plot of specific rate against $[OH^-]$ in Figure 14-9.

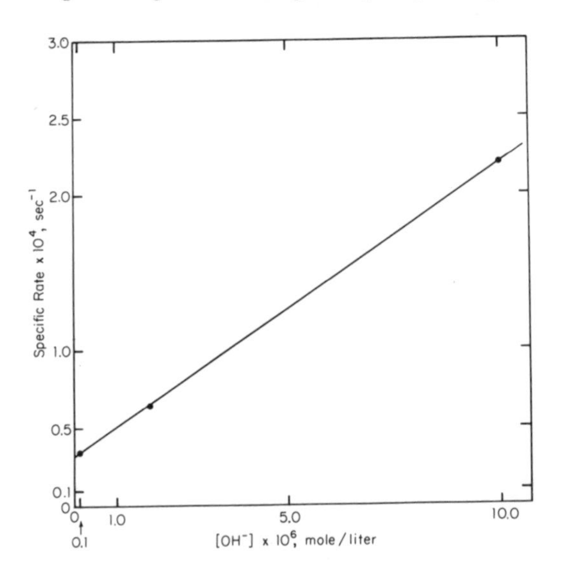

Figure 14-9. Specific rates for NCl_3 decomposition as a function of OH^- concentration at 20°C.

The decomposition of NCl_3 may then be considered either catalyzed by the presence of OH^- in accord with the reaction

$$NCl_3 + H_2O \xrightarrow{OH^-} NHCl_2 + HOCl \tag{15}$$

or a direct reaction with OH^- in accord with the equation

$$NCl_3 + OH^- \rightarrow NHCl_2 + OCl^- \tag{16}$$

During decomposition reactions spectra from 346 to 218 nm were scanned (Figure 14-10). The increases with time in absorbance near 292 nm at pH 9 indicate the accumulation of OCl^-. However, the spectra give no indication of $NHCl_2$. This can be explained by the assumption that the NCl_3 hydrolysis is the slow, rate-determining step; as soon as $NHCl_2$ is formed, it decomposes rapidly by reactions similar to those for its disappearance during breakpoint chlorination. A number of consecutive reactions are believed to be involved in the decomposition of $NHCl_2$[8] but the principal overall reaction is:

$$2\ NHCl_2 + H_2O \rightarrow N_2\uparrow + HOCl + 3\ H^+ + 3\ Cl^- \tag{17}$$

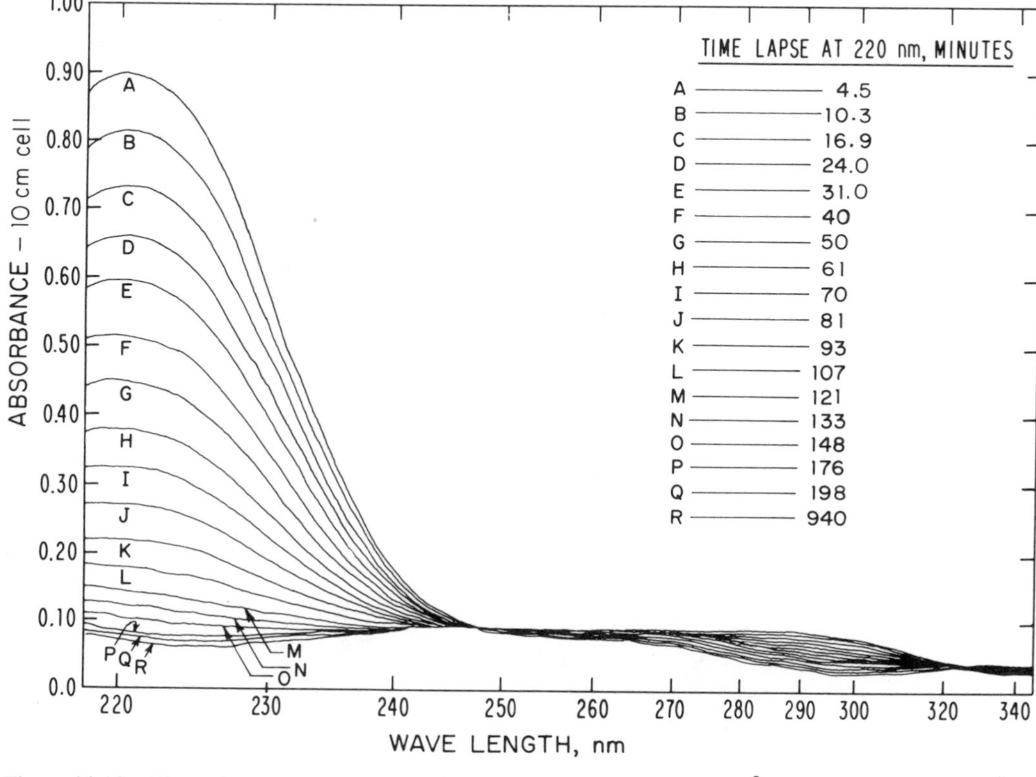

Figure 14-10. Absorption spectra during NCl_3 decomposition at pH 9 and 20°C; $(NCl_3)_o = 1.15 \times 10^{-5} M$.

In a buffered system NCl_3 decomposition is pseudo first-order; the experimental rate law may be written as

$$-(d[NCl_3]/dt) = k_3 [NCl_3] \quad (18)$$

where k_3 has the form, $k_3 = a(1 + b[OH^-])$. From the slope of the plot in Figure 14-9, a general expression for the specific rate may be written as

$$k_3 = 3.2 \times 10^{-5} (1 + 5.88 \times 10^5 [OH^-]) \quad (19)$$

It appears, then, that for the uncatalyzed decomposition, $k_3 = 3.2 \times 10^{-5}$ sec^{-1}.

Williams[1] made the observation that NCl_3 cannot persist in water with pH greater than 6.5 in the absence of HOCl, OCl^-, or both. He also claimed that NCl_3 undergoes conversion to NH_2Cl or $NHCl_2$ as soon as HOCl or OCl^- is removed from solution. The implication is that the hydrolysis of NCl_3 is a rapidly reversible reaction with its ultimate rate of decomposition governed by $NHCl_2$ decomposition. Since some HOCl was present in the stock NCl_3 solutions of these experiments, it was decided to study the effect of complete removal or absence of HOCl at the start of the reaction by adding NH_3 to convert HOCl to NH_2Cl. NH_3 was added as $(NH_4)_2SO_4$ in several concentrations to a series of reaction solutions, the reaction between HOCl and NH_3 being considered instantaneous for the pH range 7 to 9. Any contribution of a back reaction due to the accumulation of HOCl from NCl_3 hydrolysis would also be avoided by the presence of excess NH_3. A number of NH_3 concentrations were used to attempt to determine a level beyond which no further change in the initial rate of NCl_3 decomposition would be observed. However, when $[NH_3]$ was plotted against initial rate of decomposition linearity was obtained at pH 8 and 9 (Figure 14-11) even at NH_3 concentrations many times the initial NCl_3 concentration. At pH 7 the plot is linear only up to $[NH_3]$ of $5 \times 10^{-5} M$, but the tapering off is not great even at greater $[NH_3]$ and no limit is observed. A possible direct reaction between NCl_3 and NH_3 is indicated according to the overall equation,

$$NCl_3 + NH_3 \rightarrow NH_2Cl + NHCl_2$$

For the reaction at pH 9, the spectral changes (Figure 14-12) reveal a rapid decrease in NCl_3 concentration made evident by the decrease with time of the absorbance at 220 nm. Simultaneously, NH_2Cl formation is indicated by the appearance of a distinct peak at 245 nm where NH_2Cl has its maximum molar absorptivity. There is no spectrophotometric evidence for the existence of $NHCl_2$. At pH 9 $NHCl_2$ should decompose rapidly as soon as it is formed. If the assumption of direct reaction is

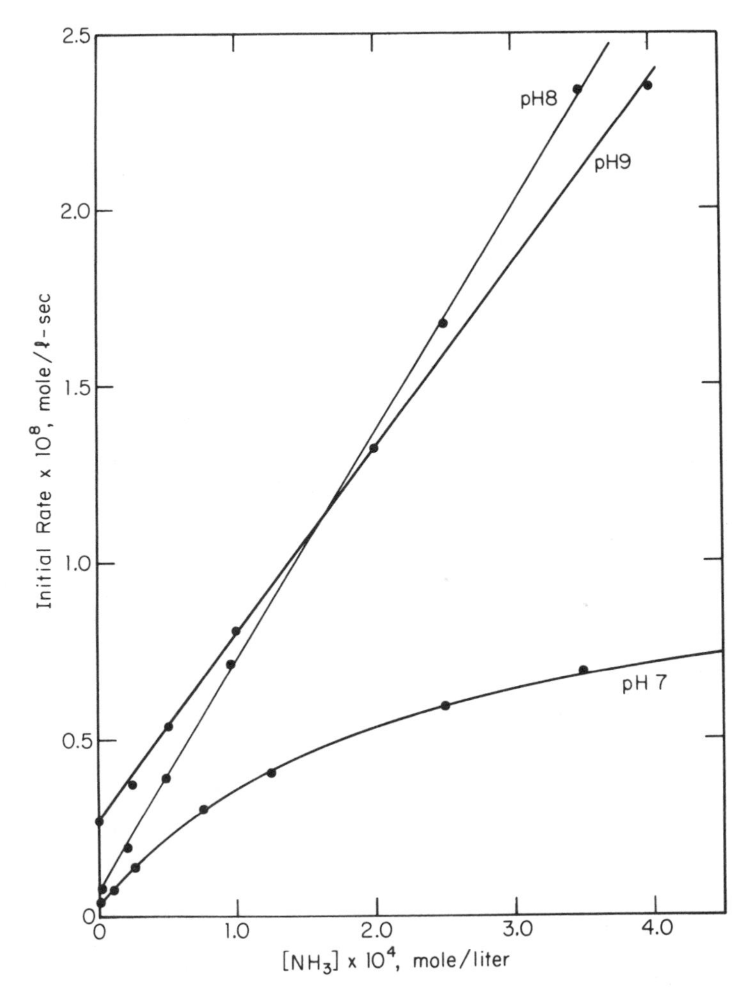

Figure 14-11. Effect of added NH_3 on initial rate of NCl_3 decomposition at 20°C; $(NCl_3)_o = 1.11 \times 10^{-5} M$ for pH 9; $(NCl_3)_o = 1.21 \times 10^{-5} M$ for pH 8; $(NCl_3)_o = 1.19 \times 10^{-5} M$ for pH 7.

correct, then the initial rate of disappearance of NCl_3 in the presence of NH_3 is the sum of the rate due to its reaction with NH_3 and the rate due to its hydrolysis, as represented by the following:

$$-(d[NCl_3]/dt)_o = k_4[NCl_3]_o[NH_3]_o + k_3[NCl_3]_o \quad (20)$$

If only $[NH_3]_o$ is varied, $k_3[NCl_3]_o$ is a constant and k_4 can be calculated from the slopes of the plots in Figure 14-11 divided by the corresponding

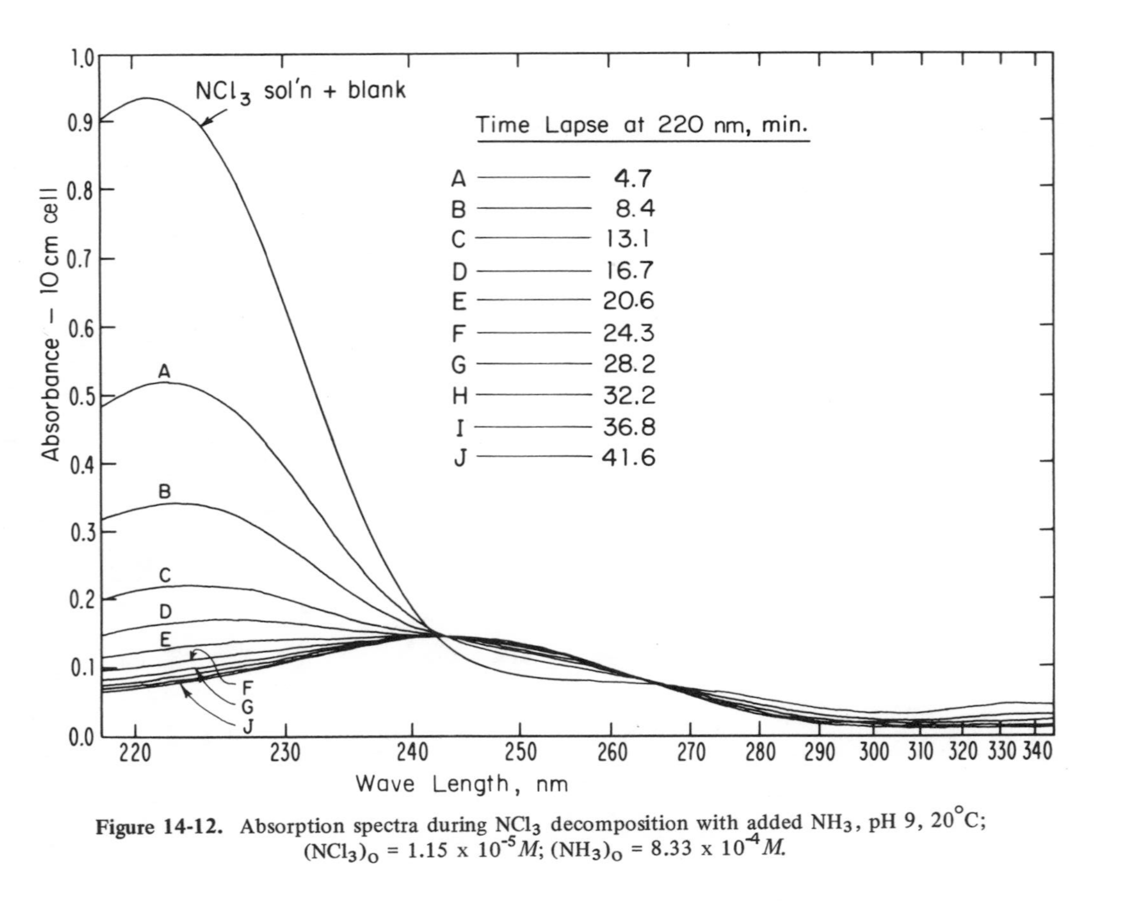

Figure 14-12. Absorption spectra during NCl_3 decomposition with added NH_3, pH 9, 20°C; $(NCl_3)_o = 1.15 \times 10^{-5}$ *M*; $(NH_3)_o = 8.33 \times 10^{-4}$ *M*.

$[NCl_3]_0$. Values of k_4 so obtained were 3.4, 5.3 and 4.3 liters per mole-sec for pH 7, 8 and 9 respectively. These values are the same within the relatively large experimental error of these experiments.

The values of k_3, the rate constant for the hydrolysis of NCl_3, can also be obtained for these runs by dividing the y-intercepts of the plots in Figure 14-11 by the initial NCl_3 concentrations, the y-intercepts being the initial rates of NCl_3 decomposition at zero NH_3 concentration. The k_3 values so evaluated agree adequately with those obtained earlier. The agreement tends to indicate that the presence of HOCl does not affect the rate of NCl_3 decomposition in the pH range 7 to 9.

SUMMARY

Dynamics of the formation and decomposition reactions of NCl_3 in dilute aqueous solutions have been studied under a variety of experimental conditions. The NCl_3 formation reaction from $NHCl_2$ and HOCl has been found to be overall second-order, first-order in each of the reactants. The intrinsic specific rate of formation has been found to be much slower than those for NH_2Cl and $NHCl_2$ and not to be strongly pH-dependent at pH greater than 3.2. At low pH, the presence of chloride ions converts a fraction of the HOCl present into aqueous Cl_2 which chlorinates $NHCl_2$ about 10^4 times as fast as HOCl.

The decomposition of NCl_3 at pH 7 to 9 is first-order with respect to NCl_3 concentration and observed to be base-catalyzed or dependent on OH^- concentration. During decomposition, it appears that NCl_3 first hydrolyzes to $NHCl_2$ in a step that is rate-determining. The $NHCl_2$ then presumably undergoes fast decomposition in a manner similar to its disappearance during breakpoint chlorination.

The presence of HOCl was observed to have no effect on the initial rate of NCl_3 decomposition in accord with the idea that simple hydrolysis is rate-determining at pH 7 to 9. From the linear relationship between the initial rate of NCl_3 disappearance and added NH_3 concentration, it is concluded that NCl_3 reacts directly with NH_3, forming NH_2Cl and $NHCl_2$ as products.

REFERENCES

1. Williams, D. B. "Elimination of Nitrogen Trichloride in Dechlorination Practice," *J. Amer. Water Works Assoc.* **58**, 248 (1966).
2. Butterfield, C. T. "Bactericidal Properties of Chloramines and Free Chlorine in Water," *U.S. Pub. Health Rep.* **63**, 934 (1948).
3. Butterfield, C. T. "Bactericidal Properties of Free and Combined Available Chlorine," *J. Amer. Water Works Assoc.* **40**, 1305 (1948).

4. Krusé, C. W., Y-C. Hsu, A. C. Griffiths and R. Stringer. "Halogen Action on Bacteria, Viruses and Protozoa," *National Specialty Conference on Disinfection* (New York: ASCE, 1970), p. 113.
5. Stringer, R. and C. W. Krusé. "Amoebic Cysticidal Properties of Halogens in Water," *J. San. Eng. Div., ASCE* **97**(SA6) 801 (1971).
6. Palin, A. T. "A Study of the Chloro Derivatives of Ammonia and Related Compounds, with Special Reference to Their Formation in the Chlorination of Natural and Polluted Waters," *Water Water Eng.* **54**, 151 (1950).
7. Morris, J. C., I. W. Weil and R. H. Culver. "Kinetic Studies on the Breakpoint Reaction with Ammonia and Glycine," Paper presented at the International Congress of Pure and Applied Chemistry, New York (1952).
8. Wei, I. "Chlorine-Ammonia Breakpoint Reactions: Kinetics and Mechanism," Ph.D. Dissertation, Harvard University (1972).
9. Pressley, T. A., D. F. Bishop, and S. G. Roan. "Ammonia-Nitrogen Removal by Breakpoint Chlorination," *Environ. Sci. Technol.* **6**, 622 (1972).
10. Putnam, D. F. "Chemical Aspects of Urine Distillation," *J. Proc. Soc. Mech. Eng.*, Paper No. 65-AV-24 (1965).
11. Weil, I. and J. C. Morris. "Kinetic Studies on the Chloramines. I. The Rates of Formation of Monochloramine, N-Chlormethylamine and N-Chlordimethylamine," *J. Amer. Chem. Soc.* **71**, 1664 (1949).
12. Morris, J. C., *et al.* "The Formation of Monochloramine and Dichloramine in Water Chlorination," Paper, 117th Meeting, ACS, Detroit (April 16-20, 1950).
13. Weil, I. and J. C. Morris. "Kinetic Studies on the Chloramines. II. The Disproportionation of Monochloramine," Paper, 116th Meeting, ACS, Atlantic City, N.J. (September 18-23, 1949).
14. Granstrom, M. L. "The Disproportionation of Monochloramine," Ph.D. Dissertation, Harvard University (1954).
15. Mellor. *A Comprehensive Treatise on Inorganic and Theoretical Chemistry, VIII* (1928), pp. 598-604.
16. Gmelin. *Handbuch der Anorganischen Chemie,* 8th ed., No. 6 (1927), pp. 410-417.
17. Berliner, J. F. T. "The Chemistry of Chloramines," *J. Amer. Water Works Assoc.* **23**, 1320 (1931).
18. Williams, D. B. "A New Method of Odor Control," *J. Amer. Water Works Assoc.* **41**, 441 (1941).
19. Williams, D. B. "The Organic Nitrogen Problem," *J. Amer. Water Works Assoc.* **43**, 837 (1951).
20. Williams, D. B. "How to Solve Odor Problems in Water Chlorination Practice," *Water Sewage Works* **99**, 358 (1952).
21. Samples, W. R. "A Study on the Chlorination of Urea," Ph.D. Dissertation, Harvard University (1959).
22. American Public Health Association. *Standard Methods for the Examination of Water and Wastewater,* 13th ed. (New York, 1971).
23. Czech, F. W., R. J. Fuchs and H. F. Antczak. "Determination of Mono-, Di-, and Trichloramine by Ultraviolet Absorption Spectrophotometry," *Anal. Chem.* **33**, 706 (1961).

24. Frost, A. A. and R. A. Pearson. *Kinetics and Mechanism,* 2nd ed. (New York: John Wiley & Sons, Inc., 1961).
25. Morris, J. C. "Kinetic Reactions Between Aqueous Chlorine and Nitrogen Compounds," *Principles and Application of Water Chemistry,* S. D. Faust and J. V. Hunter, Eds. (New York: John Wiley & Sons, Inc., 1967), pp. 23-51.
26. Connick, R. E. and Y. Chia. "The Hydrolysis of Chlorine and its Variation with Temperature," *J. Amer. Chem. Soc.* **81**, 1280 (1959).

CHAPTER 15

KINETICS OF TRIBROMAMINE DECOMPOSITION

Thomas F. LaPointe, Guy Inman,
J. Donald Johnson

Department of Environmental Sciences & Engineering
School of Public Health
University of North Carolina
Chapel Hill, North Carolina 27514

INTRODUCTION

Bromine Disinfection

Compared with chlorine, bromine, iodine and ozone have certain advantages as disinfectants. Their disadvantages are generally cost and ease of dosing. Bromine was first used as a swimming pool disinfectant in the early 1930s.[1] The absence of eye irritation among bathers and the lack of swimming pool odors associated with chlorine were then recognized as the advantages of bromine. Recent developments, such as bromine polymer complexes have facilitated dosing. The interhalogen bromine compounds (see Chapter 7), especially bromine chloride, have reduced the cost of bromine as a disinfectant. These developments have increased interest in the use of bromine as a water and wastewater disinfectant. This has, in turn, encouraged studies of the fate of bromine and such brominated compounds as might be found in wastewater and swimming pools. Some important similarities and differences exist among the ammonia derivatives of chlorine and bromine. Chlorine (Cl_2, HOCl, OCl^-) is known to react with ammonia in aqueous solution forming a group of compounds known as the chloramines: monochloramine (NH_2Cl), dichloramine ($NHCl_2$) and trichloramine (NCl_3). The formation of monochloramine the major chloramine and the subsequent formation and decomposition of dichloramine to nitrogen gas, nitrate and chloride ion, referred to as the breakpoint reaction, are of primary importance in practical application. In the 1930s Chapin[2,3] undertook initial investigations into the formation of the chloramines. Palin followed in 1949-1950 with a careful study of the breakpoint reaction and analytical methods for the determination of free chlorine and ammonia chlorine derivatives.[4,5] A final understanding of the breakpoint reactions came with the work of Morris and co-workers[6,7] in establishing rate laws and proposing mechanisms for the formation of monochloramine and the formation and decomposition of dichloramine. Chapter 14 on trichloramine completes the story on the chloramines. The biocidal action of monochloramine has been researched extensively, due largely to its prevalence and stability in neutral water solutions. Although monochloramine is a good bactericide, its virucidal properties are minimal.

The chemistries of bromine and chlorine in pure aqueous solution are similar. The primary bromine species are molecular bromine (Br_2), hypobromous acid (HOBr) and hypobromite ion (OBr^-). The major differences in the two systems are the formation of tribromide ion (Br_3^-), its chlorine analog being unimportant, and the greater predominance of Br_2 compared to the low concentrations of Cl_2 found in neutral water solutions. The following equilibrium reactions govern the behavior of bromine in aqueous solution:[8]

$$Br_2(1) + H_2O = HOBr + H^+ + Br^- \qquad K_h = 5.8 \times 10^{-9} \text{ at } 25^\circ C \qquad (1)$$

$$HOBr = OBr^- + H^+ \qquad K_a = 2 \times 10^{-9} \text{ at } 25^\circ C \qquad (2)$$

$$Br_2 + Br^- = Br_3^- \qquad K = 15.9 \text{ at } 25^\circ C \qquad (3)$$

These equations can be solved in terms of Br^- and pH, resulting in a distribution diagram, such as that shown in Figure 15-1 for 0°C. Besides the formation of primary species, bromine is also similar to chlorine in its reaction with ammonia. In 1933 and 1934 Coleman *et al.* first reported the existence of brominated ammonia derivatives, prepared in ether and referred to as the bromamines.[9,10] Using spectrophotometric analyses Johannesson confirmed the presence of monobromamine NH_2Br and dibromamine $NHBr_2$ in aqueous solution.[11,12] Tribromamine NBr_3 was later discovered spectrophotometrically.[13] A fourth bromamine specie, NH_3Br^+, reported by Johannesson[14] has never been confirmed. The reactions for the sequential formation of these compounds where hypobromous acid is the halogenating agent are:

$$NH_3 + HOBr = NH_2Br + H_2O \qquad (4)$$

$$NH_2Br + HOBr = NHBr_2 + H_2O \qquad (5)$$

$$NHBr_2 + HOBr = NBr_3 + H_2O \qquad (6)$$

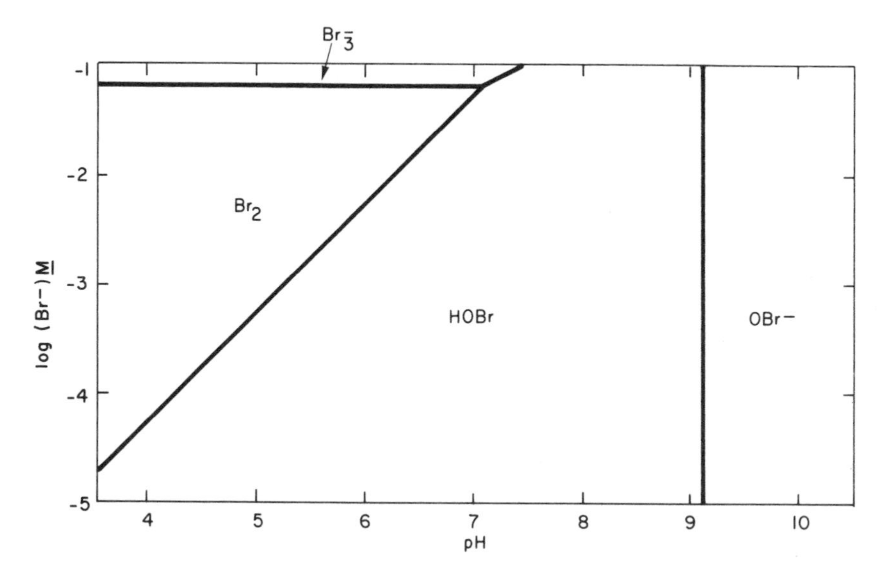

Figure 15-1. Principal species of free bromine in equilibrium with bromide ion and hydrogen ion. Lines are equal molar concentrations. Temperature is 0°C.

The bactericidal and virucidal activities of the bromamines have been studied and compared to those of the chloramines. Taylor found that tribromamine at pH 4.5 was not as virucidal as hypochlorous acid, but nevertheless possessed strong virucidal properties killing 99.99% of ϕx174 virus in 4 minutes at a concentration of 0.8 mg Br_2/l.[15] Johannesson[14] and Kruse[16] found bromamine, primarily NH_2Br, fully as bactericidal as free bromine, and Kruse further reported bromamines as potent virucides, although somewhat less so than hypobromous acid (HOBr).[16] This virucidal property of the bromamines, by contrast to the poorly virucidal chloramines, makes bromine a more attractive disinfectant in the presence of ammonia.

Chemical research on the bromamines has been faciliated by the characteristic ultraviolet absorption curves of these compounds shown in Figure 15-2. Galal-Gorchev and Morris[13] found absorption maxima at 278 nm, 232 nm and 258 nm for monobromamine, dibromamine and tribromamine respectively. Table 15-1 lists the important absorbance peaks for the major species in the chlorine- and bromine-ammonia systems. Using spectrophotometric analysis, studies were undertaken on the formation and decomposition of the bromamine species. Some very important differences from the chloramines have been noted.

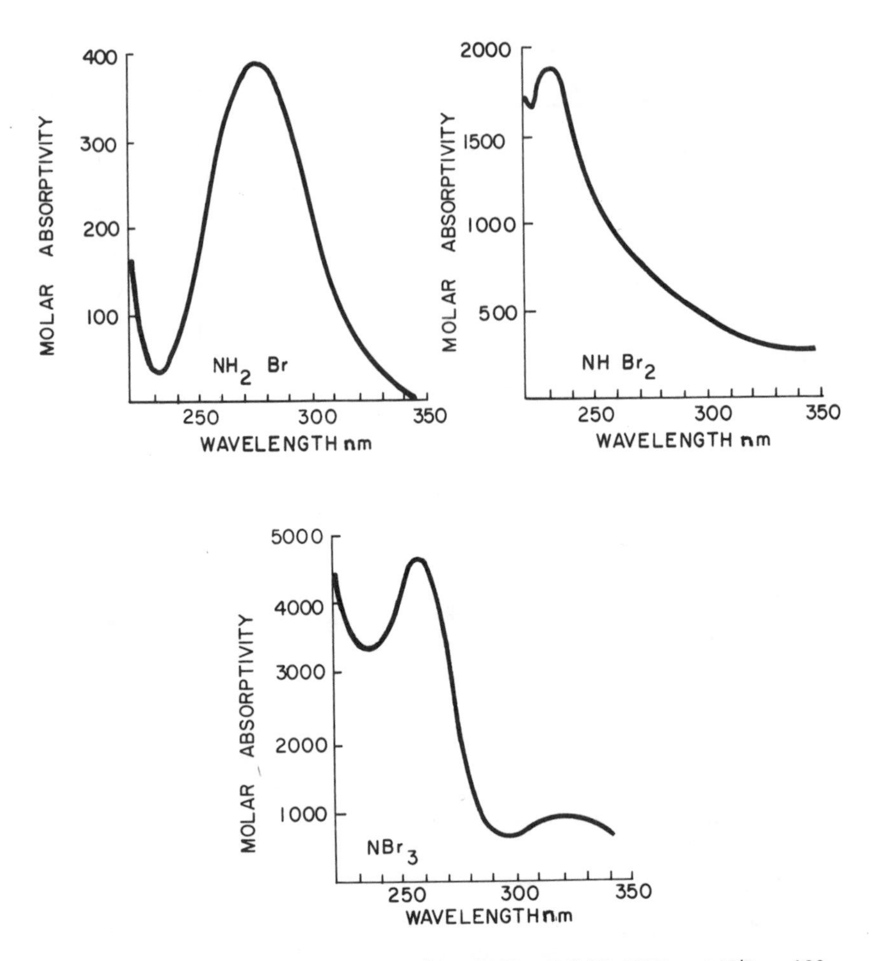

Figure 15-2. U.V.-Spectra of NH_2Br at N/Br = 1000, pH 8.25, $NHBr_2$ at N/Br = 100, pH = 5.35, NBr_3 at N/Br = 0.33, pH 6.0 versus molar absorptivity in liter $mole^{-1}$ cm^{-1}.

Comparison of Bromamine and Chloramine Systems

By contrast to the chloramines, the bromamines form very rapidly. Galal-Gorchev[17] and Johnson and Overby[18] both reported rapid formation of all bromamine species. However, once the bromamines have formed, a series of decomposition reactions takes place, in a manner similar to the chloramines.

Johnson and Overby studied the decompositions of mono-, di-, and tribromamine and observed, in regions where two of the species are present

Table 15-1. U.V. Absorption Data for Chlorine and Bromine

Compound Type	Chlorine		Bromine	
	Absorption Max, nm	Molar Absorptivity	Absorption Max, nm	Molar Absorptivity
HOX	233	100	261	93
OX^-	292	350	329	343
NH_2X	245	455	278	390
NHX_2	206 295	2100 300	232 290(?)	1900(?) —
NX_3	220 340	8100 260	258 323	4600 940

together, a simultaneous decomposition of both species, leading them to conclude that reversibility exists throughout the ammonia-bromine system.[18] By contrast with the chloramines, monobromamine is converted stepwise, through di- to tribromamine and back again, simply by changing the pH of the system. By spectrophotometric analysis they proposed a distribution diagram of bromine and bromamine species as a function of both pH and initial ammonia to bromine ratio (Figure 15-3). The concept of rapid

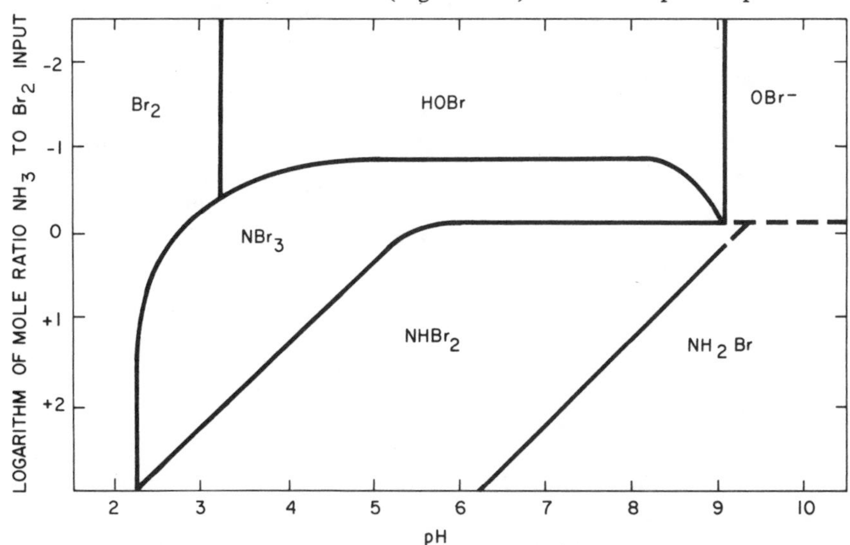

Figure 15-3. Principal species of bromine and bromamine predominating after 1-2 minutes at various pH and ammonia to bromine ratios. Lines represent equal equivalent concentrations. Hypobromous acid separation from bromine given for 10^{-3} *M* bromide.

re-equilibration is a fundamental aid to understanding the chemical characteristics of the aqueous bromamines. Another difference between the chlorine and bromine systems is that the proportion of the bromamines found (Figure 15-3) depends primarily on the ammonia-bromine ratio and pH, not on the rate of formation. In the chloramine system equilibria do not exist so the proportion of NH_2Cl, $NHCl_2$ and NCl_3 depends on the individual kinetics of formation of each specie. A third difference in the two systems can be seen by comparing the distribution diagrams for the bromamine and the chloramine systems. A distribution diagram similar to Figure 15-3 cannot be drawn for the chloramines because they are not in equilibrium with each other. For contrast, Palin's breakpoint data showing the distribution of the chloramines after 10 minutes and 2 hours are shown at pH 7.3 to 7.5 (Figures 15-4 and 15-5). In the chlorine system under conditions of moderate pH and high (>1) ammonia to chlorine mole ratios, the primary product is monochloramine; whereas the tendency in the bromine system under the same conditions is for the ammonia molecule to be brominated twice to dibromamine. Even under conditions of large excesses of ammonia (100/1 at pH 7), dibromamine is the major bromamine. This only occurs at low pH in the chlorine system. Under practical disinfecting situations pH is generally in the neutral range, and high relative molar concentrations of ammonia to the disinfectant are often present. Therefore, any analysis of disinfectant capabilities should consider the disinfectant properties of monochloramine and dibromamine.

Monobromamine, however, is only present at high pH and/or very high ammonia concentrations. The importance of dibromamine in the moderate pH range is further evidenced in bromine breakpoint studies discussed later. Another major difference in the two distribution diagrams is the appearance of tribromamine (NBr_3) at moderate pH and low ammonia to bromine ratios. At low ammonia to chlorine ratios trichloramine is formed only at low pH and in small amounts (Figures 15-4 and 15-5). Thus under the conditions of moderate pH most often encountered in disinfection, and when a high concentration of disinfectant is employed relative to the ammonia present–that is, at a low ammonia to bromine ratio, the chemistry of tribromamine is the matter of import. For chlorine, the classic breakpoint reactions of mono- and dichloramine are the determining factors.

Previous Research in Bromamine Chemistry

As noted previously each of the bromamines undergoes decomposition reactions. Monobromamine stability was studied by Galal-Gorchev[17] for pH 9 to 12 and was reported to undergo slow decomposition with the following predicted stoichiometry:

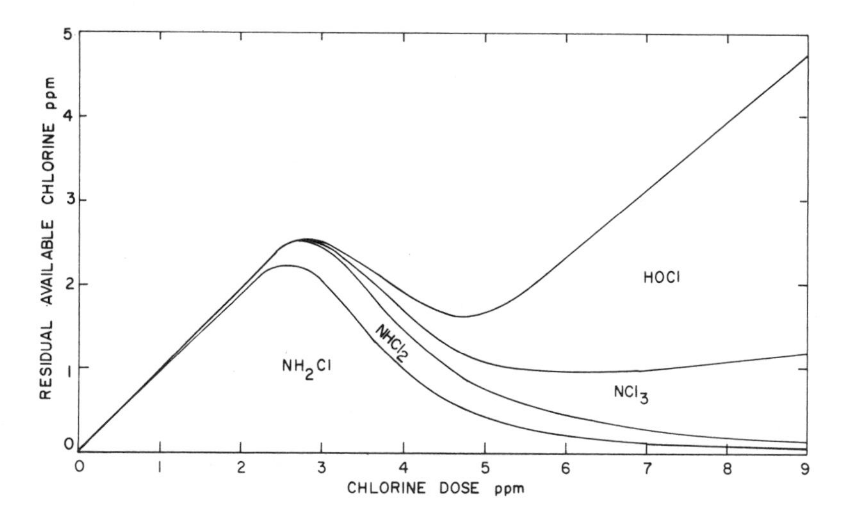

Figure 15-4. Chlorine dose-residual curve at pH 7.3-7.5 after 10 minutes contact. Initial ammonia 0.5 ppm (as N) or 3.5 micromolar. From Palin.[5]

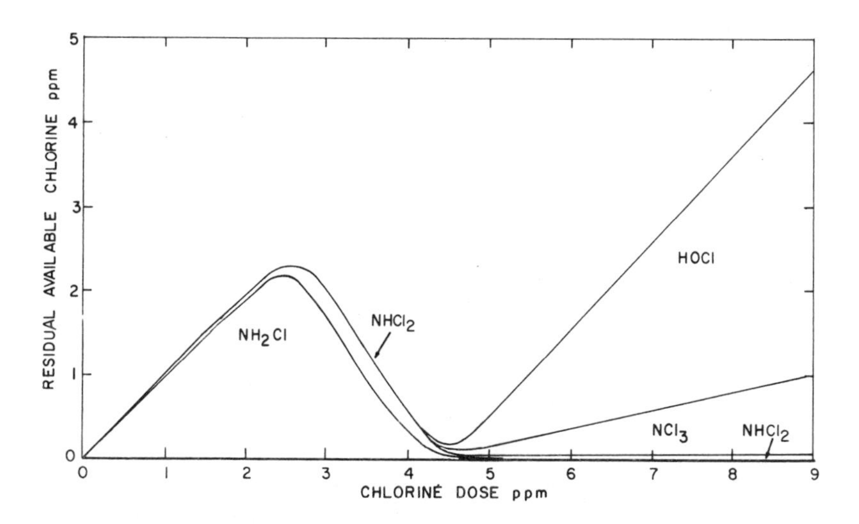

Figure 15-5. Chlorine dose-residual curve at pH 7.3-7.5 after 2 hours contact. Initial ammonia 0.5 ppm (as N) or 3.5 micromolar. From Palin.[5]

$$2NH_2Br + BrO^- + 2\ OH^- \rightarrow N_2 + 3Br^- + 3H_2O \qquad (7)$$

Low pH and high ammonia to bromine ratios were found to stabilize the compound.

Dibromamine undergoes rapid decomposition. The overall reaction stoichiometry is

$$2NH_3 + 3\ OBr^- \rightarrow N_2 + 3Br^- + 3H_2O \qquad (8)$$

Reaction 8 is quantitative in the presence of excess bromine and has been used as an analytical method for ammonia determination in alkaline solution.[19] As in the case of monobromamine, high ammonia to bromine ratios and low pH stabilize the aqueous dibromamine molecule. We attempted to explain the effect of excess ammonia by likening the dibromamine decomposition to that of dichloramine as expounded by Wei.[7] Monochloramine either reacts directly with itself to form dichloramine or hydrolyzes to HOCl.[20] The HOCl formed by hydrolysis either reacts with NH_3 to reform monochloramine or forms dichloramine by reaction with NH_2Cl. The presence of excess ammonia was found to favor the direct N-chlorination reaction by NH_2Cl by decreasing the hydrolysis of NH_2Cl to HOCl. Similarly, dibromamine could decompose through either direct reaction of $NHBr_2$ on a second $NHBr_2$ in the presence of excess ammonia, or at low concentrations of ammonia $NHBr_2$ hydrolyzes to HOBr followed by the rapid reaction of the HOBr formed with $NHBr_2$.

In any case the stoichiometry of bromamine decomposition is given in Equation 8.

A second-order observed average rate constant was calculated as 17.5 l $mole^{-1}$ sec^{-1} at pH 7 for dibromamine decomposition but there has not yet been an attempt to establish true order or propose a mechanism.[18] Johnson *et al.*[21] investigated the decomposition of the trisubstituted ammonia-bromine derivative, tribromamine, and observed a regeneration of free bromine (*i.e.*, hypobromous acid, hypobromite ion or molecular bromine) after the decomposition.[21] They proposed a reaction stoichiometry as follows:

$$2NBr_3 + 3H_2O = N_2 + 3HOBr + 3H^+ + 3Br^- \qquad (9)$$

It can be seen, however, that this stoichiometry is simply a composite of the overall decomposition stoichiometry of Equation 8 and the stoichiometry of the formation of tribromamine.

Galal-Gorchev has postulated that tribromamine decomposes directly through dibromamine.[17] This would explain the stabilizing effect of low ammonia to bromine ratios observed for tribromamine. A similar explanation might also be possible in the case of monobromamine. If, as mentioned previously, the entire bromamine system exists in a state of equilibrium, and dibromamine is the principal pathway to decomposition in regions where monobromamine is the principal specie, the likelihood of a monobromamine molecule being hydrolyzed to ammonia and free bromine is reduced if the free ammonia concentration is high. Similarly, in the case of tribromamine, the probability of formation of tribromamine is enhanced when large excesses of free bromine are present. The idea of a single set of decomposition reactions for the entire bromamine system governed by the equilibrium considerations involved cannot be proved from the rather sparse data that currently exists, but should not be dismissed.

Bromine Breakpoint Reaction

The decomposition of dibromamine like dichloramine is the basis for the bromine breakpoint reaction. A comparison between the general forms of the chlorine and bromine breakpoint curves is shown in Figures 15-4, 15-5 and 15-6.[22] While the breakpoint curve for chlorine is relatively independent of time (some changes are noticed but the general form and slope values remain the same), the bromine breakpoint curve shows marked differences for short- and long-term studies. In Figure 15-6 the pH 7 breakpoint curve for bromine is broken into three stages where I $NHBr_2$, II NBr_3, and III HOBr predominate. In the first stage (I) both 5- and 30-minute curves show little residual. This is due to the preferential formation of dibromamine and its rapid decomposition at pH 7. Point A corresponds to a Br/N molar ratio of 1.5 where the bromine is present in precisely the stoichiometric amount necessary to oxidize all ammonia to nitrogen gas, following the overall redox reaction given in Equation 8. It is interesting that this point also corresponds to the transition from dibromamine to tribromamine in the distribution diagram, Figure 15-3.

At the breakpoint a significant portion of the ammonia nitrogen is initially present as ammonium ion. This does not contradict the overall stoichiometry, however, because as has been previously emphasized, the bromamine system is governed by equilibrium by contrast to the chloramine system. Thus, as dibromamine decomposes the tribromamine and ammonia concentrations would similarly shift to maintain the same overall proportion. Any bromine to ammonia ratios greater than 1.5 would represent oxidation conditions greater than those needed to completely oxidize the ammonia, and the excess bromine would remain in the +1 oxidation state. This is precisely what happens in the second stage of the breakpoint curve represented by II. In II, during the short time curve, tribromamine is the principal species present. Because the decomposition of tribromamine is much slower than dibromamine (or perhaps because the concentration of dibromamine, the active specie in the decomposition, is so small that the decomposition rate is slowed down),

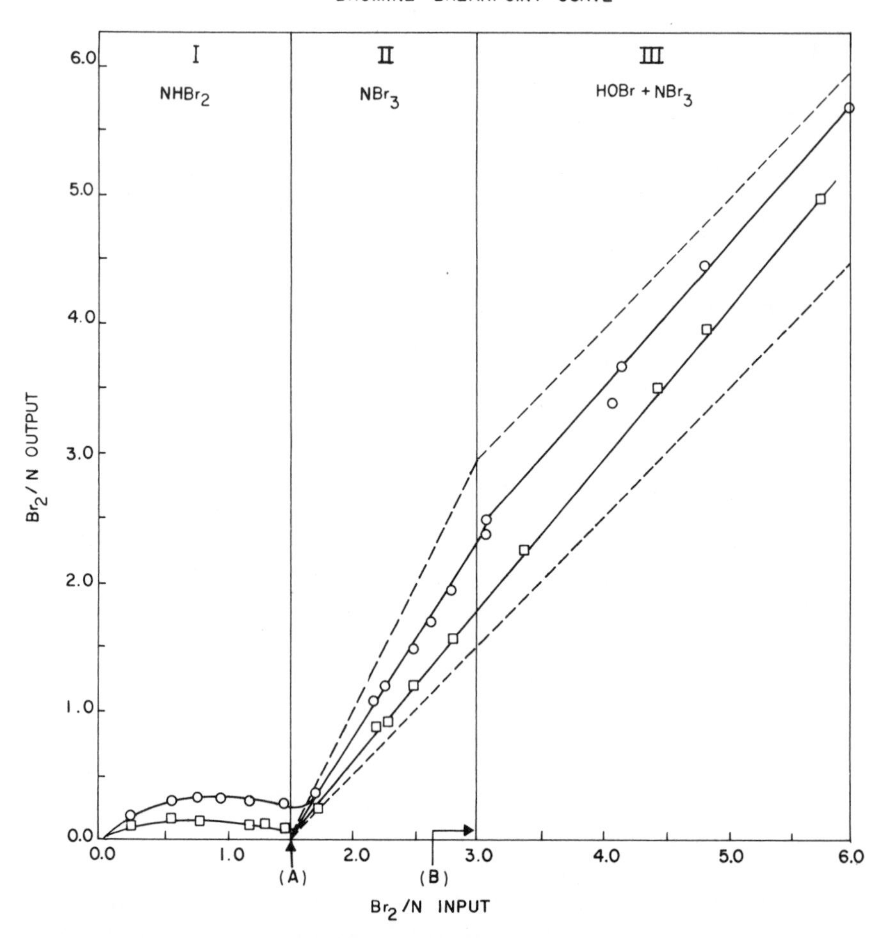

Figure 15-6. Mole ratio of bromine residual concentration to nitrogen initial concentration versus mole ratio of bromine dosed to initial nitrogen concentration. ○ is 5 minutes contact time, □ is 30 minutes contact time. pH 7, 20°C, $\mu = 0.10$, $[Br^-] < 10^{-4} M$, $[NH_3]_o = 6.67 \times 10^{-5} M$ (1.13 mg NH_3/l). Broken lines are short-time and long-time theoretical curves.

the bromine that eventually will be reduced to bromide is still in the +1 oxidation state (as bromine in NBr_3) at the time of the measurement. Thus the slope of the line between the Br/N ratios of 1.5 and 3.0 is greater than one. In more precise terms, while the ratio of Br/N dosed is changing from 1.5 to 3.0, the ratio of Br/N residual would change from 0.0 to 3.0. This is the theoretical result assuming measurements were taken immediately after addition. Greater than +1 slopes in this region have been reported by Johnson and Overby,[18] Sollo and Larson[23] and Inman and Ellmore[22] the latter's data being used for Figure 15-6. In actuality, these plots are not straight curves of constantly increasing slopes. This is due to the effect of higher Br/N ratios in the decomposition rate of tribromamine (or on the concentration of active dibromamine). Increasing Br/N ratios decrease the decomposition rate thus a greater percentage of the tribromamine will decompose to nitrogen gas, hypobromous acid and bromide ion at the lower ratios than at the higher ones. But when sufficient time is allowed for all ammonia to be oxidized (*i.e.*, tribromamine to decompose) the lower dotted line shows the results. This follows directly from the previous explanation of excess oxidizing bromine being added after the N/Br ratio of 1.5. The slope of this line is 1 and it continues to rise at this constant slope as a direct function of excess oxidizing HOBr. As again demonstrated in the figure, the difference in residuals for the short- and long-term breakpoint curves from A to B will be constantly increasing, reaching a maximum value of 1.5 at a bromine to ammonia input mole ratio of 3.0 (point B). This corresponds to the 3 moles of bromine needed to completely brominate the ammonia. From A to B there is initially no HOBr present, and the residual from the breakpoint A to B is only NBr_3. From A to B the residual which is initially NBr_3 and contains no appreciable HOBr soon produces HOBr as reaction 9 proceeds and NBr_3 decomposes into N_2 giving back half of the bromine as HOBr. After NBr_3 is formed completely addition of bromine beyond B simply remains as free bromine (*e.g.*, at pH 7, hypobromous acid). The breakpoint mole ratio for bromamines of 1.5 differs slightly from the Cl/N molar breakpoint ratio of 1.6 to 1.8 only because of the side reaction of oxidation of ammonia by chlorine to nitrate. This appears in the chlorine system but is not important for neutral bromine. In region III free bromine is present with tribromamine and, as time increases, more tribromamine decomposes (again viewed as the oxidation of NH_3) and the curve approaches the free bromine residual curve given by the rising section of the long-term breakpoint curve. The essential differences between chlorine and bromine are those occasioned by the predominant species ($NHBr_2$ and NBr_3) formed at neutral pH and the formation and subsequent decomposition of tribromamine beyond the breakpoint.

Applications of Research in Bromamine Chemistry

Determination of which bromamine specie is present and its stability under the specific conditions involved is necessary for analysis of the results of the biological studies being performed, some of which appear in Chapters 6-10.

The effect of ammonia on the devitalization of *B. subtilis* spores at pH 7,[24] was interpreted by Overby[18] utilizing the aforementioned bromamine distribution diagram and the decomposition kinetics of dibromamine. This analysis showed all the bromamine forms, NBr_3, $NHBr_2$ and NH_2Br to be as effectively sporicidal as hypobromous acid. Chapter 8 by Sollo and Larson uses a similar analysis to explain decidedly different pH effects for chlorine and bromine in treated wastewater.

Breakpoint bromination is obviously preferred in drinking water treatment and swimming pool disinfection. The above discussion has noted tribromamine as the principal specie at concentrations just beyond the breakpoint. Thus under conditions of breakpoint bromination, tribromamine stability is intimately related to the disinfection power of a given water. This study was undertaken to determine the stability of NBr_3 in the neutral pH range. Its objectives were to test the proposed decomposition stoichiometry Equation 9 and to develop a rate equation that would accurately describe the decomposition of tribromamine.

METHODOLOGY

Materials

Chlorine demand-free water was used in the experiments. All reagents were ACS grade, and glassware was presoaked in a sodium hypobromite solution of 2-4 mg/l. Sodium thiosulfate solutions were prepared and standardized according to the methods outlined in *Standard Methods.*[25] Buffers of 0.1*M* monobasic-dibasic potassium phosphate were prepared for pH 6, 7 and 8, and a 2.67 x $10^{-2}M$ ammonium chloride solution was made by directly weighing dried reagent grade ammonium chloride. Bubbling Matheson ultra-high purity gaseous chlorine through a sodium hydroxide solution for 10-15 minutes gave a sodium hypochlorite solution of 0.1-0.2*M*, which was stored in basic solution at pH 10.5-11. The concentration of this solution was then determined by the thiosulfate-iodide amperometric titration which is detailed below.

Sodium hypobromite stock solutions of reduced bromide concentration were prepared according to the work done by Farkas *et al.*[26] and Lewin and Arrahami,[27] in investigating the oxidation of bromide ion by hypochlorite ion with the subsequent formation of hypobromite ion and chloride ion.

Weighed Certified ACS grade potassium bromide was dissolved in water at pH 10.5. A stoichiometric amount of standardized sodium hypochlorite stock was subsequently added, and the pH readjusted to 11.0±0.3. This pH was chosen to minimize bromate formation (favored by low pH) and maximize reaction velocity (favored by low pH). The reaction proceeded in a brown glass stoppered bottle for 2 days, and the extent of the reaction was determined by comparison of the concentration of HOBr determined spectrophotometrically at 260 nm with the concentration of bromide initially added.

Procedure

Experiments were conducted at initial ammonia concentrations of 0.667, 1.333 and 2.667 x $10^{-4}M$ (1.13, 2.26, 4.52) mg NH_3/l respectively and henceforth will be referred to as X1, X2 and X4 powers. At each individual power, ammonia to bromine mole ratios of 1/3, 1/6 and 1/12 were run. Selection of these particular values was based on the experimental constraints in the measurement of tribromamine in the presence of free bromine. As mentioned earlier tribromamine absorbs at 258 nm, very close to an absorbance peak for hypobromous acid (HOBr, the predominant form at pH 6, 7 and 8) at 260 nm. Figure 15-7 illustrates the combined absorbances that must be considered at 258 nm. A similar absorbance problem made it impossible to follow the decomposition by spectrophotometric measurement

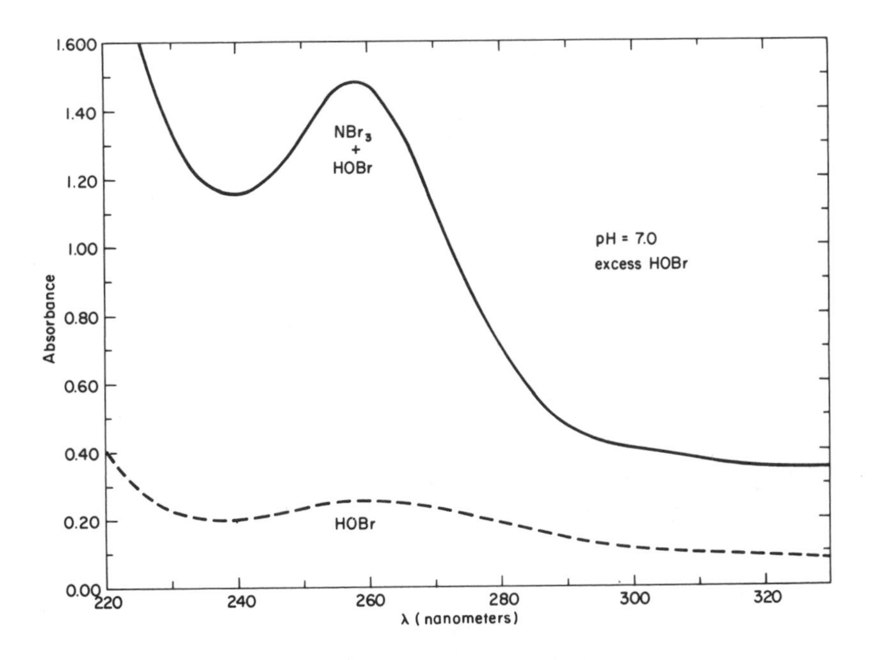

Figure 15-7. Typical U.V. spectrum of tribromamine, NBr_3, and hypobromous acid, HOBr, mixture.

alone. Because hypobromous acid is produced as tribromamine decomposes (Equation 9), a simple subtraction of hypobromous acid absorbance from the total absorbance is not possible. Owing to this, an experimental protocol was developed that involved the determination of the total absorbance at 258 nm due to tribromamine and free bromine. A Cary Model 14 spectrophotometer was used in these and the other spectrophotometric measurements.

The total residual bromine concentration was measured by a thiosulfate-iodide amperometric titration. Fifteen ml of 1.6% potassium iodide and 125 ml of deionized water were poured into a 500-ml widemouthed jar, which was then fitted to the titration apparatus consisting of a Sargent Model XV polarograph (E. H. Sargent & Co., Chicago) used in conjunction with a rotating platinum/saturated calomel electrode system. The voltage used was +0.15 volts vs SCE. Two ml concentrated acetic acid was then added to bring the pH down to about 2. The current of the electrode reaction

$$I_3^- + 2e^- \rightarrow 3I^- \qquad (10)$$

was followed and the end point taken to be that point when no current was observed. This procedure determined total reducible bromine from NBr_3, $NHBr_2$, NH_2Br, Br_2, HOBr, and OBr^- with each mole of bromine found to produce equivalent moles of iodine by the reactions:[28]

$$HOBr + 3I^- + H^+ \rightarrow I_3^- + Br^- + H_2O \qquad (11)$$

$$NBr_3 + 9I^- + 3H_4^+ \rightarrow 3I_3^- + NH_3 + NH_3 + 3Br^- \qquad (12)$$

A 4-l beaker was filled with tapewater to about the 1-l mark, and the temperature adjusted to 20 ± 0.5°C. The 1-l brown reaction bottle containing the standardized hypobromite solution adjusted to the desired concentration was placed in the 4-l beaker; both were set on a large magnetic stirrer and gentle stirring begun. The volume-adjusted ammonia solution was then poured into the reaction bottle, and two stopwatches were started simultaneously at the first instant of pouring. The reaction solution was stirred gently for one minute, then allowed to stand quiescent. Both prior to and after the cessation of the stirring aliquots of the reaction solution were withdrawn as rapidly as possible and analyzed by UV absorbance and thiosulfate-iodide amperometric titration. The time of each analysis was recorded from the stopwatches. The absorbance was measured at 258 nm, λ max for NBr_3. Thiosulfate concentrations used were such that the volume of titer needed was between 7 and 10 ml. The solution pH was measured immediately after each run.

The experimental molar absorptivity of the free bromine solution was determined prior to each experimental run. It was this molar absorptivity determination that led us to use bromide reduced stock solutions. It was discovered that the molar absorptivity of the free bromine increased with increasing concentration, quite markedly at pH 6 and less. This added absorbance was believed to originate from the tribromide ion Br_3^- which equilibrium calculations later showed would be present in low but significant concentrations at low pH considering its high molar absorptivity. Using the molar absorptivity of 38,000 of tribromide ion at 260 nm, a pH of 4.5, and a bromide concentration of $2 \times 10^{-4} M$, a solution of $6 \times 10^{-4} M$ free bromine will have an absorbance of 0.23 due to tribromide ion. Thus bromide reduced stock solutions ($< 10^{-4} M$) were used in all runs. As will be detailed later, use of these solutions, demonstrated a catalytic effect of bromide ion on the decomposition.

Calculations

Absorbance vs time (from UV data) and total bromine vs time (from titration data) plots were made of the experimental points taken during each run and smooth curves were drawn through the points. To analyze this data certain standardized times had to be chosen and individual absorbance and total bromine concentration values taken off the smoothed curves at those time points. The times chosen were 0.5, 1.0, 1.5, 2.0, 3.0, 4.0, 6.0, 8.0, 10.0, 12.0, and 15.0 minutes for pH 6 and pH 8 runs and 2.0, 3.0, 4.0, 6.0, 8.0, 10.0, 12.0, 15.0, 20.0, 25.0, 35.0, 45.0, and 60.0 minutes for pH 7 runs.

Because the absorbance at 258 nm is due to both tribromamine and hypobromous acid (Figure 15-7), and because the value of total bromine found in the titration is also due to these two compounds, simultaneous equations must be solved to determine the concentrations of tribromamine and hypobromous acid at the specific times mentioned previously.

The absorbance at 258 nm can be expressed as

$$\text{Abs.}_{258} = \epsilon_{NBr_3} \ell [NBr_3] + \epsilon_{F.B.} \ell [FB] \qquad (13)$$

where

ϵ_{NBr_3} = molar absorptivity of NBr_3 at 258 nm

ℓ = path length of cell used

$\epsilon_{F.B.}$ = experimental molar absorptivity for all free bromine species at 258 nm

The concentration of total bromine determined by titration can be expressed as:

$$[\text{Total Bromine}] = [\text{FB}] + 3[\text{NBr}_3] \tag{14}$$

where [FB] = the concentration of all free bromine species present.

In the case of our experimental parameters the free bromine forms would be predominantly hypobromous acid (HOBr) at pH 6 and pH 7, and at pH 8 hypobromous acid with some hypobromite anion (OBr^-). Forms such as molecular bromine (Br_2), and tribromide ion (Br_3^-) are present in insigificant concentrations, especially under the reduced bromide ion concentration conditions employed in these experiments.

Besides calculating the concentrations of tribromamine and the free bromine species, a stoichiometry calculation was made to verify Equation 9. The relationship used for this test was

$$\frac{[\text{Free Bromine}]_2 - [\text{Free Bromine}]_1}{[\text{NBr}_3]_1 - [\text{NBr}_3]_2} = 3/2 \tag{15}$$

This quotient of differences is calculated for the time differences between the value of 0.5 minutes for pH 6 and 8, and 2.0 minutes for pH 7, and all the other time points selected.

Individual values of tribromamine concentration given by the solutions of the simultaneous equations are then used to construct tribromamine vs time plots (Figure 15-8). These plots are then used to calculate instantaneous

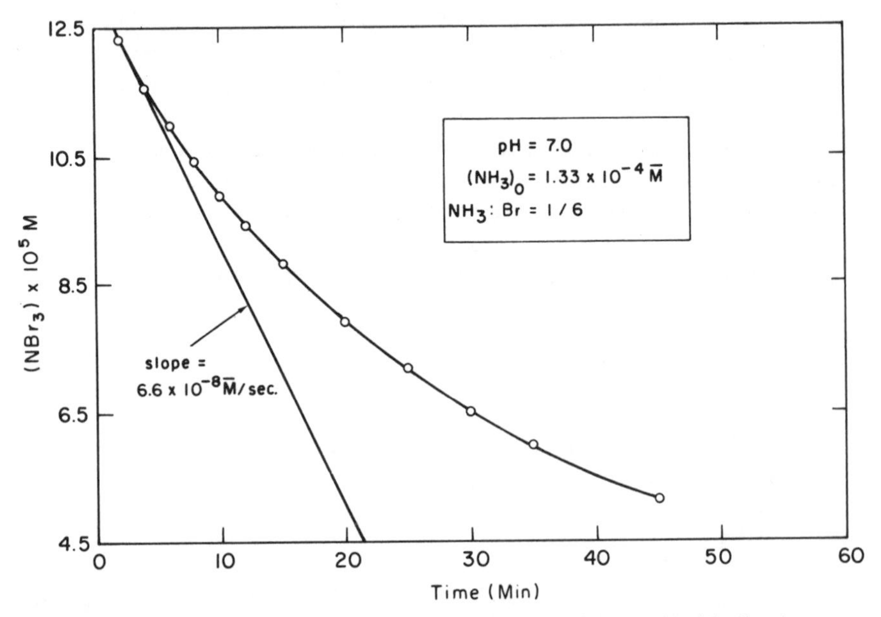

Figure 15-8. Tribromamine concentration as a function of time. Straight line is the tangent drawn to the curve in order to calculate an instantaneous rate.

rates, for van't Hoff order and rate determinations, by drawing tangents to the concentration curves. When calculating instantaneous rates, it is preferable to take the rates at a point as close to zero time as possible, because at low times the rate is the most rapid, and differences between instantaneous rates under differing conditions are greatest; also, at low times the system has gone through minimal change with minimal product formation, thus increasing the confidence in those parameters held constant.

Due to difficulty in extrapolating our data back to zero time, the two-minute instantaneous rate was chosen for pH 6 and pH 7 data. At pH 8 the decomposition rate is faster so a 30-second instantaneous rate was taken.

Stoichiometry of Tribromamine Decomposition

The stoichiometry shown in Equation 9 was originally calculated from Equation 15 for an initial time of 0.5 minutes and various final times between 1 and 60 minutes. However, at 0.5 minutes the decomposition rate is very fast. The small errors in the calculation of concentrations of tribromamine and free bromine due to the speed of the reaction at 0.5 minutes can translate to large errors in the stoichiometry. This problem is aggravated if final times are such that only a small portion of the tribromamine has decomposed. Such slight decomposition results in a ratio of small differences leading to large errors. Therefore, two initial times were chosen for comparison—zero and four minutes. At zero minutes the concentrations in the solution are very accurate because of the solution standardization prior to each run. At four minutes the reaction rate has abated enough to make fairly accurate concentration measurements. The final times selected were 15 minutes for pH 8 data and 60 minutes for pH 7 data. These final times allowed sufficient decomposition for significant differences in the concentrations, yielding a more reliable stoichiometry ratio.

Table 15-2 is a summary of the stoichiometry data calculated for these initial time conditions for pH 7 and pH 8 data. For a zero-minute initial time of 15 of 18 runs had a stoichiometry ratio between 1.3 and 1.5; for the four-minute initial time 14 of the 18 runs had a ratio between 1.3 and 1.7. Mean values were between 1.23 and 1.41 for the four groups. Table 15-3 gives the means and standard deviations for the four groups of data.

The pH 8 data calculated between 0 and 15 minutes gives the mean closest to 1.5 and the smallest standard deviation; standard deviations for the other groups are greater than a factor of two larger. This is probably due to the greater extent of reaction that takes place at pH 8 between 0 and 15 minutes. As explained below, the decomposition is hydroxide catalyzed, and at low ratios of ammonia to bromine, the reaction is slow. Thus, on Table 15-2 the ratios of 1/12 for pH 7 show the greatest deviation from 1.5 and add significantly to the standard deviation. At pH 8 after

Table 15-2

pH	Power $[NBr_3]_0$	Ratio $[Total\ Bromine]_0$	$\frac{[FB]_{t=t_f} - [FB]_{t=0}}{[NBr_3]_{t=t_f} - [NBr_3]_{t=0}}$	$\frac{[FB]_{t=t_f} - [FB]_{t=4}}{[NBr_3]_{t=t_f} - [NBr_3]_{t=4}}$
7	X4	1/12	0.8	1.1
7	X2	1/12	1.0	1.1
7	X1	1/12	1.1	1.4
7	X4	1/6	1.4	1.5
7	X2	1/6	1.5	1.6
7	X1	1/6	1.3	1.7
7	X4	1/3	1.4	1.3
7	X2	1/3	1.4	1.3
7	X1	1/3	1.4	1.4
8	X4	1/12	1.3	1.1
8	X2	1/12	1.5	1.4
8	X1	1/12	1.3	1.5
8	X2	1/6	1.5	1.7
8	X1	1/6	1.5	1.3
8	X4	1/3	1.5	1.5
8	X2	1/3	1.4	1.0
8	X1	1/3	1.4	1.3

Table 15-3

pH	Initial Time	Final Time	Stoichiometry Ratio Mean	Standard Deviation
7	0	60	1.23	0.63
8	0	15	1.41	0.26
7	4	60	1.37	0.58
8	4	15	1.38	0.61

4 minutes, much of the tribromamine has already decomposed, so the additional decomposition between 4 and 15 minutes leads to small differences in concentrations with the concomitant errors involved giving larger standard deviations. This is strong evidence that the stoichiometry presented in Equation 9 is correct for the pH range 7 to 8 under conditions of low bromide concentration and the concentration range studied in this work.

Rate and Order Determination

Data from the experimental runs were analyzed according to the method initiated by J. H. van't Hoff.

If the concentrations of all species except one are held constant and the logarithms of the instantaneous rates at that time when all the concentrations of reactants are constant are plotted against the logarithm of the concentration of the reactant varying in concentration, the slope of the line should correspond to the order (power) of that reactant in the rate equation. For a reaction proceeding by a single pathway

$$\log - (dC/dt)_{t=t_i} = \underbrace{m \log A + \log k + n \log B + \ldots}_{\text{—constant terms—}} \qquad (21)$$

Where m and n are the reaction order in A and B, and A is the initial concentrations of A chosen for a given initial concentration of B. Instantaneous rates together with the concentrations of the important species for pH 7 and pH 8 run are given in Table 15-4.

Concentration Terms in the Rate Equation

Four species were found to have an effect on the decomposition rate: free bromine (which form or forms of free bromine, *i.e.*, hypobromous acid, molecular bromine, tribromide ion, or hypobromite ion having this effect remains undertimed), hydrogen ion, and bromide ion. Chloride and sulfate ions were found to have no effect on the rate of decomposition.

Bromide Ion

Figure 15-9 shows that increasing bromide ion concentration increases decomposition rate. However, no simple relationship was found between the bromide ion concentration and the instantaneous rate. As seen in Figure 15-9, $10^{-4} M$ Br^- concentration had a very minimal effect on the initial rate and decomposition curve. However, this bromide is a significantly greater concentration than the bromide in the normal experimental situation (*i.e.*, using bromide reduced sodium hypobromite) which is estimated at between 10^{-5} and $10^{-4} M$. Therefore, experiments to determine

Table 15-4

pH	Power $[NBr_3]_0$	Ratio $[Total\ Bromine]_0$	t_i	$-(dT_r/dt)_{t=t_i}$ x 10^8 *M*/sec	$[NBr_3]_{t=t_i}$ x 10^4 *M*	$[Free\ Bromine]_{t=t_i}$ x 10^4 *M*
8	X4	1/12	0.5	99	2.42	22.6
8	X2	1/12	0.5	42	1.17	11.5
8	X1	1/12	0.5	20	0.627	5.72
8	X4	1/6	0.5	150	2.33	8.00
8	X2	1/6	0.5	66	1.12	4.24
8	X1	1/6	0.5	34	0.562	2.093
8	X4	1/3	0.5	206	1.41	2.972
8	X2	1/3	0.5	83	0.744	1.068
8	X1	1/3	0.5	50	0.368	0.636
7	X4	1/12	2.0	9.75	2.64	23.7
7	X2	1/12	2.0	4.36	1.32	11.9
7	X1	1/12	2.0	2.03	0.650	5.98
7	X4	1/6	2.0	16.3	2.41	8.13
7	X2	1/6	2.0	7.25	1.23	4.21
7	X1	1/6	2.0	4.6	0.615	2.05
7	X4	1/3	2.0	22.9	1.56	1.54
7	X2	1/3	2.0	10.2	0.810	0.791
7	X1	1/3	2.0	5.92	0.377	0.399

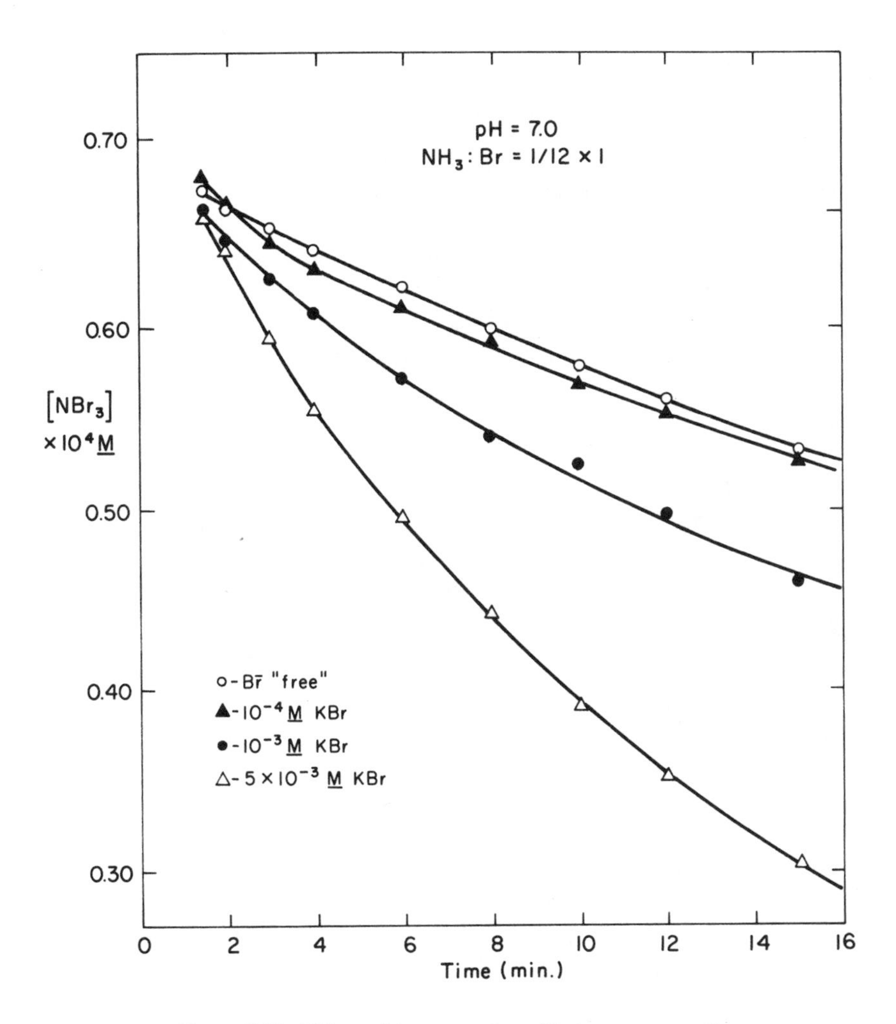

Figure 15-9. Effect of increasing bromide ion concentration on tribromamine decomposition at 20°C.

the order of the other species were done in the absence of additional bromide, that is, the only bromide present was the very minimal amount occurring in the stock bromine solution used.

Tribromamine

Ideally, to determine the order with respect to tribromamine, all other components in the rate equation would have to be kept constant. However, owing to our particular experimental constraints, due in large part to the mutual absorbance of tribromamine and hypobromous acid at 258 nm we were unable to operate under ideal van't Hoff conditions in that we could not maintain the concentration of free bromine over the wide concentration range of tribromamine studied. This wide concentration range was essential in obtaining reliable van't Hoff plots. Thus we were forced to collect data under conditions of constant ammonia to bromine ratio, where the bromine includes that portion which subsequently reacts with ammonia to form tribromamine. Figure 15-10 shows increasing tribromamine concentration increases the decomposition rate subject to the previously stated condition of a constant ammonia to bromine ratio.

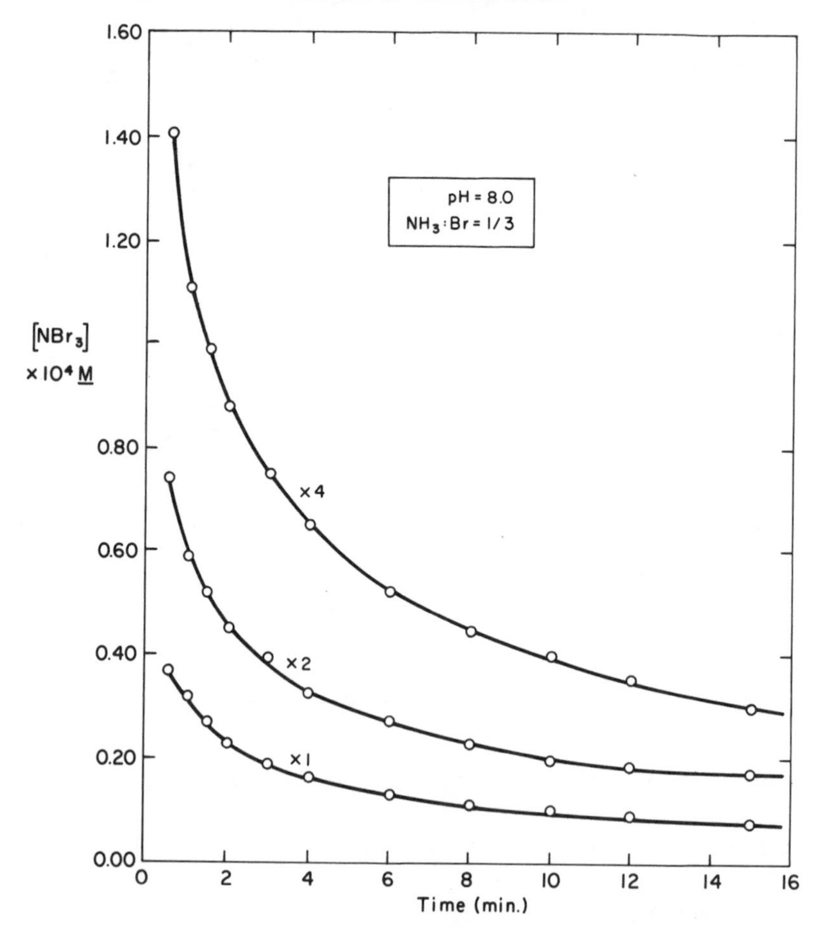

Figure 15-10. Effect of increasing tribromamine concentration on tribromamine decomposition under conditions of constant initial ammonia to bromine mole ratio, and a temperature of 20°C.

Construction of van't Hoff plots, {log (-dTr/dt) where Tr is [NBr_3] vs log [NBr_3]}, at pH of 7 and 8 (Figure 15-11) gave calculated slopes of close to 1.1 in all cases. This leads us to conclude that there is a first-order term in tribromamine in the rate equation, plus another term or terms which remain constant over varying concentrations of tribromamine, providing the initial ratio of ammonia to bromine is constant.

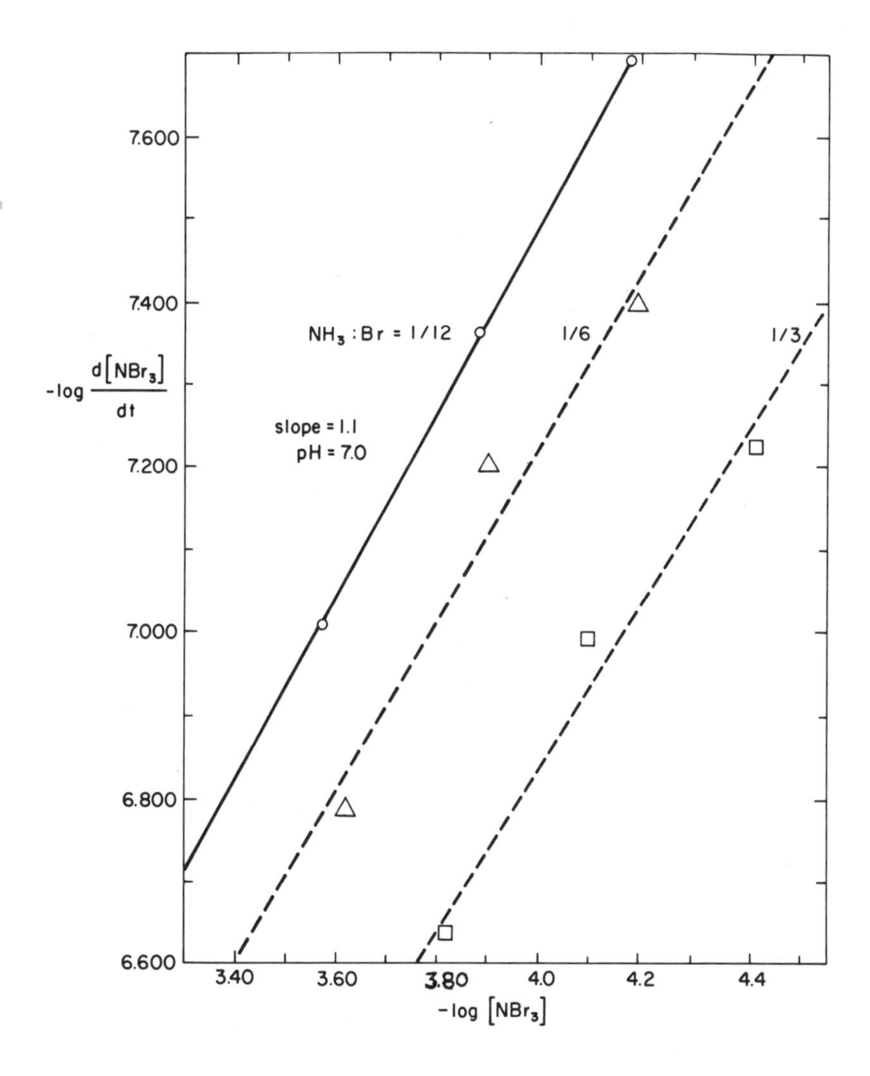

Figure 15-11a

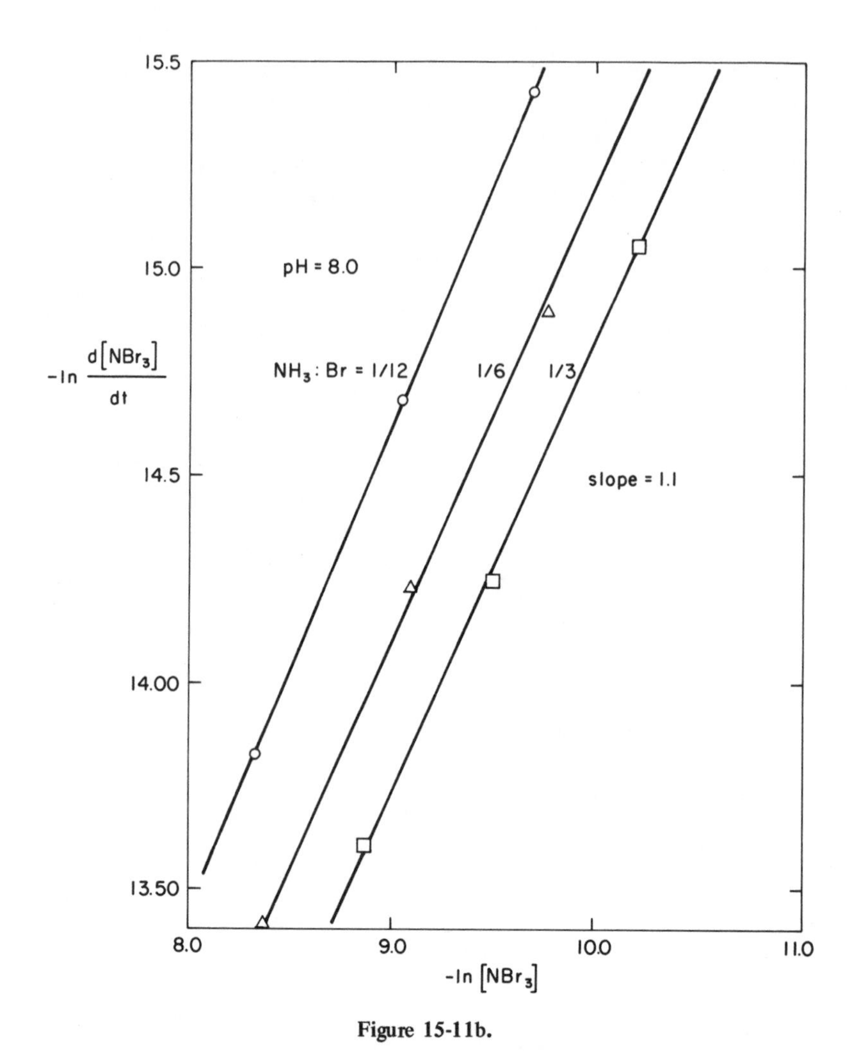

Figure 15-11b.

Figure 15-11. Van't Hoff plots of the negative logarithm of the instantaneous rate versus the negative logarithm of tribromamine concentration for pH 7 and pH 8 constant initial ammonia to bromine mole ratio and a temperature of 20°C.

Free Bromine

Identical to the problems of establishing ideal van't Hoff conditions for tribromamine, are those encountered with free bromine. Qualitatively, under conditions of constant initial tribromamine, free bromine has an inverse relationship with the decomposition rate. However, at 30 seconds and 2 minutes, enough tribromamine has decomposed to destroy the constancy needed to construct van't Hoff plots for free bromine. As mentioned previously, the rate equation should contain a term that remains relatively constant over short time intervals provided the initial ammonia to total bromine ratio was constant. One such term, is the ratio of tribromamine to *free* bromine. Van't Hoff plots of -log (-dTr/dt) versus the negative logarithm of the tribromamine to free bromine ratio were approximately linear with slopes ranging from 0.34 to 0.51, for pH's 7 and 8 (Figure 15-12), with an average slope of 0.43. In these cases the initial concentration of tribromamine is being held relatively constant as the tribromamine to free bromine ratio is varied, so slopes close to 0.5 are evidence that this ratio appears in the rate equation with an order of 1/2.

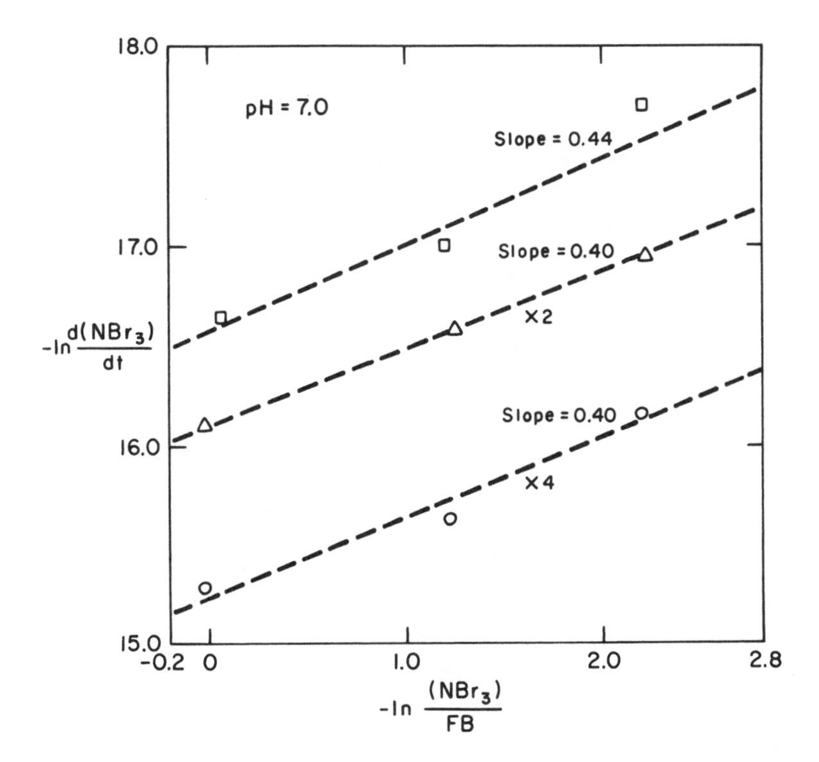

Figure 15-12. Van't Hoff plot of the negative logarithm of the instantaneous rate versus the negative logarithm of the tribromamine to free bromine ratio for pH 7, constant initial ammonia concentration and a temperature of 20°C.

Hydrogen Ion (pH)

The effect of pH on the decomposition rate is graphically given in Figure 15-13, showing hydrogen ion to have an inverse relationship to the decomposition rate. Van't Hoff plots (-log (-dTr/dt) vs. pH) for pH between 7 and 8 give a slope that is very close to +1 in all combinations studied, the range in slopes being 0.92 to 1.0 for the 9 different combinations. This suggests that the decomposition of tribromamine obeys first-order kinetics with respect to $1/H^+$ for the pH range 7 to 8.

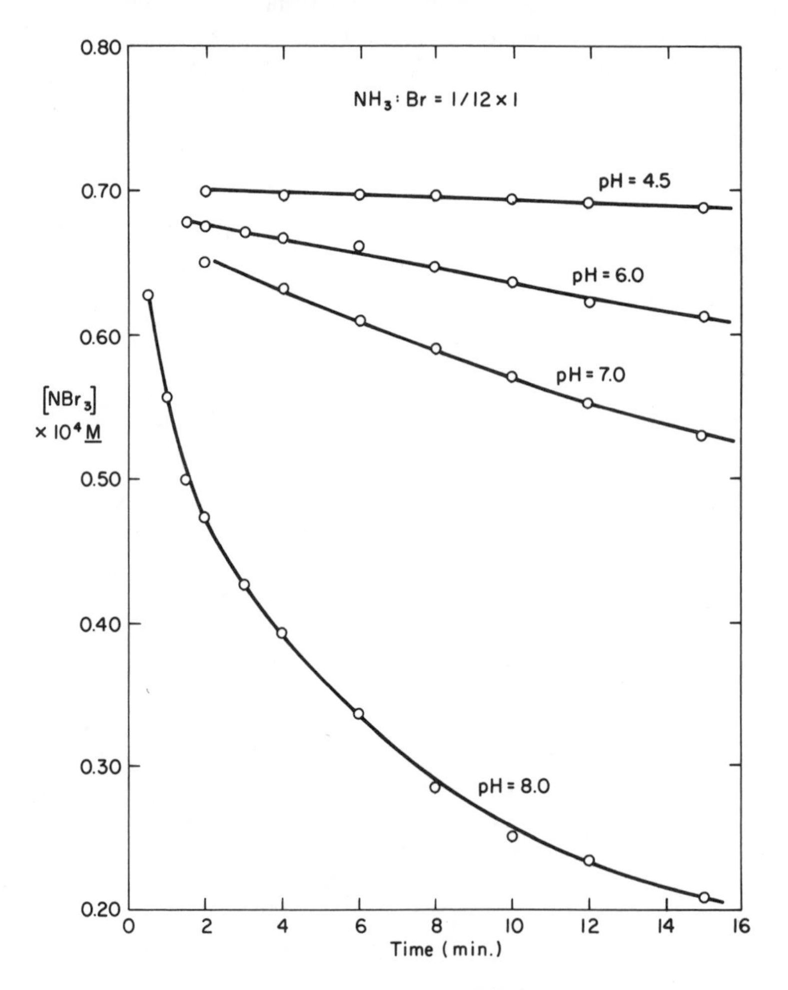

Figure 15-13. Tribromamine concentration versus time for pH's 4.5, 6.0, 7.0 and 8.0. Initial ammonia concentration = $6.67 \times 10^{-5} M$, initial bromine concentration = $8.00 \times 10^{-4} M$, temperature 0°C and bromide ion concentration $<10^{-4} M$.

Development of a Differential Rate Equation

Tribromamine and Free Bromine

The hypothesized orders of the various species in the rate equation as obtained in the previous analysis can be more rigorously tested by regarding the entire rate equation. Since our results have tentatively shown that the initial decomposition rate of tribromamine is first-order with respect to tribromamine and one-half order with respect to the ratio of tribromamine to free bromine, we may combine the appropriate terms and write the following differential rate equation at constant pH:

$$\frac{-dTr}{dt} = k' \, Tr \, (Tr/FB)^{1/2} = k' \, \frac{Tr^{3/2}}{FB^{1/2}} \tag{16}$$

where $Tr = [NBr_3]$, and FB is the concentration of free bromine at time t. This model was tested by observing the linearity of plots of (-dTr/dt) versus $(Tr^{3/2}/FB^{1/2})$. Errors in the van't Hoff plots seen previously, due to a lack of constancy in one or more of the species supposedly held constant will not effect this plot, provided pH is constant and the concentrations of all species are known, because both major species, tribromamine and free bromine, are considered together. Figure 15-14 is the plot corresponding to pH 7 data. Although there is some scatter, the points fall close enough to a straight line to indicate that the above model can account in part for the system's experimental behavior. The behavior is similar at pH 8; only the points where the initial ammonia to bromine ratio is 1/3 deviate significantly and these are the most uncertain due to the rapidity of decomposition possibly causing large errors in the calculations of the free bromine concentration, and due to the possibility of incomplete formation of NBr_3 at time zero.

Hydrogen Ion (pH)

The hydrogen ion concentration may be included as a variable in the above model by utilizing the observed order of (-1) with respect to $[H^+]$ as determined over the limited pH range 7 to 8. The resulting equation is:

$$\frac{-dTr}{dt} = \frac{k \, Tr^{3/2}}{[H^+] \, FB^{1/2}} \tag{17}$$

This expanded model can now be tested on a single graph employing all the data obtained at pH 7 and pH 8 in a manner identical to the aforementioned van't Hoff order determinations.

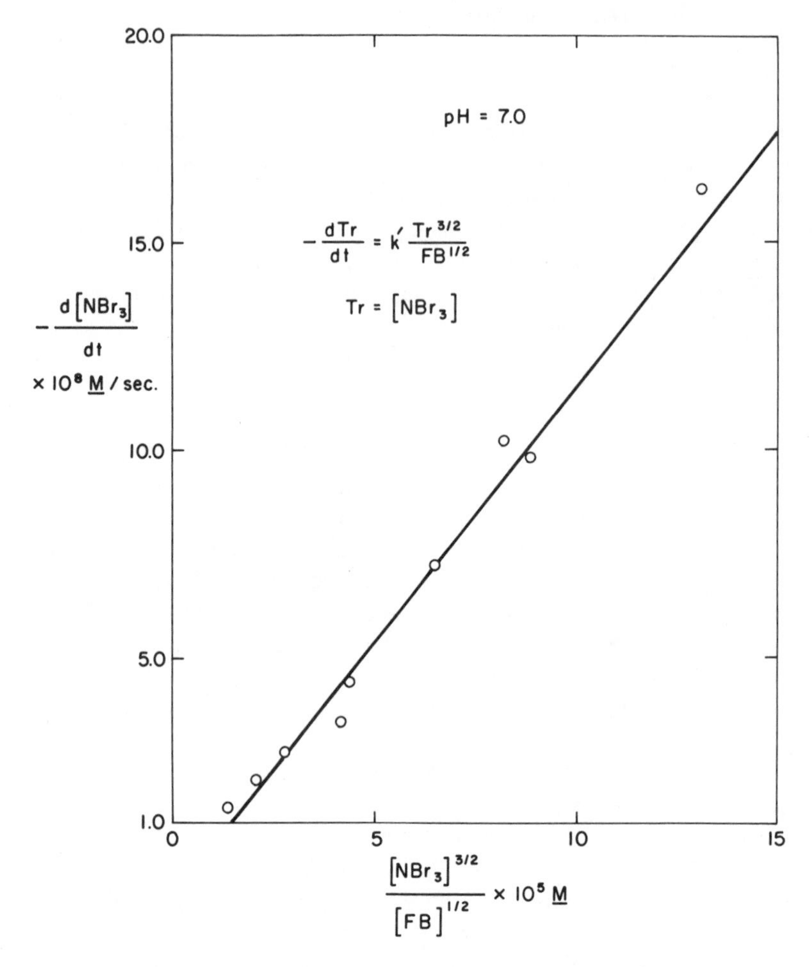

Figure 15-14. Verification of instantaneous rate, $[-d(NBr_3)/dt]$ versus $[(NBr_3)^{3/2}/(FB)^{1/2}]$ for pH 7 data.

Taking logarithms of both sides of the equation yields:

$$\ln\left(\frac{-dTr}{dt}\right) = \ln\left(\frac{Tr^{3/2}}{[H^+]\ FB^{1/2}}\right) + \ln k \qquad (18)$$

A plot was constructed, as shown in Figure 15-15, of $[-\ln(-dTr/dt)]$ versus $[\ln(Tr^{3/2}/[H^+]\ FB^{1/2}]$. In this case the slope should be a minus one (the negative log of initial rate having been plotted) and the intercept equal to

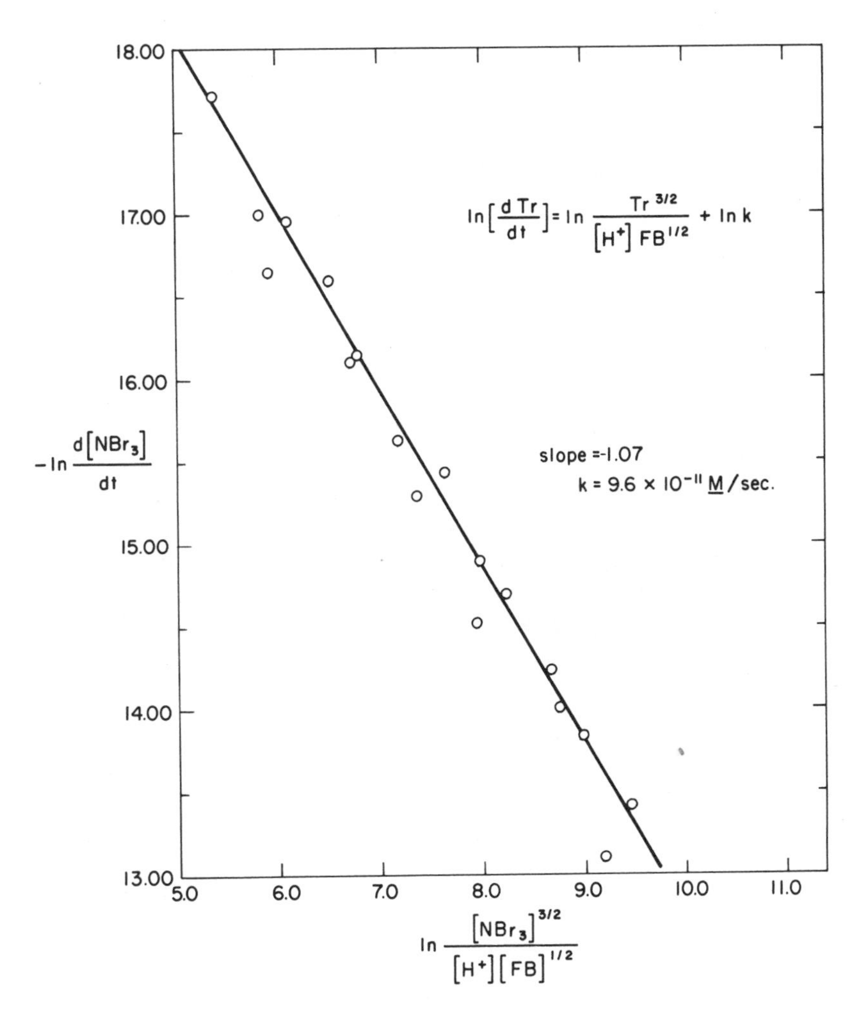

Figure 15-15. Van't Hoff plot of -ln (dTr/dt) versus $\ln([NBr_3]^{3/2}/[H^+][FB]^{1/2})$ for pH 7 and pH 8 data. Temperature = 20°C.

-ln k. The slope calculated from the logarithmic plot is -1.07 and the rate constant assumes a value of 9.6 x 10^{-11} *M*/sec. The agreement of experimental slope with the expected theoretical value indicates that the decomposition of tribromamine very closely follows a rate law such as proposed in Equation 18.

Development of an Integrated Rate Equation

For Constant Free Bromine (FB)

If this differential rate Equation 17 is integrated with respect to tribromamine for the conditions of constant free bromine and pH, the resultant integrated equation is:

$$\frac{1}{Tr^{1/2}} = \frac{kt}{2[H^+]\ FB^{1/2}} + \frac{1}{(Tr^{\circ})^{1/2}} \tag{19}$$

This integrated equation may then be compared with our kinetic results obtained over relatively long time periods.

To test this form of the rate equation, the data obtained at pH 7 was used on initial molar ratio of ammonia to bromine of 1/12, where free bromine varies only about 6% and remains in large excess throughout the measured reaction period of 60 minutes. The resulting plot is shown in Figure 15-16, with $(Tr)^{-1/2}$ plotted against time. The points form a straight line whose slope may be used to extract a k value of 9.8×10^{-11} *M*/sec, which is excellent agreement with the rate constant of 9.6×10^{-11} *M*/sec obtained from the differential analysis.

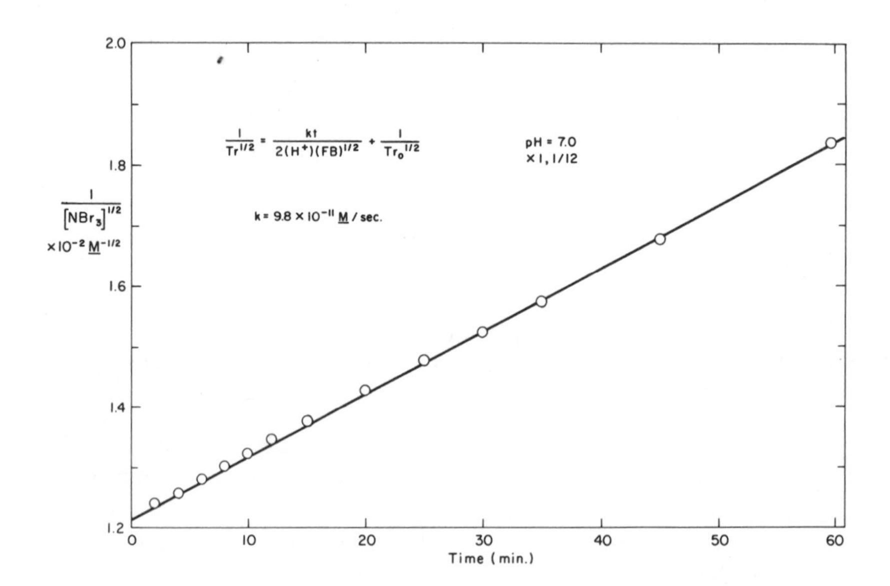

Figure 15-16. Verification of integrated rate equation for the condition of constant free bromine concentration. Temperature = 20°C.

For Variable Free Bromine

An integrated form of the rate equation, which is applicable at other initial molar ratios less than or equal to 1/3, can be obtained if FB is expressed in terms of Tr. This is done by making use of the observed decomposition stoichiometry in which three moles of free bromine (HOBr) are produced whenever two moles of tribromamine have decomposed. An expression for FB is:

$$FB = TB^{\circ} - 3Tr^{\circ} + 3/2\,(Tr^{\circ} - Tr) \tag{20}$$

where TB° and Tr° are the initial total bromine (*i.e.*, free and combined) and tribromamine concentrations respectively. The $-3Tr^{\circ}$ term accounts for the bromine which reacts with ammonia to form NBr_3, and the last term is equal to the free bromine produced during decomposition.

Substituting Equation 20 into Equation 18:

$$-\frac{1}{Tr}\left[\frac{TB^{\circ} - 3/2\,Tr^{\circ} - 3/2\,Tr}{Tr}\right]^{1/2} dTr = \frac{k}{[H^+]}\,dt \tag{21}$$

which may be integrated at constant pH to give:

$$\frac{2\,(b\,Tr - 3/2\,Tr^2)^{1/2}}{Tr} + (3/2)^{1/2}\sin^{-1}\left[\frac{3Tr}{b} - 1\right]_{Tr^{\circ}}^{Tr} = \frac{kt}{[H^+]} \tag{22}$$

where $b = TB^{\circ} - 3/2\,Tr^{\circ}$. If the limits of integration are taken as $Tr^{\circ}/2$ and $t_{1/2}$ an equation for the half-life of tribromamine is developed, this being:

$$t_{1/2} = \frac{[H^+]}{k}\left[4\left(\frac{1}{2R} - \frac{9}{8}\right)^{1/2} - 2\left(\frac{1}{R} - 3\right)^{1/2} + (3/2)^{1/2}\left[\sin^{-1}\left(\frac{6R-2}{2-3R}\right) - \sin^{-1}\left(\frac{9R-2}{2-3R}\right)\right]\right] \tag{23}$$

where $R = Tr^{\circ}/TB^{\circ}$, *i.e.*, the initial molar ratio of ammonia to bromine. Equations 22 and 23 can now be used to test the pH 7 and pH 8 data. Table 15-5 gives the correlation coefficient and computed value of k for a linear regression line of the left-hand side of Equation 29 (labeled B) plotted against time. Also shown are the rate constant and correlation coefficients for the 1.5-order approximation, good only when free bromine is relatively constant over the course of a run.

For the pH 7 runs the average value for the rate constant is 8.4×10^{-11} *M*/sec, a deviation of 14% from the van't Hoff determined value of 9.6 x

Table 15-5

pH	Ratio	Power	B Model		1.5-Order Model	
			Correlation Coefficient	Rate Constant $k \times 10^{+11}$/sec	Correlation Coefficient	Rate Constant $k \times 10^{+11}$/sec
7	1/12	X4	0.999	9.65	0.999	9.40
7	1/12	X2	0.996	7.77	0.996	7.64
7	1/12	X1	1.000	8.45	1.000	8.31
7	1/6	X4	0.998	9.48	0.997	8.82
7	1/6	X2	0.997	7.94	0.996	7.73
7	1/6	X1	0.995	8.53	0.994	8.19
7	1/3	X4	0.989	8.47	0.983	7.64
7	1/3	X1	0.991	8.81	0.985	8.26
8	1/12	X4	0.988	5.21	0.987	5.21
8	1/12	X2	0.987	5.42	0.986	5.47
8	1/12	X1	0.992	5.13	0.991	5.13
8	1/6	X4	0.989	3.92	0.988	3.89
8	1/6	X2	0.987	4.63	0.985	4.62
8	1/6	X1	0.983	4.57	0.980	4.61
8	1/3	X4	0.988	3.63	0.982	3.63
8	1/3	X2	0.975	3.48	0.967	3.47
8	1/3	X1	0.981	3.59	0.974	3.52

10^{-11} *M*/sec, with 8 of the 9 values lower than this figure. The correlation coefficients are in the range 0.989-1.000. However, when the data is viewed graphically, all the figures, both at pH 7 and pH 8, show a convex deviation from linearity.

It was thought that this deviation might be explained by either an inhibiting effect of product formation, or by competing mechanisms, with a different mechanism being the favored one at low concentrations of tribromamine. If our previously determined rate equation is correct for the start of the decomposition, then straight lines drawn through the first few points should have slopes closer to 9.6×10^{-11} *M*/sec/$[H^+]$. Such lines were drawn for all X1 runs at pH 7 and pH 8. Table 15-6 lists the rate constants so calculated.

Table 15-6

pH	Ratio	k x 10^{+11} *M*/sec
7	1/12	10.1
7	1/6	11.0
7	1/3	11.2
8	1/12	8.8
8	1/6	9.7
8	1/2	7.9

The improvement in the pH 8 rate constants is obvious. The average value from these 6 straight lines is 9.8×10^{-11} *M*/sec, however the range is rather large at 34%.

While all runs showed similar convex graphs, the degree of deviation was seen to be directly related to the initial ammonia to bromine molar ratio with smaller ratios (*i.e.*, 1/12) giving less deviation. Also related to the degree of convexity was the hydrogen ion concentration with pH 8 plots significantly more convex than pH 7 plots.

Half-Life Analysis

Since drawing straight lines through the initial few points gave results more in accord with our empirical rate equation, we hypothesized that during the period of one half-life, the integrated form of our rate equation may be valid. By using Equation 23 we were able to calculate the theoretical half-lives of our experimental ammonia to bromine ratios. We then compared these theoretical half-lives with experimental values. Table 15-7 summarizes our findings. For all the 1/12 and 1/6 ratios there is excellent agreement

Table 15-7

pH	Ratio	Power	Theoretical First Half-Life (min)	Experimental First Half-Life (min)	Per Cent Deviation	Theoretical Second Half-Life	Experimental Second Half-Life
7	1/12	X4	45.0	45.0	0		
7	1/12	X2	45.0	51.8	15		
7	1/12	X1	45.0	51.1	14		
7	1/6	X4	28.0	25.0	11		
7	1/6	X2	28.0	28.9	3		
7	1/6	X1	28.0	24.2	14		
7	1/3	X4	9.1	4.0	56	22	20.1
7	1/3	X2	9.1	5.0	45	20	26
7	1/3	X1	9.1	3.4	63	22	21.9
8	1/12	X4	4.5	4.8	6		
8	1/12	X2	4.5	4.5	0		
8	1/12	X1	4.5	6.0	33		
8	1/6	X4	2.8	3.0	7		
8	1/6	X2	2.8	2.6	7		
8	1/6	X1	2.8	2.5	11		
8	1/3	X4	0.91	0.63	30	2.3	4.5
8	1/3	X2	0.91	0.75	18	2.1	3.2
8	1/3	X1	0.91	0.64	31	1.4	3.2

at both pH 7 and pH 8 with deviations from theoretical half-lives being randomly scattered on both high and low sides and 11 of 12 results falling within 15% of the calculated half lives.

The 1/3 ratio half-lives are all much lower than what our integrated rate equation predicted. However, for the pH 7 data the second half-life gives a much closer comparison (last two columns on Table 15-7). This is not true for the pH 8 data, due probably to the severity of the curvature of the integrated plot at this high pH and low ammonia to bromine ratio.

One possible explanation for the failure of the integrated rate equation to predict the first half-life of the 1/3 ratios might be incomplete formation of tribromamine at this ratio. This is the exact stoichiometry for NBr_3 formation, so only a slight error in bromine concentration on the low side would give significant concentrations of dibromamine which undergoes very rapid decomposition. When the second half life is calculated, the ratio of tribromamine to total bromine (R) is more of the order of 1/5. Thus the presence of free bromine stabilizes the system and prevents any rapid initial formation and decomposition of dibromamine.

One other point that can be made regarding the empirical rate law concerns the term $(Tr/FB)^{1/2}$. As noted in the introduction, the bromamines appear to establish instantaneous equilibrium. Thus the reversible reaction

$$NBr_3 + H_2O = NHBr_2 + HOBr \qquad (24)$$

can be expressed in terms of the equilibrium constant shown below:

$$Keq = \frac{[NHBr_2]\ [HOBr]}{[NBr_3]} = \frac{Di\ FB}{Tr} \qquad (25)$$

The term $(Tr/FB)^{1/2}$ then equals $(Di/Keq)^{1/2}$ and the differential rate equation becomes:

$$\frac{dTr}{dt} = \frac{k}{Keq^{1/2}}\ Tr\ Di^{1/2}\ [OH^-] \qquad (26)$$

The decomposition of tribromamine through some mechanism involving dibromamine has been postulated previously by both Galal-Gorchev[17] and Hsu.[20] Galal-Gorchev proposed a decomposition directly through dibromamine, whereas Hsu hypothesized a complex mechanism involving a second-order term in tribromamine and first-order term in dibromamine.

The half-life comparison suggests that the integrated rate equation is an excellent description of our experimental system at neutral or slightly alkaline pH over a time interval of at least one half life.

The integrated rate equation, together with the ammonia concentration of a given water would allow fairly simple calculation of approximate tribromamine half-lives for different breakpoint dosages. This half-life data when coupled with bactericidal and virucidal data for free bromine and tribromamine, and some knowledge of the degree of pathogenic contamination of the water would facilitate the choice of a particular dosing concentration.

SUMMARY

Tribromamine, the major species of combined residual bromine present past the breakpoint, decomposes according to the stoichiometry:

$$2NBr_3 + 3H_2O = N_2 + 3HOBr + 3Br^- + 3H^+$$

in the pH 7-pH 8 range.

Bromide ion concentration, hydrogen ion concentration, tribromamine concentration and free residual bromine concentration, all affect the decomposition rate. A slight catalytic effect of bromide ion was found. Experiments performed to determine the decomposition kinetics of tribromamine under conditions of reduced bromide ion concentration ($<10^{-4} M$) yielded the following differential rate equation, determined by van't Hoff analysis:

$$\frac{-dTr}{dt} = \frac{k\, Tr^{3/2}}{[H^+]\, FB^{1/2}}$$

where Tr is the concentration of tribromamine and FB is the concentration of total free bromine residual. The value of the rate constant was calculated to be 9.6-9.8 x 10^{-11} *M*/sec. An integrated form of this equation, derived with the aid of the aforementioned pH 7-8 reaction stoichiometry, closely predicted the experimental rate and first half-life time (within a 15% range) for a wide range of reactant concentrations, but failed to predict accurately the tribromamine concentration at times beyond the first half-life.

ACKNOWLEDGMENT

This work was supported by the U.S. Army Medical Research and Development Command.

REFERENCES

1. Farkas-Himsley, H. "Disinfection," In *Bromine and Its Compounds,* Z. E. Jolles, Ed. (New York: Academic Press, 1966).
2. Chapin, R. M. "Dichloro-Amine," *J. Amer. Chem. Soc.* **51**, 2112 (1929).
3. Chapin, R. M. "The Influence of pH upon the Formation and Decomposition of the Chloro Derivatives of Ammonia," *J. Amer. Chem. Soc.* **53**, 912 (1931).

4. Palin, A. T. "A Study of the Chloro Derivatives of Ammonia and Related Compounds, with Special Reference to their Formation in the Chlorination of Natural and Polluted Waters," *Water Water Eng.* **54**, 151 (1950).
5. Palin, A. T. "The Estimation of Free Chlorine and Chloramine in Water," *J. Inst. Water Eng.* **3**, 100 (1949).
6. Weil, Ira and J. C. Morris. "Kinetics Studies on the Chloramines. I. The Rates of Formation of Monochloramine, N-Chloromethylamine, and N-Chlorodimethylamine," *J. Amer. Chem. Soc.* **71**, 1669 (1949).
7. Wei, I. W. "Chlorine-Ammonia Breakpoint Reactions: Kinetics and Mechanism," PhD Thesis, Harvard University, Cambridge, Mass. (1972).
8. Yaron, F. "Bromine Manufacture: Technology and Economic Aspects," In *Bromine and Its Compounds*, Z. E. Jolles, Ed. (New York: Academic Press, 1966).
9. Coleman, George H., H. Soroos and C. B. Yager. "The Reaction of Monobromamine with Grignard Reagents," *J. Amer. Chem. Soc.* **55**, 2075 (1933).
10. Coleman, George H., C. B. Yager and H. Soroos. "The Preparation of Dibromamine and Its Reaction with Grignard Reagents," *J. Amer. Chem. Soc.* **56**, 965 (1934).
11. Johannesson, J. K. "Bromamines. Part I. Mono and Dibromamine," *J. Chem. Soc. (London)* 2998 (1959).
12. Johannesson, J. K. "Anomalous Bactericidal Action of Bromamines," *Nature* **181**, 1799 (1958).
13. Galal-Gorchev, H. A. and J. C. Morris. "Formation and Stability of Bromamide, Bromimide, and Nitrogen Tribromide in Aqueous Solution," *Inorg. Chem.* **4**, 899 (1965).
14. Johannesson, J. K. "The Bromination of Swimming Pools," *Amer. J. Public Health* **50**, 1731 (1960).
15. Taylor, David G. "Kinetics of Viral Inactivation by Bromine," In *Chemistry of Water Supply, Treatment, and Distribution*, Alan J. Rubin, Ed. (Ann Arbor, Mich.: Ann Arbor Science Publishers, Inc., 1974).
16. Kruse, C. W. "Final Technical Progress Report to the Armed Forces Epidemiological Board," Commission on Environmental Hygiene, U.S. Army Medical Research and Development Command (1969).
17. Galal-Gorchev, H. A. "Bromamides: Their Formation and Occurrence in Aqueous Solution," PhD Thesis, Radcliffe University, Cambridge, Mass. (1961).
18. Johnson, J. D. and R. Overby. "Bromine and Bromamine Disinfection Chemistry," *J. San. Eng. Div. ASCE* **(SA5)**, 617 (1971).
19. Kolthoff, I. M. and E. B. Sandell, Eds. *Textbook of Quantitative Inorganic Analysis* (New York: Macmillan, Inc., 1946).
20. Morris, J. C. "Kinetics of Reactions Between Aqueous Chlorine and Nitrogen Compounds," In *Water Chemistry*, S. J. Faust and J. V. Hunter, Eds. (New York: McGraw-Hill Book Co., 1967).
21. Johnson, J. D., Y. Hsu and K. Lui. "Bromine Residual Chemistry," Annual Progress Report to the U.S. Army Research and Development Command (1972).

22. Johnson, J. D., L. Ellmore and G. Inman. Dept. of Environmental Sciences and Engineering, University of North Carolina, Unpublished data (1974).
23. Larson, T. E. and F. W. Sollo. "Determination of Free Bromine in Water," Final Technical Report to the Commission on Environmental Health of the Armed Forces Epidemiological Board. U.S. Army Medical Research and Development Command, 1970.
24. Wyss, O. and J. R. Stockton. "The Germicidal Action of Bromine," *Arch. Biochem.* **12**, 267 (1947).
25. American Public Health Association. *Standard Methods for the Examination of Water and Wastewater,* 13th ed. (New York, 1971).
26. Farkas, L., M. Lewin and R. Black. "The Reaction Between Hypochlorite and Bromides," *J. Amer. Chem. Soc.* **71**, 1987 (1949).
27. Lewin, M. and M. Avrahami. "The Decomposition of Hypochlorite-Hypobromite Mixtures in the pH Range 7-10," *J. Amer. Chem. Soc.* **77**, 4491 (1955).
28. Kolthoff, I. M. and N. H. Furman. *Potentiometric Titrations: A Theoretical and Practical Treatise* (New York: John Wiley & Sons, Inc., 1926).

CHAPTER 16

DECHLORINATION BY ACTIVATED CARBON AND OTHER REDUCING AGENTS

Vernon L. Snoeyink and Makram T. Suidan

Department of Civil Engineering
University of Illinois at Urbana-Champaign
Urbana, Illinois

INTRODUCTION

Recently more interest is being focused on dechlorination because of increased use of chlorination in wastewater treatment. A summary of the state standards for wastewater treatment shows that four states specifically require chlorination while many more require disinfection of effluents.[1] In the latter case, chlorination generally has been selected as the means of disinfection. A recently completed survey of 20 wastewater treatment plants in Illinois, a state that requires disinfection, has shown that typically 1 to 5 mg/l total chlorine residual is present in the treated effluent.[2] Many of these plants are discharging to streams that provide very little dilution. The same study showed that monochloramine, the predominant residual chlorine species in the effluents, breaks down much more slowly under the influence of ultraviolet radiation than do the other species. Under certain conditions a period of days may be required for essentially complete breakdown.

Chlorine residuals also result when the breakpoint chlorination process is used to remove ammonia from wastewater. Pressley *et al.*[3] showed that chlorine can be used to efficiently destroy ammonia; however, even after two hours of contact near the optimum pH and chlorine dose for this process, chlorine residuals near 1 to 3 mg/l remained. For conditions other than optimum, larger residuals occur.

Studies by Esvelt *et al.*[4] have shown that much of the toxicity in chlorinated waste treatment plant effluents can be attributed directly to chlorine residual. Brungs[5] made an intensive survey of the available studies of chlorine residual toxicity to fish and on the basis of this study, he suggested interim criteria for chlorine residuals in receiving waters of 0.01 mg/l where the more resistant species of fish are to be protected, and 0.002 mg/l where the more sensitive species, such as trout and salmon, are to be protected. A comparable study by Basch and Truchan[6] resulted in similar recommended criteria.

It is apparent from the studies of Basch and Truchan,[6] Brungs[5] and Snoeyink and Markus[2] that if chlorination is the process selected for disinfection, dechlorination will be necessary in many instances to protect fish life in the receiving water.

Dechlorination might also be necessary or desirable in the treatment of industrial or potable water supplies. According to White,[7] dechlorination might be required before demineralizers, for boiler makeup water, certain food plant operations or in the beverage industry. A process to remove excess chlorine might also be required when superchlorination is used to destroy odor-causing compounds.

This chapter presents a detailed review of what is known about dechlorination by activated carbon and compares this process with alternative means of dechlorination. Proper understanding of the various processes depends on knowledge of the aqueous chemistry of chlorine, particularly when ammonia is present (see Chapter 1 for a review of this subject). In the present chapter, dechlorination by sulfur compounds and other reducing agents is first reviewed then followed by an extensive discussion of the reactions of free and combined chlorine residual with activated carbon, and of the procedures used in the design of carbon dechlorination beds. In the latter discussion, emphasis is placed on the use of carbon for ammonia removal and the adsorption of organic compounds in conjunction with dechlorination. Because of the infrequent use of this process to date, much remains to be learned about the process and about how it can be designed and operated to best perform these various functions.

DECHLORINATION WITH SULFUR COMPOUNDS

The most common procedure to eliminate or reduce chlorine residual involves the use of sulfur compounds with sulfur in the +IV oxidation state. Sulfur dioxide is the most popular among the S(+IV) species, the major reason for which appears to be the cost of using it. Dean[9] has estimated that the chlorination cost of a secondary effluent for disinfection followed by dechlorination with SO_2 will be on the order of 1.2 to 1.3 times as costly as chlorination alone.

Sulfur dioxide is generally purchased as a liquid which is then converted to a gas in preparation for adding it to water. The gas dissolves readily in water forming sulfurous acid

$$SO_2 + H_2O \rightleftharpoons H_2SO_3 \quad (1)$$

The H_2SO_3 partially ionizes to give HSO_3^- and $SO_3^=$, with the relative concentrations of these species being dependent on pH. Sodium bisulfite ($NaHSO_3$) and sodium sulfite (Na_2SO_3) are salts that are also used as a source of S(+IV) for dechlorination, but these are generally more expensive and less stable than SO_2.[10] A major advantage in using SO_2 is that the same equipment is used for feeding it as for dosing chlorine.[7] This similarity serves to cut down on equipment variability in a plant, thus reducing operational difficulties.

According to White,[7] hypochlorous acid and monochloramine, respectively, react with SO_2 as follows,

$$SO_2 + H_2O + HOCl \rightarrow 3\,H^+ + SO_4^= + Cl^- \quad (2)$$

$$SO_2 + 2H_2O + NH_2Cl \rightarrow NH_4^+ + 2\,H^+ + SO_4^= + Cl^- \quad (3)$$

with similar reactions being applicable when the other S(+IV) species are used and when the other chlorine residual species are reduced. Using Equation (2) as a basis for calculation, it can be shown that 0.9 mg of SO_2 is required per mg of chlorine as Cl_2 reduced. Similarly, 2.1 mg/l of alkalinity as $CaCO_3$ is required to react with the H^+ produced by the reaction. If the H^+ resulting from hydrolysis of gaseous Cl_2 when it is added to the water is included, a total of 2.8 mg of alkalinity are required per mg of chlorine. It is important to note from Equation (3) that S(+IV) cannot be used for ammonia removal because the chloramino-N is reconverted to ammonia. Also, there is some evidence that the required S(+IV) dose is a function of the composition of the water.[11]

The kinetics of S(+IV) dechlorination are very fast. White[7] states that it reacts nearly instantaneously with free chlorine while monochloramine, dichloramine, nitrogen trichloride and poly-N-chlor compounds each react somewhat more slowly than the preceding compound, but the total reaction time of all compounds constituting measurable chlorine residual is never more than a few minutes. Because of the very rapid reaction kinetics, mixing is the most important parameter to be considered when S(+IV) compounds are to be used for dechlorination.

The reaction between S(+IV) and oxygen

$$O_2 + 2SO_2 + 2H_2O \rightarrow 4H^+ + 2SO_4^= \quad (4)$$

is important and merits careful consideration. The extent to which it proceeds will increase the S(+IV) dose, but, more importantly, depletion of dissolved oxygen may necessitate a costly reaeration step if dechlorination is being applied prior to discharge of a treated wastewater to receiving waters. It also indicates the importance of avoiding any overdose of S(+IV), which may be difficult in certain situations. This reaction is commonly used in the study of gas transfer.[12]

This reaction is catalyzed and inhibited by several elements and compounds commonly present in water. Traces of Cu, Co, Fe, Ce, Mn and O_3 increase the rate of reaction while ethyl alcohol, glycerol and mannitol have been found to decrease it.[13] Westerterp *et al.* state that at S(+IV) concentrations below 0.02*M*, the conversion rate is first-order with respect to S(+IV) concentration and zero-order with respect to oxygen concentration.[13] Reinders *et al.*[14] reported experimental data that were used to calculate a rate constant on the order of 0.005 sec^{-1} for a system in which no catalysts or inhibitors were present; the pH was 7.35, and the S(+IV) concentration 0.01*M*. The rate of reaction increased markedly with increasing pH. For concentrations of S(+IV) above 0.01*M*, the reaction was first-order with respect to oxygen concentration and independent of the S(+IV) concentration.[13] These same authors found that Co had a very significant catalytic effect, although they did not determine rate constants applicable to concentration levels of importance in dechlorination.

If the chemical composition of the water being dechlorinated is such that a significant amount of oxygen reacts with the S(+IV) during the dechlorination step, additional S(+IV) will be required, and reaeration of the water subsequent to dechlorination will be necessary. However, the reaction would have to be catalyzed to proceed faster than reported by Reinders *et al.*[14] if this were to be the case. This reaction should be studied to determine the effect of various constituents and if, under any circumstances, the reaction does proceed at a significant rate.

Sodium thiosulfate ($Na_2S_2O_3$) can also be used in dechlorination, but reacts more slowly than the S(+IV) compounds and can contribute to sensory effects involved with taste and odor.[15] It is usually employed for dechlorinating water samples prior to bacteriological analysis.

The stoichiometry of the reaction between thiosulfate and residual chlorine is not well defined. Reactions are given in which $S_4O_6^=$,[8,16,17] $SO_4^=$, and S and $SO_4^=$[18] are the end products of thiosulfate oxidation with chlorine

$$2\ Na_2S_2O_3 + H^+ + HOCl \rightarrow S_4O_6^= + 4\ Na^+ + Cl^- + H_2O \qquad (6)$$

$$Na_2S_2O_3 + 4\ HOCl + H_2O \rightarrow 6\ H^+ + 2\ SO_4^= + 4\ Cl^- + 2\ Na^+ \qquad (7)$$

$$Na_2S_2O_3 + HOCl \rightarrow 2\ Na^+ + SO_4^{=} + S + H^+ + Cl^- \quad (7)$$

Other reactions in which the ratio of the amount of S to $SO_4^=$ formed is different than given in Equation (7) could also be written. On the basis of Equation (5), 44.45 parts by weight $Na_2S_2O_3$ are required for each part of free chlorine as Cl_2. No experimental data justifying this reaction could be found, however. Alterman found the experimentally determined weight ratio of $Na_2S_2O_3$ to Cl_2 to vary from 1.9 at pH less than 6.2 to 1.02 at pH 9, and to 0.56 at very alkaline pH.[18] The weight ratio of 0.56 is that which is expected if $SO_4^=$ is the only sulfur end product in accordance with Equation (6). If S and $SO_4^=$ are formed as given by Equation (7), a weight ratio of 2.2 is expected, which is slightly higher than that observed by Alterman for low pH. This could be explained if higher proportions of $SO_4^=$ to S were formed than are given in Equation (7), but it should be noted that the endproducts of the reaction were not measured by Alterman and that alternative explanations might be possible.

It is apparent that more remains to be learned about chlorine-thiosulfate reactions if thiosulfate is to be used extensively. Thiosulfate is present in some industrial wastes, such as those from the paper and pulp industry, and when such a source is readily available, its use could become more attractive.

DECHLORINATION WITH HYDROGEN PEROXIDE

Hydrogen peroxide has not been extensively used for dechlorination although it may have some potential in this respect. According to Mischenko *et al.*,[19] the following reaction will take place

$$HOCl + H_2O_2 \rightarrow O_2\ (aq) + H^+ + Cl^- + H_2O \quad (8)$$

This reaction is not thermodynamically limited (ΔG° = -36.3 kcal at 25°C), but the kinetics of the reaction with both free and combined chlorine have not been well defined. According to Equation (8), each mg of chlorine as Cl_2 would require 0.48 mg of H_2O_2. The current cost of H_2O_2 is approximately 49.3¢/lb,* plus freight charges. The chemical cost only to remove 1 mg/l of Cl_2 would be 0.4¢/1000 gal.

A major advantage in using H_2O_2 for removal of chlorine from treated wastewater is that one of the endproducts of the reaction is O_2 [1 mg of H_2O_2 will yield 0.94 mg of O_2 per Equation (8)]. The use of H_2O_2 could

*McKesson Chemical Co., Decatur, Illinois. The 49.3¢/lb for pure H_2O_2 is based upon the cost for 48-500-lb drums of technical grade, 35% pure H_2O_2, of 17.25¢/lb. The cost of the chemical increases as the size of the order decreases. One 500-lb drum, for example, costs 70% more on a per-pound basis than the 48-500-lb drums.

make a reaeration step prior to discharge of the wastewater unnecessary, which is a very important consideration in view of the cost of reaeration. Some additional disinfection could also result. Potential problems of its use include the possibility that it will be reduced by organic matter. If this reaction were to proceed rapidly, relative to the rate at which it reduces chlorine, excessively high doses would be required making its application too costly. There is some indication, though, that the rate of reaction between H_2O_2 and chlorine residual is too slow for the reaction to be useful.[20] In view of the lack of information about the use of this compound for dechlorination, it is obvious that much more remains to be learned about it to assess fully the feasibility of using it.

DECHLORINATION WITH AMMONIA AND FERROUS SULFATE

According to White[7] as well as others, ammonia and ferrous sulfate can be used as dechlorinating agents. Ammonia can be used to eliminate free chlorine via the breakpoint reaction,

$$2\ NH_3 + 3\ HOCl \rightarrow N_2 + 3\ H^+ + 3\ Cl^- + 3\ H_2O \qquad (9)$$

or it can be used to convert free chlorine to combined chlorine. A very long reaction time, approximately 20 minutes in the pH range 7 to 7.5, must be allotted for the breakpoint reaction, however.[7] Ammonia may be used in this manner in water treatment plants to eliminate the troublesome NCl_3 sometimes present with free chlorine.

Ferrous ion (Fe^{+2}) can also be used as a dechlorinating agent, but its use is limited to situations where dechlorination is followed by a solids removal step. When oxidized by chlorine, Fe^{+2} is converted to Fe^{+3} which is very insoluble and which serves as a good coagulant. Sedimentation and/or filtration would be required to remove it from the water. According to White,[7] Fe^{+2} reacts readily with free chlorine and monochloramine, but much more slowly with dichloramine.

DECHLORINATION WITH ACTIVATED CARBON

Free Chlorine-Activated Carbon Reactions

Activated carbon has been used for dechlorination for some time. The process was first installed in a municipal water treatment plant at Reading, England in 1910. It appears to have been used predominantly for the removal of free chlorine, although reference has been made to the fact that it will also remove combined chlorine.[21] Magee was the first to conduct an intensive study of this process, but despite the advancement which he and others following him have made, much remains to be learned about it.[21]

Investigators have studied the reaction between free chlorine and carbon and according to their findings,[21,22] it can be described in the following equation:

$$C^* + HOCl \rightarrow C^*O + H^+ + Cl^- \qquad (10)$$

where C* represents the activated carbon and C*O represents a surface oxide on the carbon. Magee showed that when free chlorine is initially contacted with activated carbon, there is an initial buildup of Cl-containing species on the carbon surface.[21] After a period of time, however, the chlorides produced in the reaction are stoichiometrically equal to the free chlorine removed from the aqueous solution in the reactor. Snoeyink *et al.*[23] also observed similar results.

The hydrogen ion produced in accordance with Equation (10) is important because it might necessitate a pH adjustment step subsequent to dechlorination. The problem of a pH decrease becomes more severe if chlorine gas is used as the source of chlorine. Chlorine gas hydrolyzes to produce H^+ when added to water. The stoichiometry of the reaction in Equation (10) requires further study with respect to production of H^+, however. Based on this equation, it is predicted that no H^+ will be produced if OCl^- reacts in place of HOCl. However, Olsen and Binning[24] and Snoeyink *et al.*[23] noticed a pH drop when OCl^- reacted, the reason for which was not determined. It could be attributable to the formation of a surface oxide in the form of a carboxyl group which subsequently ionizes to produce the H^+, however.[24]

The production of surface oxides via Equation (10) is important because of their effect on the dechlorination reaction as well as on the adsorption of organic compounds by the carbon. Magee[27] attributed the gradual reduction in efficiency of dechlorination of a carbon bed to the gradual poisoning of the carbon surface with these oxides. He also observed that these oxides were unstable to a certain degree because CO and CO_2 were released to the solution. That this occurs is fortunate because it probably results in extended life of the carbon for dechlorination.

Using the analytical procedure of Boehm,[25] Snoeyink *et al.*[23] studied the accumulation of acidic surface oxides as a function of the amount of free chlorine reacted. As the amount of chlorine reacted increases, the surface concentration of NaOH titratable oxides reached a plateau as can be seen in Figure 16-1. At the plateau value, approximately 15 mmoles free chlorine (1.06 g as Cl_2) have reacted per gram of carbon. Many of the titratable oxides were relatively volatile in nature and could be removed by drying at 105-110°C or, alternatively, by drying and outgassing, as shown

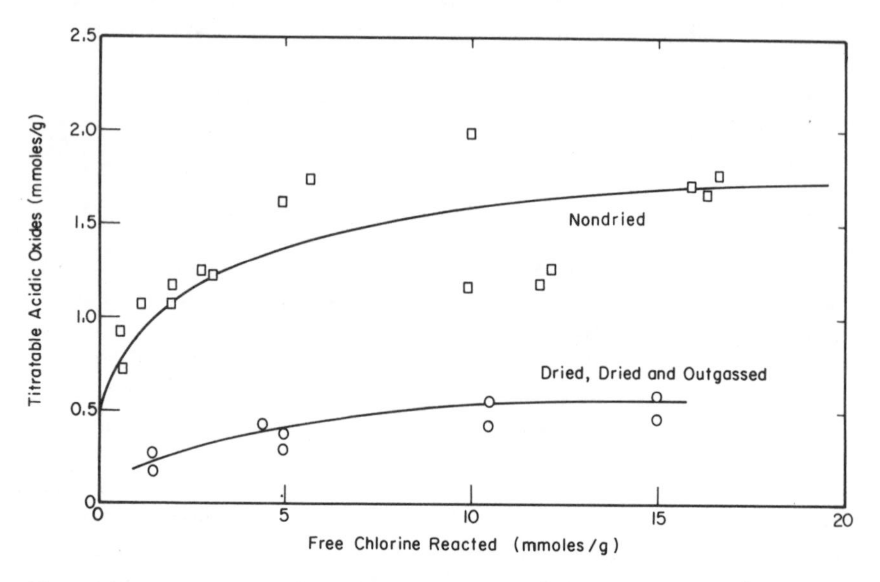

Figure 16-1. NaOH titratable surface oxides produced by the reaction of free chlorine with activated carbon (after Snoeyink *et al.*[23])

in Figure 16-1. The same study showed that the oxides removed by this procedure had little effect on adsorption of phenolic compounds.

Only a small fraction of the oxides produced via Equation (10) were titratable with NaOH. On the basis of the reaction stoichiometry, a one-to-one correspondence between oxides produced and chlorine reacted is expected. Such was not the case, however. Only 1.5 to 2 mmoles of titratable oxides were formed for each 15 mmoles reacted per gram; also, the increase in oxides was not linear but reached a plateau. Some of the oxides were probably evolved while others were not titratable with NaOH. It was also found possible to nearly double the concentration of NaOH titratable oxides by using three sequential treatments of the carbon with 15 mmoles chlorine per gram, with each treatment being followed by drying of the carbon.

The surface oxides which result from reaction with chlorine also affect adsorption of organic compounds. Figure 16-2 shows the decrease in capacity of carbon for *p*-nitrophenol as a function of the amount of chlorine reacted. Each treatment with chlorine was followed by drying of the carbon at 105-110°C prior to the adsorption studies. The value at 60 mmoles/g was obtained by using carbon that had been reacted with 15 mmoles chlorine per gram four times and dried after each treatment. The

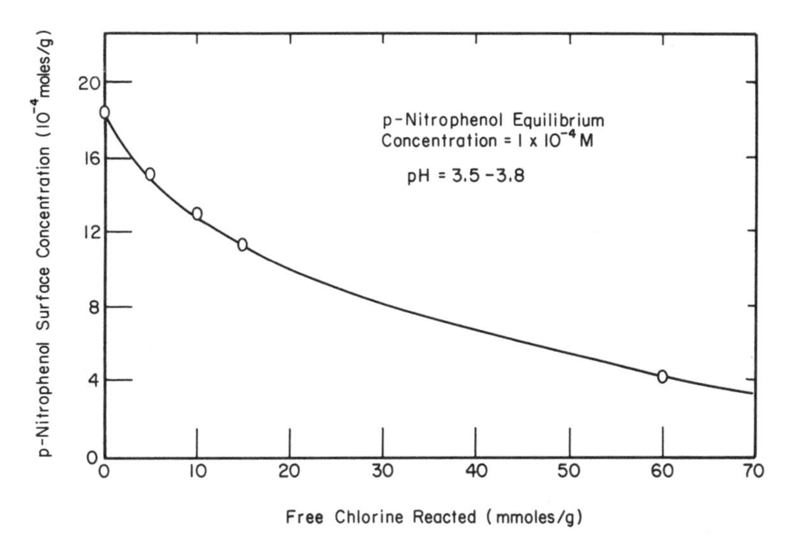

Figure 16-2. Reduction in *p*-nitrophenol adsorption capacity as a function of the amount of free chlorine reacted (after Snoeyink *et al.*[23]).

decrease in capacity could possibly be attributed to destruction of the carbonyl functional groups or to pore blockage by the surface oxides. These results are generally consistent with those of Coughlin *et al.*[26] The effect on adsorption of other types of organic compounds may well be different than was observed for *p*-nitrophenol, however, especially if the compound will react specifically with the surface oxides formed.

In batch systems, a distinct brown color was found in solution after reaction of approximately 15 mmoles chlorine per gram carbon (Snoeyink *et al.*[23]). Boehm[25] similarly found brown colloidal matter in suspension after extensive oxidation of carbon. Olsen and Binning[24] observed a similar material and noted that a certain portion of it was not adsorbable by carbon. Recent observations in the authors' laboratory showed that this color appeared in the effluent from a carbon dechlorination bed after 3 to 5 g free chlorine as Cl_2 had reacted per gram of carbon in the bed. It should be noted, however, that the amount of chlorine which must react before the color is produced is near the amount that reacts during the life of a typical dechlorination bed as given by Hagar and Flentje.[27] If alternative designs of the bed are used, the production of color by the carbon bed, rather than the appearance of chlorine in the effluent of the bed, might signal the need for regeneration or replacement of the carbon.

Dechlorination Bed Life and Regeneration

There is little quantitative information available on the life of dechlorination beds or on procedures that will allow the prediction of bed life under different conditions. One exception is the set of curves presented by Hagar and Flentje[27] as reproduced in Figure 16-3, although no information is

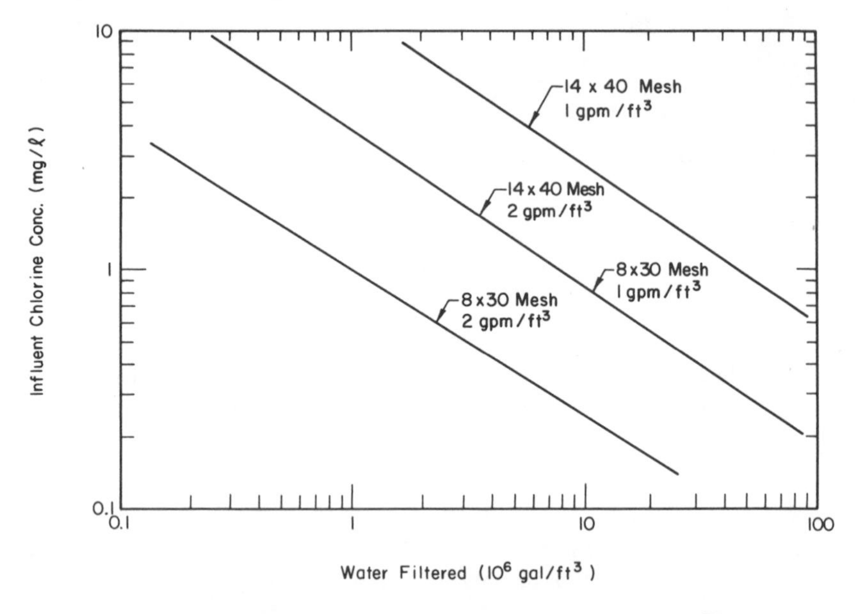

Figure 16-3. Dechlorination bed life (after Hagar and Flentje[27]); reprinted by permission of *J. Amer. Water Works Assoc.*

given on the procedures which were used to develop these curves. Using a concentration of 0.01 mg/l chlorine to define breakthrough, they found bed life to be a significant function of influent chlorine concentration, hydraulic loading rate, temperature, pH, and particle size of the activated carbon. Calculations made from these curves show that at an influent concentration of 1 mg/l of free chlorine, 1 g of 8 x 30 U.S. standard mesh carbon will react with 0.28 and 2.1 g chlorine when loaded at rates of 2 and 1 gpm/ft^3, respectively, while 1 g of 12 x 40 mesh carbon will react with 2.1 and 12.5 g chlorine when hydraulic loading rates of 2 and 1 gpm/ft^3 are used, respectively. If 2.1 g chlorine react per gram of carbon at the 1 gpm/ft^3 loading rate, a bed life of 12.5 years is predicted for a 2.5-ft bed depth. This period of time is very long, however, and it is expected that adsorption of organics, that would cause a decrease in dechlorination efficiency, would necessitate regeneration much before the 12.5 years.

Care must be exercised in using Figure 16-3. If Equation (10) correctly represents the reaction, only 5.9 g free chlorine can react per gram of carbon before each carbon atom has combined with an oxygen atom if activated carbon is assumed to be 100% carbon. But if each carbon atom can accept 2 oxygen atoms, then 17.5 g chlorine/g carbon represents the maximum. The type of oxide formed is important in this respect. If the reaction were to proceed to these limits, all the carbon would have been converted to CO or CO_2. Prior to complete conversion, however, the carbon particles would fragment with the smaller particles escaping from the bed and the dark color, as noted above, would also be produced. It is not apparent that Figure 16-3 takes these latter phenomena into account.

It is not apparent at the present time why the capacity of the dechlorination bed should be such a significant function of carbon particle size. Current evidence points toward the surface reaction as being the rate-limiting step[21] and unless surface area is vastly different for the two particle sizes, as is generally not the case, the differences observed by Hagar and Flentje[27] are not expected. Various aspects of this reaction are currently under study by the authors of this chapter.

One possible mechanism of reaction which could significantly affect bed life and which merits discussion concerns the catalytic destruction of HOCl by carbon as follows:

$$2\ HOCl \rightarrow 2\ H^+ + 2\ Cl^- + O_2 \qquad (11)$$

This reaction is thermodynamically favored as can be shown by free energy calculations. In our laboratory, preliminary monitoring of dissolved oxygen during dechlorination has shown no significant change in concentration, however, thus reducing the probability that this reaction is important. Magee[21] similarly found no increase in dissolved oxygen concentration during dechlorination.

The original activated carbon dechlorination efficiency can be regenerated by heating the carbon to temperatures in excess of 500-700°C.[21] Boehm[25] and Puri[22] noted that carbon must be heated in an inert atmosphere to a temperature exceeding 1000°C to evolve all oxides, thus indicating that elimination of all oxides is not necessary to reestablish the original dechlorination efficiency.

A large weight loss may be noted during regeneration. The very extensive surface area of carbon can hold a large number of oxides and the carbon evolved with the oxides can be significant. If Equation (10) is correct, for example, each part by weight of free chlorine requires 0.17 parts by weight of carbon. Because many of the oxides formed as products are attached to the carbon, a very significant portion of the weight loss will take place during the regeneration step.

Stuchlik[28] reports that carbon in pressure dechlorinators can be regenerated by steam treatment. The exhausted bed is first washed with a basic solution to remove the acid formed during the dechlorination reaction. Then 105-110°C steam, 1 atm gauge pressure, is introduced at the top of the bed. Several hours are required for the steam to heat the bed; steaming is continued for about one hour after steam first appears in the effluent of the bed. After steaming, the carbon is rinsed and returned to operation. Stuchlik indicates that regeneration treatments after the first regeneration are necessary after successively shorter time intervals and, after a time, the carbon will require replacement. He also notes that in gravity feed dechlorinators, some recovery of capacity can be achieved by washing with caustic solution.

Combined Chlorine-Activated Carbon Reactions

A study by Lawrence *et al.*[10] showed that chloramines were reduced by activated carbon and that a portion of the chloramino-N was not reconverted to NH_3. D'Agostaro[29] further examined the reaction products and found the principal nitrogen product other than ammonia to be N_2. No oxides of nitrogen were found with the exception of a trace of NO_3^-. Atkins *et al.*[30] also noted the oxidation of some of the chloramino-N at chlorine doses significantly less than the breakpoint dose when the chloramines were contacted with carbon. That such oxidation is possible is important because it indicates that carbon can be used to effect partial removal of NH_3 from water in addition to serving as a dechlorinating agent.

Bauer and Snoeyink studied the reaction of monochloramine with carbon in a pH 7.4 batch system in which monochloramine was the only chloramine species[31] (Figure 16-4). As can be observed, the carbon readily reduced the monochloramine concentration. Also, the nitrogen in the form of NH_3 and chloramines initially remained constant, thus indicating reconversion of NH_2Cl-N to NH_3. A reaction describing this observation is given in the following equation:[31]

$$NH_2Cl + H_2O + C^* \rightarrow NH_3 + H^+ + Cl^- + C^*O \qquad (12)$$

After approximately 1.42 mmoles NH_2Cl/g carbon had reacted, the chloramino-plus NH_3-N in the sample began to decrease. The following equation was proposed to account for this decrease.[31]

$$2\ NH_2Cl + C^*O \rightarrow N_2 + H_2O + 2\ Cl^- + C^* \qquad (13)$$

Similar tests were conducted with samples of carbon which had been reacted with 3, 7 and 10 mmoles of free chlorine per gram to place oxides on the surface in accordance with Equation (10). As NH_2Cl reacted with these

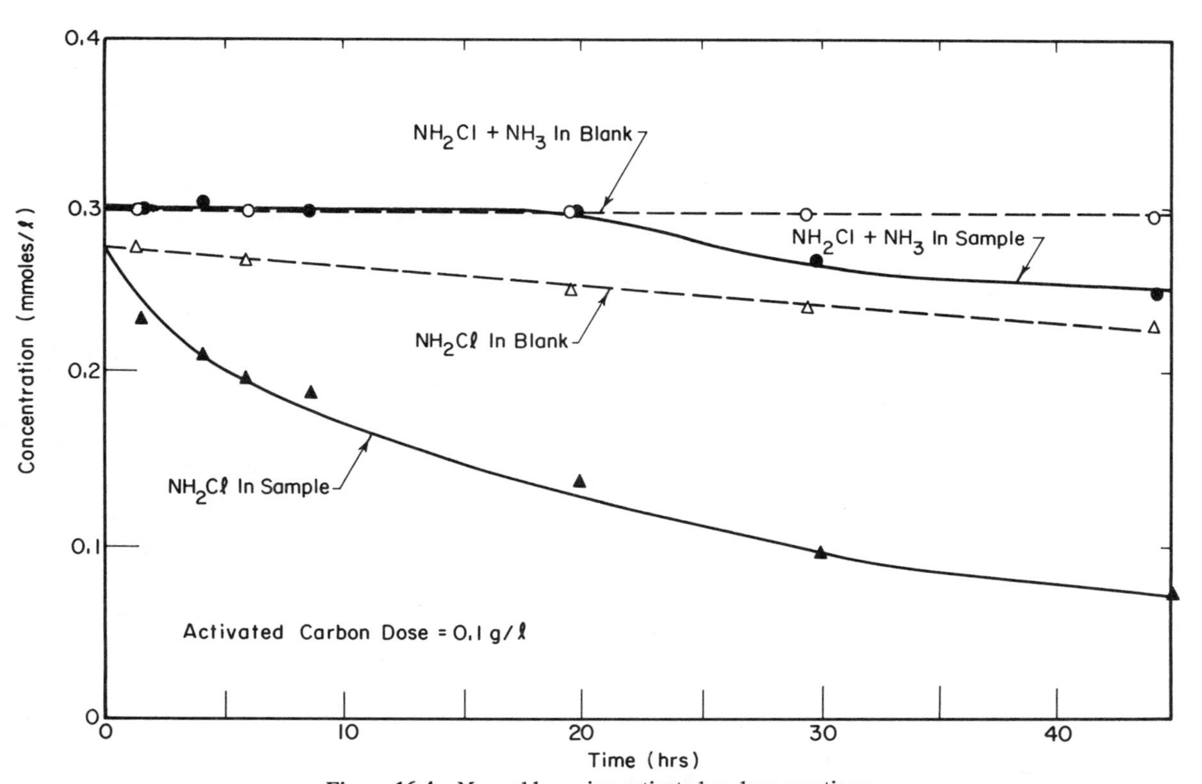

Figure 16-4. Monochloramine-activated carbon reactions (after Bauer and Snoeyink,[31] reprinted by permission of *J. Water Pollution Control Federation*).

carbons, immediate oxidation of NH_2Cl-N was observed in accordance with Equation (13), consistent with the hypothesis that some minimum level of surface oxides is required for NH_2Cl-N oxidation. The ratio of the moles of NH_2Cl removed to NH_2Cl-N oxidized varied from 1:1 to 3:1, however, and the reason for this variation is not apparent. Many aspects of these reactions require further study including the stoichiometry and the exact role of the surface oxides.

Other batch tests were performed at a pH of 4.3 where the predominant chloramine species was dichloramine, $NHCl_2$,[31] and the results are shown in Figure 16-5. As can be seen, the decrease in chloramino-plus NH_3-N paralleled the decrease in $NHCl_2$ concentration. All nitrogen originally in the form of $NHCl_2$ was converted to another oxidation state. Based on the findings of D'Agostaro[29] who used a similar system and found that N_2 was a predominant end product, the following equation was proposed,

$$2\ NHCl_2 + H_2O + C^* \rightarrow N_2 + 4\ H^+ + 4\ Cl^- + C^*O \qquad (14)$$

The fact that all eliminated $NHCl_2$-N was oxidized is of particular importance because of the implication that pH control during chlorination to cause $NHCl_2$ formation, and subsequent contact of the $NHCl_2$ with carbon can yield essentially any desired degree of NH_3 removal.

The reaction of NCl_3 with carbon has not been studied extensively but the work of Williams[32] and Atkins *et al.*[30] indicates that it is readily removed by carbon.

Atkins *et al.*[30] used a pilot plant to study ammonia removal by chlorination of wastewater followed by contact of the wastewater with activated carbon. Wastewater clarified with lime was filtered and passed through an activated carbon bed for removal of dissolved organic compounds before it entered the chlorine contact basin. Gaseous chlorine was fed at a controlled rate; a contact time of 15 minutes was used. Chlorination was followed by 10 minutes empty bed contact time with carbon in two expanded bed upflow contactors in series. Because of the low alkalinity of the water after coagulation with lime, the pH decreased during chlorination so that $NHCl_2$ concentration was significant. In general, their data were consistent with conversion of $NHCl_2$-N to NH_3 by the carbon. No attempt was made to optimize design of the process, however.

Atkins *et al.* found nearly 50% ammonia removal was possible when a chlorine dose to NH_3-N weight ratio of approximately 4:1 was used.[30] They concluded that essentially complete removal was possible at a weight ratio of 9:1. This is similar to the breakpoint ratio in the absence of carbon. Control of pH did not appear important in controlling efficiency at the higher dosages. They also found that the carbon reduced all chloramines and free chlorine and left the effluent essentially void of residual chlorine.

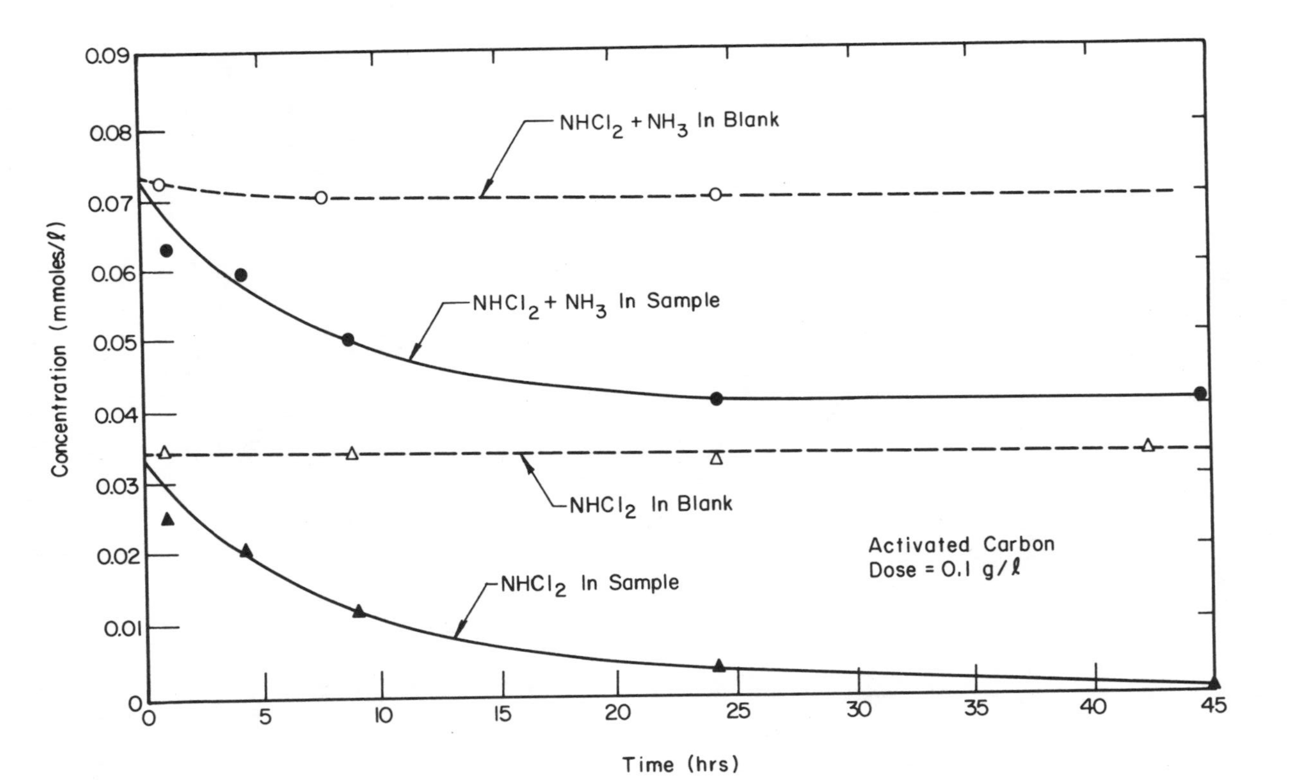

Figure 16-5. Dichloramine-activated carbon reactions (after Bauer and Snoeyink,[31] reprinted by permission of *J. Water Pollution Control Federation*).

The increase in chloride concentration in mg/l resulting from ammonia oxidation was found to be equal to the concentration of gaseous chlorine in mg/l applied to the wastewater. The cost for removal of 10 mg/l NH_3-N to meet an effluent standard of 5 mg/l NH_3-N was 6.8¢/1000 gal for a 6 MGD plant (1972), including all operation, maintenance and capital costs. The chemical costs were approximately one-half of the total cost for this removal.

RATE OF REACTION AND MODELING OF THE DECHLORINATION BED

Magee[21] studied the rate of reaction of free chlorine with carbon in a column apparatus and found results generally consistent with the rate of surface reaction being the rate-limiting step. He observed that an initial stage, which he called a diffusion controlled stage, was in effect prior to initiation of the surface reaction rate-controlled step. The initial phase was observed to last approximately 1000-bed volumes when inlet concentration of free chlorine was 30 mg/l as Cl_2. Subsequent to this initial period, he found that the removal rate could be described by the first-order rate equation

$$dC/dt = -kC \qquad (15)$$

which, for the initial conditions of concentration $C = C_o$ at time $t = 0$ integrates to

$$C = C_o e^{-kt} \qquad (16)$$

The time parameter, t, in this case represents the time of contact of the chlorinated water with carbon.

Magee found the rate constant to be an important function of pH, with the value at pH 4-5.5 being four times greater than the value at pH 8.5-10; marked differences in the efficiency of different types of carbon were also observed.[21] His model does not account for the poisoning, or reduction in reaction rate, which occurs as the extent of the reaction increases. The data on which he based his model were taken after a significant amount of chlorine had reacted with the carbon, and it was assumed that a further reduction in rate as the reaction proceeded would not occur. Equation (16) thus predicts that a dechlorination bed operating under a given set of conditions will produce a given quality of effluent indefinitely; this prediction is contrary to observed results.

The magnitude of the poisoning effect is shown by the results of some recent research conducted in the authors' laboratory wherein the reduction in rate of reaction was determined as a function of the amount of chlorine reacted. The experiment was conducted in a rapidly stirred batch reactor to which a measured amount of 60 x 80 U.S. Standard Mesh carbon was added. The pH of the solution was maintained at 4 and the HOCl concentration was allowed to vary between $C_o + \Delta C$ and $C_o - \Delta C$. The average concentration, C_o, was 5 mg/l as Cl_2 and ΔC was approximately 0.5 mg/l. Reaction rate was then measured by determining the time required for the concentration to change from $C_o + \Delta C$ to $C_o - \Delta C$. The measured rate is plotted as a function of the amount of chlorine reacted per gram of carbon in Figure 16-6.

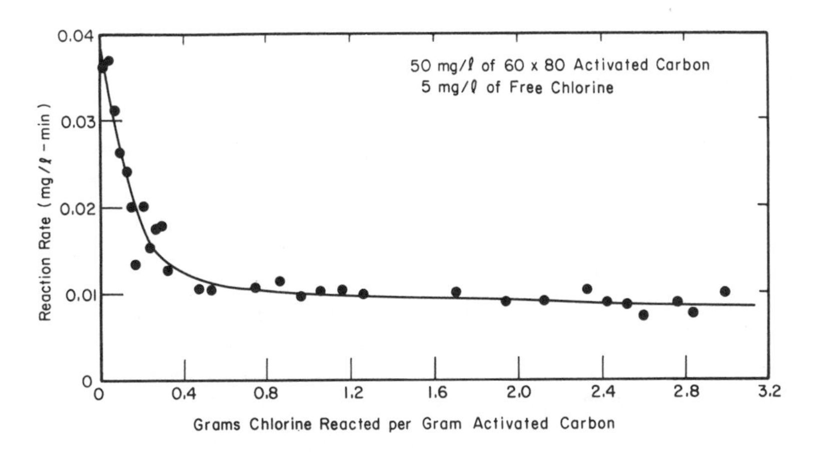

Figure 16-6. Activated carbon-free chlorine reaction rate as a function of the amount of chlorine reacted.

This curve is of interest because it shows a very significant drop in rate as the reaction proceeds and, at very large amounts of chlorine reacted per gram, it appears to reach a plateau. Research is continuing in this area to determine the effect of chlorine concentration, type of residual, particle size of carbon, pH, and adsorbed organic matter on such curves and to use these data to develop a predictive model for the dechlorination bed.

Although Magee's model is based upon the surface reaction being the rate-limiting step, it is not apparent that the diffusion-controlled step can be eliminated from an accurate model that describes the reaction. He found that the reaction rate constant was inversely proportional to the particle size of carbon—an observation inconsistent with the surface reaction being rate-controlling because total surface area of a porous adsorbent generally varies very little with particle size. Additionally, Magee reported that the reaction rate doubled for a temperature increase of 20°C for the range 0-90°C.[21] This temperature effect corresponds to an activation energy of 5.5 to 8.6 kcal/mol, values that correspond to those generally attributed to a diffusion controlled reaction rate.[33]

In other research in the authors' laboratory, it was observed that in general, for the various chlorine species studied, $NHCl_2$ will react most rapidly with carbon followed in order by HOCl, OCl^-, and NH_2Cl. The rate of dechlorination was reduced by the presence of adsorbed *p*-nitrophenol and by adsorbed organic matter from sewage. There was good indication, however, that even though the rate was decreased by adsorption of organics, the carbon beds could serve to dechlorinate a water such as a chlorinated secondary effluent.

CONCLUSIONS

Review and analysis of the state-of-knowledge of dechlorination by activated carbon has revealed it is a viable alternative for dechlorinating water and wastewater. Free and combined chlorine are readily reduced by carbon with the monochloramine reaction being the slowest of all the species. Under controlled conditions, carbon may serve as an integral part of an ammonia removal process because it will convert chloramino-N to N_2. The dual usage of dechlorination beds as adsorbers for organic compounds appears possible, but is an area that requires much study. Although reaction of carbon with chlorine residual affects the ability of the carbon to adsorb organic compounds, this effect does not appear serious enough to eliminate the possible conjunctive use of carbon for adsorption and dechlorination. Thus the adsorption of undesirable chlororganics would occur along with the removal of inorganic chlorine.

Sulfur compounds such as SO_2 and $S_2O_3^{=}$, can also be used for dechlorination. These compounds effectively reduce the residual, but they cannot be used simultaneously to remove organic compounds and ammonia, as can activated carbon. They may well be the most economical chemicals, however, if the only desired effect is that of removing chlorine residual.

Hydrogen peroxide has been used as a dechlorinating agent and it appears especially desirable because the reduction reaction results in the production of oxygen. The studies of this reaction are very few in number, however, and potential interferences with the reduction reaction exist.

ACKNOWLEDGMENT

The partial support of this study by the Office of Water Research and Technology is gratefully acknowledged.

REFERENCES

1. Environmental Protection Agency. *Water Quality Standards Criteria Digest*. Washington, D.C. (1972).

2. Snoeyink, V. L. and F. I. Markus. "Chlorine Residuals in Treated Effluents," *Water Sewage Works* **121**, 35 (1974).
3. Pressley, T. A., D. F. Bishop and S. G. Roan. "Ammonia-Nitrogen Removal by Breakpoint Chlorination," *Environ. Sci. Tech.* **6**, 622 (1972).
4. Esvelt, L. A., W. J. Kaufman and R. E. Selleck. "Toxicity Assessment of Treated Municipal Wastewaters," *J. Water Pollution Control Fed.* **7**, 1558 (1973).
5. Brungs, W. A. "Effects of Residual Chlorine on Aquatic Life," *J. Water Pollution Control Fed.* **45**, 2180 (1973).
6. Basch, R. E. and J. G. Truchan. "Calculated Residual Chlorine Concentrations Safe for Fish," Interim Report, Michigan Water Resources Commission, Department of Natural Resources, Lansing, Michigan (1973).
7. White, G. C. *Handbook of Chlorination* (Cincinnati, Ohio: Van Nostrand Reinhold Co., 1972).
8. Laubusch, E. J. In *Water Quality and Treatment*, 3rd ed. (New York: McGraw-Hill Book Co., 1971).
9. Dean, R. B. "Toxicity of Wastewater Disinfectants," In *News of Environmental Research in Cincinnati,* (Cincinnati, Ohio: U.S. Environmental Protection Agency, July 5, 1974).
10. Lawrence, A. W., W. S. Howard and A. K. Rubin. "Ammonia-Nitrogen Removal from Wastewater Effluents by Chlorination," 4th Mid-Atlantic Industrial Waste Conference, University of Delaware (November, 1970).
11. Gagen, O. M. "Chlorination of Water with Large Doses and Subsequent Dechlorination," *Vodosnahzhenie Sanit. Tekh.* **16**, 27 (1941), In *Chem. Abstr.* **35**, 7597 (1941).
12. Eckenfelder, W. W. and D. J. O'Connor. *Biological Waste Treatment* (New York: Pergamon Press, 1961).
13. Westerterp, K. P., L. L. Van Dierendonck and J. A. DeKraa. "Interfacial Areas in Agitated Gas-Liquid Contactors," *Chem. Eng. Sci.* **18**, 157 (1963).
14. Reinders, W. and P. Dingemans. "Die Oxydationsgeschwindigkeit von Hydrochinon mit Sauerstoff," *Rec. Trav. Chim.* **52**, 231 (1934) (German).
15. Baker, R. A. "Declorination and Sensory Controls," *J. Amer. Water Works Assoc.* **56**, 1578 (1964).
16. Fair, G. M., J. C. Geyer and D. A. Okun. *Water and Wastewater Engineering,* Vol. 2 (New York: John Wiley & Sons, Inc., 1968).
17. Weber, W. J., Jr. *Physicochemical Processes* (New York: Wiley-Interscience, 1972).
18. Alterman, N. A. "Dose of Thiosulfate Needed in Dechlorination of Water," *Gigiena i Sanit.* **23**, 66 (1958) (Russian).
19. Mishchenko, K. P., I. E. Flis and V. A. Kastodina. "Thermodynamic Characteristics of Aqueous Solutions of Hydrogen Peroxide and its Reaction with Chlorine at Different Temperatures," *Zhur. Priklad. Khim.* **33**, 2671 (1960), In *Chem. Abstr.* **55**, 9023 (1961).
20. Connick, R. E. "The Interaction of Hydrogen Peroxide and Hypochlorous Acid in Acidic Solutions Containing Chloride Ion," *J. Amer. Chem. Soc.* **69**, 1509 (1947).

21. Magee, V. "The Application of Granular Active Carbon for Dechlorination of Water Supplies," *Proc. Soc. Water Treat. Exam.* **5**, 17 (1956).
22. Puri, B. R. "Surface Complexes on Carbons," In *Chemistry and Physics of Carbon,* Vol. VI, P. L. Walker, Ed. (New York: Marcel Dekker, Inc., 1970), pp. 191-282.
23. Snoeyink, V. L., H. T. Lai, J. H. Johnson and J. F. Young. "Active Carbon: Dechlorination and the Adsorption of Organic Compounds," In *Chemistry of Water Supply, Treatment and Distribution,* A. Rubin, Ed. (Ann Arbor, Michigan: Ann Arbor Science Publishers, Inc., 1974).
24. Olson, L. L. and C. D. Binning. "Interaction of Aqueous Chlorine with Activated Carbon," In *Chemistry of Water Supply Treatment a and Distribution*, A. Rubin, Ed. (Ann Arbor, Michigan: Ann Arbor Science Publishers, Inc., 1974).
25. Boehm, H. P. "Chemical Identification of Surface Groups," *Adv. in Catalysis*, D. D. Eley *et al.*, Eds., Vol. 16 (New York: Academic Press, 1966), pp. 179-222.
26. Coughlin, R. W., F. Ezra and R. N. Tan. "Influence of Chemisorbed Oxygen in Adsorption into Carbon from Aqueous Solution," *J. Coll. Interface Sci.* **28**, 386 (1968).
27. Hagar, D. G. and M. W. Flentje. "Removal of Organic Contaminants by Granular Carbon Filtration," *J. Amer. Water Works Assoc.* **57**, 1440 (1965).
28. Stuchlik, H. *Active Carbon: Manufacture, Properties and Applications* (New York: Elsevier Publishing Co., 1970).
29. D'Agostaro, R. A. "An Investigation into the Mechanism of Nitrogen Removal from Aqueous Solutions Utilizing Chlorination Followed by Activated Carbon Contact," M.Sc. Thesis, School of Civil Engineering, Cornell University, Ithaca, New York (1972).
30. Atkins, P. F., Jr., D. A. Scherger, R. A. Barnes and F. L. Evans, III. "Ammonia Removal by Physical-Chemical Treatment," *J. Water Pollution Control Fed.* **45**, 2372 (1973).
31. Bauer, R. C. and V. L. Snoeyink. "Reactions of Chloramines with Active Carbon," *J. Water Pollution Control Fed.* **45**, 2290 (1973).
32. Williams, D. B. "Elimination of Nitrogen Trichloride in Dechlorination Practice," *J. Amer. Water Works Assoc.* **58**, 248 (1966).

CHAPTER 17

COMPARATIVE INACTIVATION OF BACTERIA AND VIRUSES IN TERTIARY–TREATED WASTEWATER BY CHLORINATION

Peter P. Ludovici, Robert A. Phillips, Wayburn S. Jeter

Departments of Microbiology & Medical Technology
and Civil Engineering
University of Arizona
Tucson, Arizona 85721

INTRODUCTION

For many years water has been known to play a role in the transmission of virus diseases. Over 50 documented waterborne epidemics of infectious hepatitis have been recorded.[1] In addition water has been implicated in the transmission of poliomyelitis and viral gastroenteritis. Epidemiological evidence for the latter diseases is inconclusive, however. As pointed out by Melnick,[2] it will be virtually impossible to obtain such evidence in modern communities; as the longer and more variable incubation period for enteric viral diseases, with their greater ratio of inapparent-to-clinical-disease, makes their transmission more difficult to trace. Yet the sheer number of cases makes this an important problem. More than 18,000 cases of gastroenteritis and related diarrheal diseases frequently attributed to enteric viruses were reported in 142 outbreaks over a 16-year period from 1946-1960 in the United States.[3] Currently, an average of 14 known waterborne-disease outbreaks occur each year in the United States causing an estimated 1600 illnesses and one death per year.[4]

Furthermore, the massive waterborne outbreak of infectious hepatitis in New Delhi, India, in 1955 in which 30,000 individuals became infected by drinking water from the municipal supply system which had been treated by chlorination and had met the accepted coliform bacteria standards, points up the necessity for developing newer and more sensitive methods to assess the microbiological safety of water. Such studies are of utmost importance now that our waterways are becoming more heavily polluted and our recycling needs increase.

It was because of such predicted needs that the city of Tucson constructed a sand and gravel pilot filter for the tertiary treatment of wastewater in 1966. The pilot filter was loaded with secondary treated effluent (activated sludge) from Tucson's sewage plant. After approximately 150 feet (45 m) of horizontal and 15 feet (4.5 m) of vertical filtration, this wastewater emerged as pilot filter effluent (PFE) or tertiary-treated water which was used to fill four experimental fish ponds. Reclaimed water of this type will be used in recreational lakes for boating, fishing, swimming and ultimately, when the need arises, as potable water.

The authors have reported their efforts to find an efficient and economical method to determine the effectiveness of this pilot filter in removing enteroviruses from wastewater.[5] This paper presents some of our studies on the comparative effectiveness of chlorine disinfection in eliminating bacteria and enteroviruses from such tertiary-treated wastewater and, at the same time, reviews the general area of bacteriological and virological disinfection of water by chlorination. Sections on factors believed to influence the results of such comparative studies are included, but no attempt is made to present an inclusive review–there are many excellent reviews on similar subjects (see Chapter 7).[6-9]

BACTERIAL ENUMERATION METHODS

Standard bacterial enumeration procedures are documented in *Standard Methods.*[10,11]

VIRAL CONCENTRATION PROCEDURES

Detection of viruses in water is similar to their detection in other sources. With heavily polluted wastewater (like sewage), viruses can be detected by testing water directly. However, when potable and treated waters are tested it is necessary to concentrate viruses from such waters before detection and enumeration. The field of virus concentration from water has advanced so rapidly that it has been impossible to adequately test one method before a newer method has replaced it. The method of virus concentration used to isolate viruses from a water sample will affect the number and types of viruses obtained with and without disinfection. A brief review of the major methods used to concentrate viruses, with their advantages and disadvantages, is presented below. Valid comparisons of data cannot be made between studies using a laboratory model that does not require virus concentration, and studies that require virus concentration under natural conditions.

Gauze Pad Method

This was the original method used to concentrate viruses in water, adapted from the Moore[12] swab procedure for adsorbing bacteria from water. Gauze or cotton filled pads were suspended for one to several days in water sites in order to trap or adsorb viruses from the water flowing through the gauze. Pads were then collected in plastic bags and treated with sodium hydroxide to bring the pH to 8.0 which facilitated virus elution from the pads.[13] Liquid expressed from the pads was tested directly from viruses or further concentrated by ultracentrifugation or other procedures. While this method is far superior to the common grab sample procedure for detecting viruses in water, a major disadvantage is that it is not quantitative. This problem may be overcome by suspending the gauze pad in a closed container and pumping a known volume of water sample through the pad per unit time. However, the pad technique was abandoned once simpler virus concentration procedures were developed.

Ultracentrifugation

Liquids containing viruses ranging in size from 10-200 nm are ultracentrifuged at approximately 120,000 g for 2-4 hr. The ultracentrifugation method is used routinely to harvest and concentrate viruses from small volumes of cell culture fluids and from water expressed from gauze pad samples. The expressed fluid is centrifuged first at low speeds to sediment the bacteria and larger particles. Gelatin is added to the supernatant fluid to make a 2% solution, and the sample is centrifuged at high speed to trap the viruses in a pellet on the bottom of the tube. The pellet is resuspended in a small volume of cell culture medium and tested on cell cultures directly. Sample concentrations on the order of 10-100 times are possible by this method.

Unfortunately, the size of the samples is limited to 100 ml per run unless an expensive continuous flow apparatus is used. This operational disadvantage has prevented extensive use of the method, especially for detecting viruses in large volumes of water.

Membrane Filter Adsorption

This was the basis for the first quantitative virus concentration procedure. Cliver[14] found that enteroviruses adsorb to membrane filters even when the pore diameter is considerably larger than the virus. Using Coxsackievirus

A9 in tap water and 0.45 nm porosity membranes, Cliver showed there is a 50% chance of detecting virus at a concentration of two plaque-forming units per liter. Wallis and Melnick[15-17] further developed the technique to concentrate enteroviruses in various types of waters approximately 1000-fold.

The major disadvantage is that filters clog when large volumes of water are assayed. Another is loss of virus adsorptivity when other components in wastewater coat the membrane. To adapt the procedure to large samples of water (100-400 gal/hr) in the field, a series of depth filters (textiles, orlon and cotton) are used to clarify the water and remove particulates and complexed metals, with final adsorption of the virus on a 10-inch fiberglass disc or a cellulose acetate depth filter.[18]

Dialysis

A sample of water may be concentrated approximately 100 times by placing it in a cellulose dialyzing bag immersed in polyethylene glycol.[19-21]

Ultrafiltration

Viruses are adsorbed on a soluble aluminum alginate filter which is then dissolved with a small volume of 3.8% citrate. The solution containing the viruses is inoculated directly into cell cultures.[24] It is possible to recover 10 $TCID_{50}$ of virus from 10 liters of water if the water is not too turbid. Recoverability of Poliovirus ranged from 25-100%.[22]

Recently Nupen and Stander[23] concentrated viruses from 10 liters of water using a stirred ultrafiltration cell (Amicon high performance cell model 2000) with a Diaflo P.M. 30 non-cellulose membrane. The thin skin produced a combination of selectivity, high throughput and resistance to clogging by rejecting substances at the surface. Viruses were recovered by flooding the membrane with 10% fetal calf serum, 0.5% lactalbumin hydrolysate in Hanks' balanced salt solution. This method gives a 70% recovery of virus and permits the processing of 10 liters of clear or slightly turbid water in a matter of hours.

The disadvantage of these ultrafiltration methods is the high possibility of damaging the delicate membranes before or during filtration.

Phase Separation

This method is based on Albertsson's finding that mixtures made from two polymers, like dextran sulfate and polyethylene glycol, in an aqueous solution leads to the formation of a two-phase system.[24] Viruses, depending on their size and surface properties, concentrate in one phase of the system,[25,26] and can be concentrated 100-200 times depending upon the kind of phase system used. Shuval *et al.*[20,27] using a two-step phase separation technique, showed that as few as 1-2 viruses per liter of water can be detected; they achieved a concentration factor of 500. The method is simple and inexpensive; its chief disadvantages are the limited sample volume and the requirement of two overnight incubations.

Adsorption to Particulate Substances

Particulate materials to which viruses have adsorbed are collected by sedimentation or filtration, and the viruses are eluted using a small volume of fluid. A number of substances have been used including: (1) precipitates of inorganic salts like aluminum hydroxide,[16,28] calcium phosphate,[16,18,28,29] ammonium sulfate[16] and cobalt chloride;[30] (2) vegetable flocs;[31] (3) protamine sulfate;[32] (4) ion exchange resins;[33,34] (5) passive hemagglutination;[19,35] and (6) insoluble polyelectrolytes.[36-38] The adsorbent is added to the water sample and magnetically stirred ½-2 hr, followed by the separation of the adsorbent-virus complex by sedimentation or filtration, and finally elution of the viruses in a small volume of fluid. Concentration factors as high as 100,000 times can be achieved by several of the newer methods. These techniques show considerable promise since they are relatively simple and inexpensive and highly efficient.

COMPARISON OF VIRUS CONCENTRATION METHODS

Only a few comparative studies of these different virus concentration procedures for the detection and enumeration of viruses has been done. Gibbs and Cliver[19] compared the dialysis, ultracentrifugation and passive hemagglutination procedures. They found that the passive hemagglutination technique was specific for Reovirus but ineffective for others, whereas ultracentrifugation was superior to demonstrate viruses in food extracts. The dialysis procedure yielded a concentrate which was toxic to the cell cultures.

Shuval *et al.*,[20] in comparing the phase separation (PS) method with dialysis concentration, found the PS procedure 10 times more sensitive in detecting enteroviruses in several types of laboratory prepared water as well as with clarified sewage. Moreover, the PS method showed an efficiency of virus recovery of approximately 100% compared to only 40-50% for dialysis.

Lund and Hedstrom[39] compared the PS procedure with an ammonium sulfate-ultracentrifugation method and found both techniques equally successful in detecting enteroviruses in sewage.

Gravelle and Chin[34] collected gauze pad samples in sewage and compared three methods of processing the samples before inoculations into tissue culture. One sample was inoculated without concentration and the other two were concentrated by ultracentrifugation or resin precipitation. The ultracentrifugation procedure was superior to the others and the use of a combination of methods increased the number of isolates.

Moore, Ludovici and Jeter[5] compared the membrane filtration (using protamine sulfate or anion exchange resin to remove membrane coating components) with adsorption on aluminum hydroxide and found the aluminum hydroxide technique far superior–not only for seeded Poliovirus but also for detecting low levels of indigenous viruses in wastewater. Previous work (Ludovici & Jeter, unpublished data) had shown the superiority of the membrane filtration procedure over the ultracentrifugation technique in detecting enteroviruses from wastewater.

Wallis *et al.*[37] compared the adsorption technique of aluminum hydroxide with insoluble polyelectrolytes and reported that the polyelectrolyte procedure was more efficient in detecting enteroviruses in sewage.

England[32] compared a protamine sulfate (salmine) method to several adsorption procedures including polyelectrolyte, aluminum hydroxide, and $CaHPO_4$ for preferentially concentrating Reoviruses and Adenoviruses over enteroviruses in several types of wastewater. The reason for such a study is that more rapidly growing enteroviruses tend to overgrow the slower growing Reoviruses and Adenoviruses and thus mask their detection in water. She found that recoveries of Reoviruses and Adenoviruses by the salmine adsorption method exceeded recoveries by polyelectrolyte or $CaHPO_4$, and was at least comparable to that recovered by $Al(OH)_3$. This success of the $Al(OH)_3$ method in concentrating Reoviruses differs from the finding of Wallis and Melnick[16] who reported that Reovirus did not adsorb to $Al(OH)_3$.

This discrepancy points out that our lack of information on the specificity of the various concentration methods for detecting the different types and strains of viruses present in wastewaters. Obviously, such information is important in evaluating the effectiveness of each concentration method in detecting viruses before as well as after chlorination. For example, Grindrod and Cliver[40] showed that the phase separation method was highly efficient in the recovery of 3 types of Poliovirus and Coxsackieviruses A-9 and B-3, but inhibitory for Coxsackievirus B-2, Echovirus 6 and Influenza virus. Berg[7] experienced similar selectivity of virus recoveries with the insoluble polyelectrolyte adsorption method in which Poliovirus I recoveries were less than 30 and 20%, respectively. Nevertheless, Berg[7] believed that the insoluble polyelectrolyte method was the most sensitive method available for large volumes of water, despite its erratic efficiency.

Another example is Gibbs and Cliver's[19] finding that the passive hemagglutination was specific only for Reoviruses. Further work needs to be done in this important area of differential selectivity in concentrating viruses from natural waters. Perhaps no one method will suffice and viral concentration procedures will have to be used in tandem to assess properly the viral population present in water.[7]

QUANTITATION OF VIRUSES

Two methods are used for the quantitative determination of viruses in a concentrate prepared by one of the previously described concentration procedures. The method used will depend on the viruses and cell cultures available to the laboratory. When unknown viruses are being isolated, it is sometimes best to use both methods; or if only one method can be used, it may be preferable to use the test tube technique discussed below.

Test Tube Assay Method

Serial dilutions of the concentrate are inoculated, usually 0.1 ml per tube containing a monolayer of virus-susceptible cells (primary rhesus or African green monkey kidney cells, human amnion, or VERO, WI38, HEp2, HeLa, etc.). The tubes are incubated at 37°C and examined daily for cytopathogenicity (CPE). The lowest dilution of the concentrate causing CPE in 50% of the tubes is known as the $TCID_{50}$ (tissue culture infectious dose - 50%). Also, the most probable number (MPN) of infectious virus particles can be estimated based on the number of tubes showing CPE as calculated from standard statistical tables.[41]

Plaque Assay Method

Serial dilutions of the concentrate are inoculated usually 0.1-1 ml per petri dish (60 x 15 mm) or 1-4 oz prescription bottle containing a monolayer of the same virus-susceptible cells used for the test tube method. After 1 hr adsorption at 37°C, the monolayer is usually washed once with medium containing 2% fetal calf serum and antibiotics before an agar overlay is added. The agar overlay medium varies according to the cell type used.

After 24-48 hr incubation at 37°C in a standard incubator for sealed bottles, or a 5% CO_2 humid atmosphere for petri dishes and unsealed bottles, the cultures are examined for plaques. The number of plaque-forming units (PFU) of infectious virus is calculated based on the dilution, and reported as PFU per unit volume of water.

Plaque assay is a more precise estimate than either the $TCID_{50}$ or MPN concentration obtained by the test tube assay technique. Nevertheless, Ludovici and Riggin (unpublished data) have noted that natural viruses in wastewater will often produce CPE in cell monolayers of tubes where they fail to form plaques by the plaque assay method. England[32] reported similar observations for Adenovirus recovered from raw sewage, in which plaques were not produced in rhesus monkey kidney, whereas CPE was produced in tube cultures of human embryonic kidney. For this reason it is best to use both assay procedures to detect and enumerate unknown natural viruses in water. If only one technique is used, it should be the test tube assay procedure. Perhaps both methods should be used in studies of chlorination effect on known (seed) viruses, since partially inactivated viruses may cause CPE in tube monolayers where they are not able to produce plaques under agar.

CHLORINE DISINFECTION

In its gaseous form, chlorine reacts with water to form hypochlorous acid (HOCl), the most effective germicidal form of chlorine known. At pH values above 7.0, HOCl ionizes to form hypochlorite ion (OCl^-) which has been shown by a number of investigators to be a considerably less effective germicide.[42-45] Both HOCl and OCl^- are known as free chlorine. Chlorine reacts with ammonia and nitrogenous substances usually found in natural waters to form ammonia chloramines or organic chloramines which are even less effective disinfectants than OCl^-.[46,47] It has been estimated that *Escherichia coli* is killed 60-70 times faster with HOCl than with OCl^- whereas chloramines act 270 times slower than HOCl.[45,48] A 10°C increase in temperature increases the rate of virus inactivation 200-300%[46] and the rate of inactivation is a function of concentration and pH.[42]

Pure Water Studies

Wattie and Butterfield[49] found that a 0.6 ppm residual chlorine and 1-hr contact time was required to obtain 100% kill of *E. coli* and *Eberthella typhosa* at pH 7.0 using pure water. From their data,[50] Weidenkopf,[42] Clarke and Kabler[51] and Clarke, Stevenson and Kabler,[52] Clarke *et al.*[45] constructed death rate curves of *E. coli* and three enteric viruses exposed to HOCl in pure water. These curves showed that at 0-6°C Poliovirus I and Coxsackie A-2 are more resistant than *E. coli* but Adenovirus 3 was more sensitive; *e.g.*, at a 99% destruction level, 0.1 ppm HOCl took 99 seconds for *E. coli*, 33 seconds for Adenovirus, 8-15 minutes for Poliovirus, and more than 40 minutes for Coxsackievirus A-2.

Data published by Kelly and Sanderson[43] demonstrated a similar resistance pattern to HOCl for several enteroviruses in pure water. They reported that inactivation of viruses by chlorine was affected by pH, chlorine concentration, exposure time and temperature as well as virus strain. Poliovirus I (strain MK500) was most resistant while Coxsackie B-5 was most sensitive. Poliovirus I (Mahoney strain), Poliovirus II and III, and Coxsackie B-1 virus were intermediate in resistance to free chlorine. Their data suggested that at a pH of 7 and 25°C, a minimum free chlorine residual of 0.3 ppm and a contact period of at least 30 minutes was needed to inactivate enteroviruses. It is apparent from these early studies that some viruses are considerably more resistant to HOCl than the coliform bacteria and that enteric viruses *per se* differ in their resistance.

Scarpino *et al.*[53] demonstrated the effectiveness of HOCl and OCl^- as disinfectants for *E. coli* and Poliovirus I at pH 6 and 10 at 5°C. At pH 6, Poliovirus I was 130 times more resistant to HOCl than *E. coli*. At pH 10, *E. coli* was 3 times more resistant to OCl^- than Poliovirus I. These authors reported that, overall, OCl^- was 7 times as effective against Poliovirus I as HOCl which, in turn, was 50 times as effective in destroying *E. coli* as OCl^-. Unexplained salt effects were present in this data.

Shah and McCamish[54] compared the resistance of Poliovirus I and *E. coli* bacteriophages f_2 (RNA type) and T_2 (DNA types) to 4 mg/l combined chlorine. The f_2 phage was markedly more resistant to combined chlorine in pure water than Poliovirus I and T_2 phage. Poliovirus I was more resistant than T_2 phage. They suggest that this greater chlorine resistance makes the f_2 an eminent candidate for the role of indicator virus in water and wastewater.

Dahling *et al.*[55] compared the resistance of two enteric viruses (Poliovirus I and Coxsackie A-9), two DNA bacteriophages (T_2 and T_5), two RNA bacteriophages (f_2 and MS2), and *E. coli* ATCC 11229 to free chlorine in demand-free water at pH 6.0. The enteric viruses were most resistant to free chlorine followed by the RNA bacteriophages, *E. coli*, and the DNA bacteriophages.

Unfortunately, the resistance of Infectious Hepatitis virus, the most important virus in wastewater disinfection, is still unknown because this virus cannot yet be grown in tissue culture. The finding of Neefe *et al.*[56] that at least a 0.4 ppm chlorine residual (primarily HOCl and chloramine) in distilled water containing 40-50 ppm of infective feces was needed to destroy infectivity of Infectious Hepatitis virus for human volunteers in 30-35 min suggests a fairly high resistance of this virus to chlorination.

Kelly and Sanderson[44] noted that 1 ppm of monochloramine in clean water required 3 hours at pH of 7 and 25°C to destroy both Polio- and Coxsackie-viruses. As the pH increased to 9.0 there was a corresponding increase in contact time to 5 or 6 hours for inactivation of the viruses. Increasing the concentration of combined chlorine decreased contact time.

Overall, these studies in pure water suggest that the susceptibility of various microorganisms to chlorine will vary depending on the type of chlorine, whether it is free or combined, the temperature, and the pH of the water. Under optimum conditions at approximately a neutral pH where HOCl would predominate, the descending order of resistance to free chlorine would be Polio I, Coxsackie A-9, f_2 phage, MS2 phage, *E. coli*, T_2 phage and T_5 phage. Conversely, when the chlorine in pure water is in the combined form the f_2 phage appears to be more resistant than the enteroviruses. Additional comparative studies are needed to substantiate the above conclusions. The results of Cramer and Kruse[57] and Olivieri *et al.* (see Chapter 7) are useful. These articles point out the previously unrecognized importance of mixing and physical state of the virus.

Raw Sewage Studies

Not as much work has been done with raw sewage as with clean water models or secondary effluents. Wyckoff[58] reported that in sewage treatment a chlorine residual greater than 0.5 ppm resulted in a median kill of coliform of 80% after one minute contact.

McKee showed that a combined chlorine dosage of 8 mg/l to freshly settled sewage having a starch-iodine chlorine demand of only 3 mg/l after 15 minutes contact produced a 99.995% kill of coliforms.[59]

Lothrop and Sproul compared the inactivation by chlorine of T_2 bacteriophage with Poliovirus I in settled wastewater having BOD of 153 mg/l, suspended solids of 110 mg/l, and an average NH_3-N 21.3 mg/l.[60] A 99.99% kill required doses of 10-50 mg/l chlorine for T_2 and 10-70 mg/l for Poliovirus I using a contact time of 30 minutes. The chlorine residual to achieve the above results was 28 mg/l for T_2 versus 33-43 mg/l for Poliovirus I.

More recently Cramer and Kruse[57] compared the inactivation of Poliovirus III (Leon) and f_2 bacteriophage with chlorine in buffered sewage. Both viruses were treated in the same reaction flask, thus eliminating any possible inherent errors in preparation or replication. At pH 6.0 and 10.0 with a 30 mg/l dose of chlorine under prereacted (viruses added after chlorine is allowed to react with sewage) and dynamic (viruses added before chlorine is allowed to react with sewage) conditions, f_2 bacteriophage in both cases was at least as resistant or more resistant to chlorine than Poliovirus III.

Although the data for raw sewage is sparse, the studies suggest that f_2 phage (RNA type) is as resistant or more resistant to combined chlorine as the enteric viruses, Polio I and III. The T_2 phage (DNA type) is less resistant than the enteroviruses. No comparative studies were done with coliform bacteria so the bacteria cannot be ranked in the sewage studies.

Secondary Effluent Studies

Until recently no data had been published on the effect of chlorine on viruses in wastewater effluents. In fact, Clarke's discussion of the work of Shuval *et al.*,[61] stated: "There is essentially no published information detailing the conditions and parameters necessary to ensure destruction of enteroviruses in sewage by use of chlorine." Shuval *et al.*[61] reported the inactivation by chlorine of enteric viruses seeded in the effluent from the Haifa wastewater treatment plant, a high rate biofiltration plant with an average effluent BOD of 45 mg/l, average suspended solids of 50 mg/l, and concentrations of NH_3 ranging from 5-20 mg/l. Chlorine dosages applied to this effluent with a pH of 7.7-7.8 at 20°C were 3.6 to 11 mg/l, and contact periods up to six hours were used. In 80% of samples using 8 mg/l of chlorine after 2-hr contact, the coliform concentration of the sewage effluent was less than 100 per ml. In 50% of samples this level of coliform inactivation (less than 100 per 100 ml) was achieved with the same dosage (8 mg/l) but with only 30-minutes contact. Reducing the chlorine dose to 5 mg/l achieved the same level of coliform inactivation after 1½ hr contact. These data for coliform compare with the more resistant Poliovirus I data in which only 50 and 90% inactivation occurred when 5 and 11 mg/l chlorine doses were used, respectively. Conversely, the Echovirus strain in this study had approximately the same susceptibility to chlorine as the coliform organisms, since it showed a 99% reduction in 30 minutes and a 99.39% inactivation in 6-hr contact, with only a 3 mg/l dose of chlorine. This was one of the first studies to show that the standard coliform index would not give a true picture of the presence of more resistant strains of enteroviruses like Poliovirus I. Furthermore, in all cases the residual chlorine levels remaining after 6-hr contact, as measured by Shuval *et al.*,[61] varied but was never less than 0.3 mg/l (all combined chlorine).

In a later study Shuval[27] showed essentially the same results except that Echovirus 9 used was more resistant. For example, this author demonstrated that to achieve a 99.9% inactivation with a one-hour contact time, 2 mg/l of chlorine for coliforms, 8 mg/l for Echovirus 9, and 20 mg/l for Poliovirus I were required. Thus it required a chlorine dose approximately 10 times greater to inactivate Poliovirus I than it does to inactivate coliforms.

Marais *et al.*[62] investigated the efficiency of chlorination in secondary effluents with varying amounts of nitrogenous organic compounds and ammonia. From their data, they concluded that it would be necessary to chlorinate to the breakpoint, which in all cases immediately inactivated both *E. coli* and Poliovirus I (attenuated and virulent). It is of interest that two of these authors, Nupen and Stander,[23] arrived at the same conclusion for tertiary treated wastewater from the Windhoek reclamation project (see tertiary-treated effluent).

Warriner reported on field trials for the inactivation of Poliovirus III by chlorinating a trickling filter effluent. A 75% inactivation of Poliovirus III was achieved with an average detention time of 30 minutes in the chlorine contact chamber and a 4 mg/l combined amperometric residual.[63] A 99% inactivation occurred with a 10-minute contact time and combined amperometric residual of 21 mg/l. Burns and Sproul reported for trickling filter and settled wastewater effluents using 8 and 10 mg/l chlorine dosages that a 99% reduction of T_2 reduction in one and a half hours.[64] Coliform bacteria were inactivated both with lower chlorine dosages and at shorter detention times than the T_2 bacteriophage. These data are interesting because they demonstrate the greater chlorine resistance of the T_2 bacteriophage (a naturally occurring bacterial virus that attacks *E. coli*) with the naturally occurring indicator organisms (coliform bacteria) under field conditions in secondary effluents. These investigators believe, in agreement with the results of other workers, that increased contact time is more important than increased chlorine dosage in disinfection of viruses and bacteria in wastewater. Furthermore, they did not find that orthotolidine-arsenite residuals showed any correlation with virus inactivation, but that amperometric residuals provided some indication of virucidal action with chlorinated trickling filter effluents.

Lothrop and Sproul extended the study to compare the chlorine inactivation of T_2 bacteriophage with Poliovirus I in secondary effluent (extended aeration), 10 and 20% storm water overflow, and settled wastewater.[60] With the secondary effluent (average NH_3-N concentration of 0.97 mg/l), a 100% inactivation was achieved using a chlorine dose of 4-6 mg/l for T_2 bacteriophage as compared with 11-16 mg/l required for Poliovirus I with a contact time of 30 minutes and approximately 0.2-0.5 mg/l free chlorine residual. Inactivation of poliovirus in both the 10 and 20% storm water overflows was achieved by chlorinating to the breakpoint to produce a small free chlorine residual. The 10% storm water overflow (NH_3-N of 2.8 mg/l) required a dosage of 20 mg/l to produce free chlorine residuals, whereas the 20% storm water overflow (NH_3-N of 3.65 mg/l) required a dosage of 35 mg/l. The free chlorine residuals are based on the forward titration amperometric method, whereas the combined residuals are based on the iodometric amperometric back titration procedure. This study demonstrates the comparative efficiency of free versus combined chlorine residuals and, significantly, the possibility of obtaining small free chlorine residuals even in wastewaters that have a low level of nitrification.

The overall data on chlorination studies of secondary effluents suggest that the T_2 (DNA type) phage is more resistant to chlorine (combined) than the coliform bacteria, but T_2 is probably more sensitive to chlorine than Polio I and other enteroviruses.

Estuarine Water Studies

Liu *et al.*[65] found a wide range of free-chlorine-susceptibility in 20 strains of human enteroviruses suspended in Potomac estuarine water without using virus concentration procedures. The viruses studied included the three strains of Poliovirus and Reoviruses; Adenoviruses 3, 7a, and 12; Echo 1, 7, 9, 11, 12 and 20; Coxsackie B1, B3, B5; and Coxsackie A5 and A9. The most resistant was Poliovirus II requiring 40 minutes for a 99.99% inactivation using 0.5 mg/l free chlorine whereas the most sensitive was Reovirus 1 requiring only 2.7 minutes for the same per cent inactivation. The remaining 18 enteroviruses fell between these extremes in a rather haphazard manner, except for the three Reoviruses which were the most sensitive. Their results indicate that the degree of chlorine resistance of any given virus is unpredictable. Their results are open to some question because of variable concentrations and controls.

Tertiary Treated Effluent Studies

England *et al.*,[66] in studies relating to the Santee water reclamation project, showed that 19% of the specimens contained viruses after detention in an oxidation pond as a tertiary treatment. However, less than 10% of the chlorinated effluents from the oxidation pond yielded viruses. Percolation through a natural sand and gravel layer as tertiary treatment removed seeded attenuated Poliovirus and, over a 33-month period, no viruses were recovered from recreational lake samples.

Nupen and Stander,[25] using breakpoint chlorination to achieve a free residual chlorine of 0.5 mg/l as a standard requirement for disinfection, reported that on routine monitoring of the final tertiary treated effluent of the Windhoek Waste Water Reclamation Plant, no enteroviruses could be detected in 10 liters of water samples using the ultrafiltration method. These authors concluded that the water was safe for drinking purposes, since the *European Standards for Drinking-Water*[67] recommended the examination of 10-liter samples of water and stated that less than one plaque-forming unit of virus per liter of water provided a reasonable assumption that the water was safe for drinking. Melnick,[2] however, proposed minimum standards of not more than one infectious virus unit per 10 gallons (37.85 liters) of recreational water used for swimming, and not more than one infectious virus unit per 100 gallons (378.5 liters) of water used for drinking or food preparations. Berg[7] stated that no virus should be present in a 100-gallon sample of disinfected, renovated, and other potable waters and that 100 gallons should be the minimum sample size for the detection of low level virus contamination.

It is clear that there is some confusion and disagreement as to what sample size is adequate as well as the maximum proposed number of viruses allowable. As stated by Scarpino, in discussing Nupen and Stander's paper,[25] not only must a standard procedure be developed for detecting low numbers of virus in large volumes of water, but experience must be gained on the size of sample that would provide the necessary confidence that the water is microbiologically safe for human use. The epidemiological information gained from Nupen and Stander's study[25] indicated that the incidences of typhoid, other *Salmonella*, *Shigella*, and infectious hepatitis did not markedly change after the introduction of reclaimed water to the community in 1968. Further epidemiological data of this type are needed before realistic microbiological safety standards on viruses can be established.

EXPERIMENTAL

Wastewater Samples and Chlorine Treatment

Our studies were done to learn how effective chlorine treatment would be to eliminate enteroviruses and remaining bacteria from Tucson's tertiary treated wastewater (pilot filter effluent, PFE).

Table 17-1 illustrates the experimental design of the study. Five-gallon (20 liter) samples of PFE were collected in stainless steel pressure vessels (Millipore) at the pilot filter site and brought to the laboratory for immediate processing. The sample was divided into 3 portions: 4 liters processed for indigenous enteroviruses, 1 liter processed for bacteria, and 15 liters "spiked" sample was processed for recoverability of the added virus, and the remaining 11 liters were divided into 2 portions and treated with either 2 mg or 4 mg/l of standard chlorine solution prepared from household bleach and given a contact time of 30 minutes. Samples were dechlorinated with thiosulfate after removing a portion for total residual chlorine determinations by the starch iodine procedure.[10] Four- and one-liter samples of chlorine treated water were assayed for enteroviruses and bacteria, respectively.

Bacteriological Enumeration Procedures

The enumerations performed for total coliforms and fecal streptococci in our studies were made by the membrane filtration procedures as described in *Standard Methods*[10] using m-Endo Broth (Difco) and m-Enterococcus Agar (BBL), respectively.

Table 17-1. Experimental Design

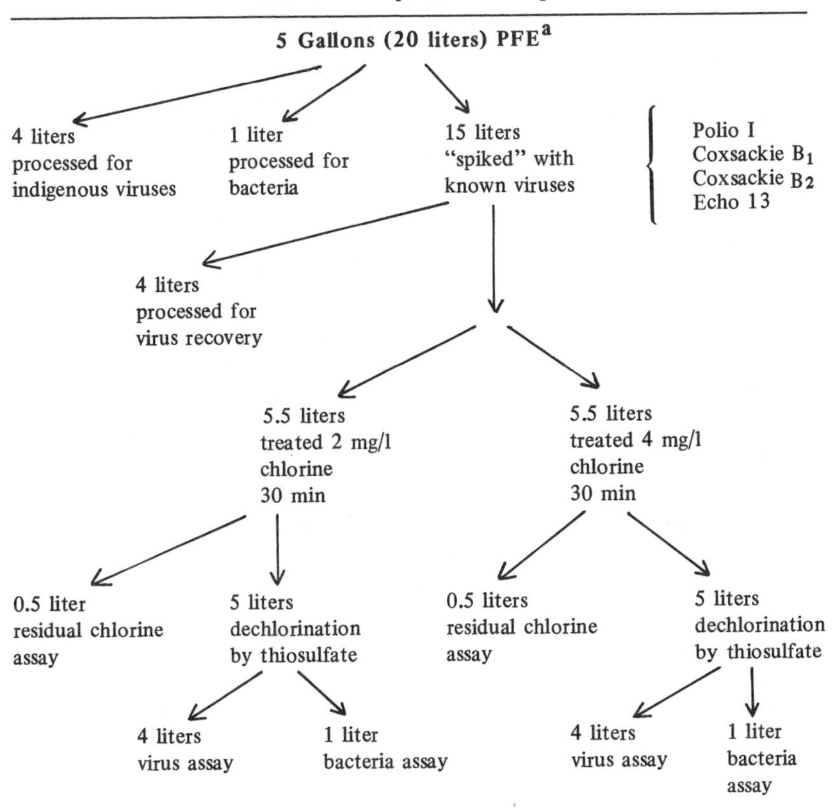

[a]PFE, pilot filter effluent (tertiary treated)

Virus Concentration Procedure

The virus concentration procedure used with our tertiary treated wastewater samples was the aluminum hydroxide adsorption procedure described by Wallis and Melnick[16,17] and modified by Moore, Ludovici and Jeter.[5]

Test Tube Assay of MPN for Coxsackie B1 and 2, and Echovirus 13

Monolayers of VERO in 16 x 125 mm test tubes were washed once with balanced salt solution (BSS) and inoculated in groups of 10 with 0.1 ml of several dilutions (undiluted, 10^{-1}, 10^{-2}) of the water concentrates. After adsorption at 37°C for one hour, the cells were washed once with

Eagles basal medium (EBM) containing 2% fetal calf serum (FCS) and 100 units of penicillin, 50 μg of Streptomycin, and 2.5 μg of fungizone per ml. The tubes were examined microscopically daily for cytopathic effect. The MPN of infectious virus particles was estimated based on the number of positive tube cultures as calculated from standard statistical tables.[41] Values are expressed on the basis of MPN per 4 liters of water.

Plaque Assay for Poliovirus I

Monolayers of HEp-2 in one ounce bottles were washed once with BSS and inoculated in duplicate with 0.1 ml of several dilutions (undiluted, 1:5, 1:50) of the water concentrates. After adsorption at 37°C for one hour, the cells were washed once with EBM containing 2% FCS and 100 units of penicillin, 50 μg of streptomycin, and 2.5 μg of fungizone per ml. The monolayers were covered with a 5-ml agar overlay medium consisting of Earles balanced salts, 1:60,000 neutral red, 0.1% skim milk, 1.5% Bacto-agar, 0.78% $NaHCO_3$, 25 m*M* $MgCl_2$, and 2 m*M* L-glutamine in addition to 100 units of penicillin and 50 μg of streptomycin per ml. Poliovirus I plaques were read at 48 hr after inoculation, and the results expressed in plaque forming units per 4 liters (PFU/l).

Tertiary Treated Wastewater Sample Quality

The effluent sample quality varied somewhat according to the pilot filter loading procedures. Generally, the pH ranged from 7.3-7.6, NH_3-N from 2.8-5.5, organic nitrogen 1.4-5.6, BOD 1.5-8.5, and COD 15-33.

SURVIVAL OF SEEDED POLIOVIRUS IN PFE AFTER 30-MINUTE CHLORINE CONTACT TIME

Table 17-2 presents the survival data for Poliovirus I seeding experiments. It is clear that as the number of plaque-forming units of virus present decreases from a maximum of 21,000 to a minimum of 700 per 4 liters of water, the percentage of Poliovirus reduction with chlorine treatment increases to a maximum of 100%. The range of Poliovirus reduction was 92-100% with a mean of 96.19% for 2 mg/l chlorine treatment compared with 96-100% with a mean of 99.3% for 4 mg/l.

The chlorine residuals (Table 17-2) ranged from 0.31 to 1.2 mg/l and 0.99 to 2.76 mg/l for the 2 and 4 mg/l chlorine doses, respectively.

Table 17-3 presents the comparative data for total coliform and fecal streptococci in these same water samples. The range of total coliform reduction was 89.76-99.98% with a mean of 98.77% for 2 mg/l chlorine treatment, compared with 99.86-99.99% with a mean of 99.91% for 4 mg/l.

Figure 17-2. Survival of Seeded Poliovirus in PFE[a] after 30-Minute Chlorine Contact Time[b]

Experiment Number	Poliovirus Seeded	Chlorine Dosage					
		2 mg/l			4 mg/l		
		Chlorine Residual (mg/l)	Poliovirus Surviving	Poliovirus Reduction (%)	Chlorine Residual (mg/l)	Poliovirus Surviving	Poliovirus Reduction (%)
1	13,500	1.2	752	95	ND[c]	ND[c]	
2	6,300	0.9	137	98	ND[c]	ND[c]	
3	21,000	1.1	560	97	2.1	47	99.7
4	9,000	0.34	560	94	2.24	340	96
5	10,000	0.31	80	99.0	2.04	0	100
6	5,550	0.37	370	94	0.99	50	99.1
7	12,000	0.39	930	92	1.29	130	99.0
8	14,000	0.75	282	98	1.77	160	99.0
9	910	0.63	26	97	2.1	0	100
10	5,250	0.78	27	99.4	2.41	0	100
11	900	0.99	0	100	2.41	0	100
12	700	0.92	0	100	2.76	0	100
$\bar{X}$				96.9			99.3

[a]PFE, pilot filter effluent (tertiary treated).

[b]Indigenous virus concentration of PFE was 1 virus infective unit or less for these 12 samples (March 19 to July 17, 1968).

[c]ND = Not Done.

Table 17-3. Survival of Bacteria in PFE[a] after 30-Minute Chlorine Contact Time (PFE Seeded with Poliovirus 1)

	Chlorine Dosage									
	0 mg/l		2 mg/l				4 mg/l			
Experiment Number	Total Coliform (No/100 ml)	Fecal Streptococci (No/100 ml)	Total Coliform (No/100 ml)	Coliform Reduction (%)	Fecal Streptococci (No/100 ml)	Streptococci Reduction (%)	Total Coliform (No/100 ml)	Coliform Reduction (%)	Fecal Streptococci (No/100 ml)	Streptococci Reduction (%)
1	1270	69	130	89.76	9	86.96	ND	ND	ND	ND
2	83	28	< 1	> 98.80	0.4	98.57	ND	ND	ND	ND
3	1440	68	16	98.89	< 0.4	> 99.41	1.1	99.99	<< 0.4	> 99.41
4	10,700	115	14	99.87	1.5	98.70	8.4	99.92	< 0.4	99.65
5	1800	49	< 0.4	> 99.98	< 0.4	> 99.18	< 0.4	> 99.98	< 0.4	> 99.18
6	500	58	1.5	99.70	< 0.4	> 99.31	< 0.4	> 99.92	< 0.4	> 99.31
7	1050	248	< 0.4	> 99.96	< 0.4	> 99.84	< 0.4	> 99.96	< 0.4	> 99.84
8	163	165	< 0.9	> 99.75	< 0.4	> 99.76	< 0.4	> 99.75	< 0.4	> 99.76
9	280	ND	2.9	98.96	ND	ND	0.4	99.86	ND	ND
10	163	143	< 0.4	> 99.75	< 0.4	> 99.72	< 0.4	> 99.75	0.4	99.72
11	1400	61	< 0.4	> 99.97	3	95.08	< 0.4	> 99.97	< 0.4	> 99.34
12	1920	180	2.9	99.85	< 0.4	> 99.78	< 0.4	> 99.97	< 0.4	> 99.78
$\bar{X}$				99.77		97.84		99.91		99.55

[a]PFE, pilot filter effluent (tertiary treated).

Fecal streptococcus reduction ranged from 86.96-99.78% with a mean of 97.84% for 2 mg/l chlorine treatment, compared with 99.18-99.84% with a mean of 99.55% for 4 mg/l chlorine treated samples tested.

When the percentage reduction figures of Tables 17-2 and 17-3 are compared, it is clear that overall Poliovirus is most resistant followed by total coliforms, and then fecal streptococci to the effects of chlorine.

Survival of Seeded Coxsackie B_2 in PFE After 30-Minute Chlorine Contact Time

Table 17-4 presents the survival data for the Coxsackie B_2 seeding experiments. This virus is apparently more susceptible to chlorine treatment than Poliovirus I for there was a 100% reduction of virus in all cases except experiment 23, where 5120 virus particles were seeded and 80 were recovered with 2 mg/l chlorine treatment of the water.

The chlorine residuals as shown in Table 17-4 ranged from 0.43-1.03 mg/l and 1.5-2.94 mg/l for the 2 and 4 mg/l chlorine doses, respectively.

Table 17-5 presents the comparative data for total coliform and fecal streptococci in these water samples. In this series of experiments, treatment of water with chlorine at the 4 mg/l level resulted in most bacteria being eliminated. Chlorine treatment with 2 mg/l again was less effective, but the resistant organisms in this series were fecal streptococci instead of total coliform. The mean values were: 99.8% reduction for total coliform, compared with only 97.3% reduction for fecal streptoccus in the 2 mg/l chlorine treatment group. Similarly, for 4 mg/l chlorine treatment, the fecal streptococcus were more resistant giving a mean value of 98.9% reduction to 99.8% for total coliforms.

When the percentage reduction figures of Tables 17-4 and 17-5 are compared, it is evident that overall Coxsackie B_2 is most sensitive to chlorine followed by coliforms and fecal streptococci.

Survival of Seeded Coxsackie B_1 in PFE After 30-Minute Contact Time

Table 17-6 presents the survival data for the Coxsackie B_1 seeding experiments. Coxsackie B_1 like B_2 virus was fairly susceptible to chlorine treatment. Even at high levels of input virus, 97-99% was inactivated with 2 mg/l chlorine and 99.3-100% was inactivated with 4 mg/l chlorine. The mean values were 98.6% reduction for 2 mg/l chlorine treatment to 99.8% reduction for 4 mg/l.

The chlorine residuals as shown in Table 17-6 ranged from 0.21-1.4 mg/l and 0.96-3.31 mg/l for the 2 and 4 mg/l chlorine doses, respectively.

Table 17-7 presents the comparative data for the survival of coliforms and fecal streptococci in these water samples. The results were similar

Table 17-4. Survival of Seeded Coxsackie B_2 in PFE[a] after 30-Minute Chlorine Contact Time[b]

		Chlorine Dosage					
		2 mg/l			4 mg/l		
Experiment Number	Coxsackie B_2 Seeded (MPN/4 liter)	Chlorine Residual (mg/l)	Coxsackie B_2 Surviving (MPN/4 liter)	Coxsackie B_2 Reduction (%)	Chlorine Residual (mg/l)	Coxsackie B_2 Surviving (MPN/4 liter)	Coxsackie B_2 Reduction (%)
14	40	1.03	0	100	2.94	0	100
15	40	0.78	0	100	2.86	0	100
16	40	0.92	0	100	2.32	0	100
18	40	0.57	0	100	1.96	0	100
19	484	0.64	0	100	2.77	0	100
20	400	0.96	0	100	2.16	0	100
21	40	0.53	0	100	1.78	0	100
22	2060	0.92	0	100	2.38	0	100
23	5120	0.96	80	98	2.56	0	100
24	1040	0.96	0	100	2.53	0	100
25	3020	0.43	0	100	1.50	0	100
$\bar{X}$				99.8			100

[a]PFE, pilot filter effluent (tertiary treated)

[b]Indigenous virus concentration of PFE was 4 virus infective units per 4 liters or less for these 11 samples (August 7-November 20, 1968).

Table 17-5. Survival of Bacteria[a] in PFE after 30-Minute Chlorine Contact Time (PFE Seeded with Coxsackie B_2)

	Chlorine Dosage									
	0 mg/l		2 mg/l				4 mg/l			
Experiment Number	Total Coliform (No/100 ml)	Fecal Streptococci (No/100 ml)	Total Coliform (No/100 ml)	Coliform Reduction (%)	Fecal Streptococci (No/100 ml)	Streptococci Reduction (%)	Total Coliform (No/100 ml)	Coliform Reduction (%)	Fecal Streptococci (No/100 ml)	Streptococci Reduction (%)
14	290	192	< 0.4	>99.86	< 0.4	> 99.79	< 0.4	> 99.86	< 0.4	> 99.79
15	120	82	< 0.4	>99.67	1.1	98.66	< 0.4	> 99.67	0.7	99.15
16	200	71	0.4	99.80	< 0.4	> 99.44	< 0.4	> 99.80	< 0.4	> 99.44
17	330	87	< 0.4	>99.88	1.5	98.28	< 0.4	> 99.88	0.4	99.54
18	290	71	< 0.4	>99.86	1.1	98.45	<< 0.4	> 99.86	0.4	99.44
19	263	41	< 0.4	>99.85	2.2	94.63	<< 0.4	> 99.85	< 0.4	> 99.02
20	318	49	< 0.4	>99.87	0.4	99.18	<< 0.4	> 99.87	< 0.4	> 99.18
21	118	49	0.4	99.66	0.4	99.18	< 0.4	> 99.66	< 0.4	> 99.18
22	127	79	< 0.4	>99.69	9.8	87.59	< 0.4	> 99.69	< 0.4	> 99.49
23	4700	46	< 0.4	>99.99	< 0.4	> 99.13	< 0.4	> 99.99	< 0.4	> 99.13
24	4830	18	< 0.4	>99.99	< 0.4	> 97.78	< 0.4	> 99.99	< 0.4	> 97.78
25	2540	46	13	99.49	2.2	95.22	< 0.4	> 99.98	1.1	97.61
$\bar{X}$				99.8		97.3		99.8		98.9

[a]PFE, pilot filter effluent (tertiary treated)

Table 17-6. Survival of Seeded Coxsackie B_1 in PFE[a] after 30-Minute Chlorine Contact Time[b]

		Chlorine Dosage					
		2 mg/l			4 mg/l		
Experiment Number	Coxsackie B_1 Seeded (MPN/4 liter)	Chlorine Residual (mg/l)	Coxsackie B_1 Surviving (MPN/4 liter)	Coxsackie B_1 Reduction (%)	Chlorine Residual (mg/l)	Coxsackie B_1 Surviving (MPN/4 liter)	Coxsackie B_1 Reduction (%)
26	98,200	0.28	2,000	98	1.53	400	99.7
27	88.000	0.21	1,600	98	1.31	394	99.5
28	24,400	0.67	800	97	0.96	160	99.3
29	22,800	0.96	240	99	2.66	8	99.9
30	920	0.78	0	100	1.84	0	100
31	27,600	1.40	240	99	3.31	0	100
32	27,600	1.27	560	98	2.96	0	100
33	4,722	1.30	32	99	3.28	0	100
34	4,722	1.23	16	99	2.76	0	100
35	12,600	1.24	80	99	2.82	0	100
$\bar{X}$				98.6			99.8

[a]PFE, pilot filter effluent (tertiary treated)

[b]Indigenous virus concentration of PFE was 0-30 virus infective units per 4 liters for these 10 samples (December 4, 1968-March 4, 1969).

Table 17-7. Survival of Bacteria in PFE[a] after 30-minute Chlorine Contact Time (PFE Seeded with Coxsackie B_1)

	Chlorine Dosage									
	0 mg/l		2 mg/l				4 mg/l			
Experiment Number	Total Coliform (No/100 ml)	Fecal Streptococci (No/100 ml)	Total Coliform (No/100 ml)	Coliform Reduction (%)	Fecal Streptococci (No/100 ml)	Streptococci Reduction (%)	Total Coliform (No/100 ml)	Coliform Reduction (%)	Fecal Streptococci (No/100 ml)	Streptococci Reduction (%)
26	6350	181	13	99.80	23	87.29	<0.4	>99.99	8	95.58
27	11,700	36	1.8	99.98	5.1	87.2	<0.4	>99.99	<0.4	>98.89
28	19,100	250	6.4	99.97	4.8	98.08	1.1	99.99	<0.4	>99.84
29	6900	200	1.1	99.98	25	87.50	<0.4	>99.99	4.3	97.85
30	4900	190	0.6	99.99	6.0	96.84	<0.3	>99.99	<0.3	>99.84
31	10,600	200	7.7	99.93	1.7	99.15	<0.3	>99.99	<0.3	>99.85
32	12,300	350	1.7	99.99	1.1	99.69	<0.3	>99.99	0.9	99.74
33	8720	27	2.3	99.97	2.3	91.48	<0.3	>99.99	0.9	96.67
34	2640	9	5.7	99.78	0.9	90.00	<0.3	>99.99	<0.3	96.67
35	1360	12	0.3	99.98	0.3	97.50	<0.3	>99.98	<0.3	>97.50
$\bar{X}$				99.9		93.5		99.9		98.2

[a]PFE, pilot filter effluent (tertiary treated).

to those observed in the previous series, namely fecal streptococci were more resistant to chlorine than coliforms and the higher concentration (4 mg/l) was necessary to eliminate most bacteria from the PFE. The mean values were 99.9% reduction for total coliform, compared with only 93.5% reduction for fecal streptococcus in the 2 mg/l chlorine treatment group. For 4 mg/l chlorine treatment, the total coliform mean value was 99.9% reduction, compared with 98.2% for fecal streptococci.

From a comparative standpoint in this series, the fecal streptococci were most resistant to chlorine, and the total coliforms and Coxsackie B_1 type viruses were approximately equal in susceptibility to chlorine inactivation.

Survival of Seeded Echovirus 13 in PFE After 30-Minute Contact Time

Table 17-8 presents the survival data for the echovirus type 13 seeding experiments. The data indicate that increasing the chlorine dosage from 2-4 mg/l had an increased destructive effect on echovirus 13 in one sample but no effect in two other samples. This result apparently was not affected by virus input level. The percentage reduction value of 99.9% remained the same for both 2 and 4 mg/l chlorine treatment groups.

The chlorine residuals as shown in Table 17-8 ranged from 0.7 to 1.33 mg/l and 1.5 to 3.2 mg/l for the 2 and 4 mg/l chlorine doses, respectively.

Table 17-9 presents the corresponding survival data for coliforms and fecal streptococci in these water samples. The mean values were 99.7% reduction for total coliforms, compared to only 89% for fecal streptococcus in the 2 mg/l chlorine treated group; and 99.8% reduction for total coliforms, compared to 96% for fecal streptococcus in the 4 mg/l chlorine treated group.

The comparative data for these four experiments suggest that the fecal streptococci were again most resistant to chlorine treatment while the total coliforms were slightly more resistant than echo 13 viruses to chlorine inactivation.

Comparative Survival of Seeded Enteroviruses and Bacteria in PFE after 30-Minute Chlorine Contact Time

Table 17-10 summarizes the data for our study. Data for the total coliforms or fecal streptococcus groups were averaged for all four experiments. Overall the fecal streptococci were the most resistant microorganisms in the study. The two viruses, Polio I and Coxsackie B_1 were next in resistance to chlorine inactivation, followed closely by the total coliforms. The viruses Coxsackie B_2 and Echo 13 appeared to be most sensitive. However, Echo 13 showed no increased destruction when the dose of chlorine was

Table 17-8. Survival of Seeded Echo 13 in PFE[a] after 30-Minute Chlorine Contact Time[b]

Experiment Number	Echo 13 Seeded (MPN/4 liter)	Chlorine Dosage					
		2 mg/l			4 mg/l		
		Chlorine Residual (mg/l)	Echo 13 Surviving (MPN/4 liter)	Echo 13 Reduction (%)	Chlorine Residual (mg/l)	Echo 13 Surviving (MPN/4 liter)	Echo 13 Reduction (%)
36	16,000	1.33	8	99.9	2.51	8	99.9
37	16,000	1.26	8	99.9	2.78	0	100.0
38	8,000	1.2	0	100.0	3.2	0	100.0
39	9,200	0.7	8	99.9	1.5	8	99.9
$\bar{X}$				99.9			99.9

[a]PFE, pilot filter effluent (tertiary treated)

[b]Indigenous virus concentration of PFE was less than 1 virus infective unit per 4 liters for these four samples (March 11-April 8, 1969).

Table 17-9. Survival of Bacteria in PFE[a] after 30-Minute Chlorine Contact Time (PFE Seeded with Echovirus 13)

	Chlorine Dosage									
	0 mg/l		2 mg/l				.4 mg/l			
Experiment Number	Total Coliform (No/100 ml)	Fecal Streptococci (No/100 ml)	Total Coliform (No/100 ml)	Coliform Reduction (%)	Fecal Streptococci (No/100 ml)	Streptococci Reduction (%)	Total Coliform (No/100 ml)	Coliform Reduction (%)	Fecal Streptococci (No/100 ml)	Streptococci Reduction (%)
36	820	36	<0.3	>99.96	0.9	97.50	<0.3	>99.96	0.6	98.33
37	140	10	<0.3	>99.79	2.3	77.00	<0.3	>99.79	0.9	91.00
38	110	9	1.0	99.09	1.0	88.89	<0.3	>99.73	<0.3	96.67
39	1090	15	<0.3	>99.97	1.1	92.67	<0.3	>99.97	<0.3	98.00
$\bar{X}$				99.7		89.0		99.8		96.0

[a]PFE, pilot filter effluent (tertiary treated).

Table 17-10. Comparative Survival of Enteroviruses and Indigenous Bacteria in Tertiary Treated Wastewater (PFE) after 30-Minute Chlorine Contact Time

	Mean Percentage Reduction	
	2 mg/l Cl	4 mg/l Cl
Fecal Streptococci	94.4	98.2
Polio I	96.9	99.3
Coxsackie B_1	98.6	99.8
Total coliform	99.6	99.9
Coxsackie B_2	99.8	100.0
Echo 13	99.9	99.9

increased from 2-4 mg/l and this explains the only discrepancy in the ranking of the organisms for 2-4 mg/l in the summary of Table 17-10.

Overall, the data indicate that chlorine treatment at 2 and 4 mg/l for 30 minutes contact further reduced the viral and bacterial levels in Tucson's tertiary treated wastewater (PFE). However, it is clear that if the concentration of these enteroviruses was high enough they could survive such chlorine treatment at levels which presumably could cause diseases.

Our results with Coxsackie virus B_2 probably can explain why we have never isolated Coxsackieviruses from Tucson's wastewater, and the data for the remaining organisms, except for fecal streptococci, appear to be in general accord with results of other workers. For example, the data of investigators mentioned previously, plotted by Clarke *et al.* as the chlorine concentration required to produce 99% destruction of *E. coli* and several viruses in several types of water, indicated that Poliovirus I was more resistant than *E, coli* (the predominant organism in the coliform group) while Adenovirus 3 was less resistant.[45] The plot for Coxsackie A_2 virus showed even greater resistance to chlorine than Poliovirus. We worked with Coxsackie B viruses rather than the A types, and different types of water were involved, so that the results are not comparable. Our data, comparable with that of Liu *et al.*,[65] who also studied enteroviruses treated in natural waters, show that Coxsackie B_1 virus was more sensitive to chlorine inactivation than Poliovirus I. Unfortunately, Liu *et al.* did not include Coxsackie B_2 virus and Echovirus 13 in their studies.

Burns and Sproul demonstrated that coliphage T_2 was more resistant to chlorine inactivation than coliform organisms and they questioned, as others have done, the reliability of the coliform test as an indicator of the effectiveness of virus chlorination.[64] Lothrop and Sproul showed that although coliphage T_2 was much less sensitive to chlorine than the coliform organism, it was consistently more sensitive than Poliovirus I.[62] Recently, Shah and McCamish showed that phage f_2 (an RNA virus) was markedly more resistant to chlorine than either Poliovirus or coliphage T_2 (a DNA virus) and they suggested its possible use as an indicator of viral pollution of wastewater.[54]

We have been unable to find any similar comparative studies of fecal streptococci with viruses. Geldreich and Kenner have pointed out the limited sanitary significance of fecal streptococci as indicators of stream pollution since *Streptococcus faecalis* var. liquifaciens is a ubiquitous organism which persists in such things as storm wastes and river water.[68] They recommend, however, the use of fecal coliform to fecal streptococci ratios to define possible sources of fecal discharge into streams. Our findings, that the fecal streptococcus group overall was more resistant to chlorine than the enteroviruses studied, are significant in the search for a reliable indicator microorganism to assess the microbiological safety of chlorine treated wastewater. In this respect, the development of a rapid fluorescent antibody identification method for fecal streptococci by Pavlova *et al.*,[69] which could lend itself to automation for continuous wastewater monitoring, may be extremely useful.

From the data presented in this chapter, fecal streptococci would certainly be a more reliable indicator of the microbiological safety of wastewater than the coliform bacteria presently used. However, we believe that reclaimed wastewater or potentially sewage-contaminated surface waters should be subjected to a direct viral assay for enteric viruses *per se* before they are considered microbiologically safe for human use. Dependence on a bacterial indicator organism, especially coliform bacteria, to judge the total efficiency of water treatment, disinfection and the microbiological safety of wastewater is not only unjustified with our present knowledge, but is also potentially dangerous.

Because of its simplicity, economy and usefulness for so many years, it is clear that the coliform bacteria index will not be easily abandoned. However, concentrated and intensified research efforts must be made to develop more reliable and sensitive indicator systems.

REFERENCES

1. Mosley, J. W. "Transmission of Viral Diseases by Drinking Water," In *Transmission of Viruses by the Water Route,* G. Berg, Ed. (New York: Wiley-Interscience, 1967), pp. 5-24.
2. Melnick, J. L. "Detection of Virus Spread by the Water Route," In *Proc. 13th Water Qual. Conf.*, Virus and Water Quality: Occurrence and Control, University of Illinois, Urbana (1971), pp. 114-124.
3. Weibel, S. R., F. R. Dixon, R. B. Weidner and L. J. McCabe. "Waterborne-Disease Outbreaks, 1946-60," *J. Amer. Water Works Assoc.* **56**, 947-957 (1964).

4. Craun, G. F. and L. J. McCabe. "Review of the Causes of Water-borne-Disease Outbreaks," *J. Amer. Water Works Assoc.* **65**, 74-84 (1973).
5. Moore, M. L., P. P. Ludovici and W. S. Jeter. "Quantitative Methods for the Concentration of Viruses in Wastewater," *J. Water Pollution Control Fed.* **42**, R21-R28 (1970).
6. Grabow, W. O. K. "The Virology of Wastewater Treatment," *Water Res.* **2**, 675-701 (1968).
7. Berg, G. "Integrated Approach to Problem of Viruses in Water," *J. San. Eng.* **97**, 867-882 (1971).
8. Chambers, C. W. "Chlorination for Control of Bacteria and Viruses in Treatment Plant Effluents," *J. Water Pollution Control Fed.* **43**, 228-241 (1971).
9. Shuval, H. I. and E. Katzenelson. "The Detection of Enteric Viruses in the Water Environment," In *Water Pollution Microbiology*, R. Mitchell, Ed. (New York: Wiley-Interscience, 1972), pp. 347-361.
10. American Public Health Association. *Standard Methods for the Examination of Water and Wastewater*, 12th ed. (New York: 1965).
11. American Public Health Association. *Standard Methods for the Examination of Water and Wastewater*, 13th ed. (New York: 1971).
12. Moore, B. "The Detection of Paratyphoid Carriers in Towns by Means of Sewage Examination," *Monthly Bull. Med. Res. Council (Great Britain)* **8**, 241 (1948).
13. Melnick, J. L., J. Emmons, E. M. Opton and J. H. Coffey. "Coxsackie Viruses from Sewage Methodology Including an Evaluation of the Grab Sample and Gauze Pad Collection Procedure," *Amer. J. Hyg.* **59**, 185-195 (1954).
14. Cliver, D. O. "Enterovirus Detection by Membrane Chromatography," In *Transmission of Viruses by the Water Route*, G. Berg, Ed. (New York: Wiley-Interscience, 1967), pp. 139-141.
15. Wallis, C. and J. L. Melnick. "Concentration of Viruses from Sewage by Adsorption on Millipore Membranes," *Bull. WHO* **36**, 219 (1967).
16. Wallis, C. and J. L. Melnick. "Concentration of Viruses on Aluminum Phosphate and Aluminum Hydroxide Precipitates," In *Transmission of Viruses by the Water Route*, G. Berg, Ed. (New York: Wiley-Interscience, 1967), pp. 129-138.
17. Wallis, C. and J. L. Melnick. "Concentration of Viruses on Aluminum and Calcium Salts," *Amer. J. Epidemiol.* **85**, 459-468 (1967).
18. Wallis, C. and J. L. Melnick. "A Portable Virus Concentration for Use in the Field," *Adv. Water Pollution Res.*, Proc. 6th Internat. Conf. 119-131 (1972).
19. Gibbs, T. and D. O. Cliver. "Methods for Detecting Minimal Contamination with Reovirus," *Health Lab. Sci.* **2**, 81-88 (1965).
20. Shuval, H. I., S. Cymbalista, B. Fattal and N. Goldblum. "The Concentration of Enteric Viruses in Water by Hydroextraction and Two-Phase Separation," In *Transmission of Viruses by the Water Route*, G. Berg, Ed. (New York: Wiley-Interscience, 1967), pp. 45-56.
21. Cliver, D. O. "Detection of Enteric Viruses by Concentration with Polyethylene Glycol," In *Transmission of Viruses by the Water Route*, G. Berg, Ed. (New York: Wiley-Interscience, 1967), pp. 109-120.

22. Gartner, H. "Retention and Recovery of Polioviruses on a Soluble Ultrafilter," In *Transmission of Viruses by the Water Route,* G. Berg, Ed. (New York: Wiley-Interscience, 1967), pp. 121-128.
23. Nupen, E. M. and G. J. Stander. "The Virus Problem in the Windhoek Waste Water Reclamation Project," *Adv. Water Pollution Res.,* Proc. 6th Internat. Conf. (1972), pp. 133-145.
24. Albertsson, P. A. "Partition of Proteins in Liquid Polymer-Polymer Two-Phase Systems," *Nature* **182**, 709-711 (1958).
25. Albertsson, P. A. "Particle Fractionation in Liquid Two-Phase Systems. The Composition of Some Phase Systems and the Behaviour of Some Model Particles in Them. Application to the Isolation of Cell Walls from Microorganisms," *Biochim. Biophys. Acta* **27**, 373-395 (1958).
26. Frick, G. and P. A. Albertsson. "Bacteriophage Enrichment in a Two-Phase System with Subsequent Treatment with 'Freon' 113," *Nature* **183**, 1070-1072 (1959).
27. Shuval, H. I., B. Fattal, S. Cymbalista and N. Goldblum. "The Phase-Separation Method for the Concentration and Detection of Viruses in Water," *Water Res.* **3**, 225-240 (1969).
28. Taverne, J., I. H. Marshall and R. Fulton. "The Purification and Concentration of Viruses and Virus Soluble Antigens on Calcium Phosphate," *J. Gen. Microbiol.* **19**, 451-461 (1958).
29. Salk, J. E. "Partial Purification of the Virus of Epidemic Influenza by Absorption on Calcium Phosphate," *Proc. Soc. Exp. Biol. Med.* **46**, 709-712 (1941).
30. Grossowitz, N., A. Mercado and N. Goldblum. "A Simple Method for Concentration of Live and Formaldehyde-Inactivated Poliovirus," *Proc. Soc. Exp. Biol. Med.* **103**, 872-877 (1960).
31. Konowalchuk, J. and J. I. Speirs. "Enterovirus Recovery with Vegetable Floc," **26**, 505-507 (1973).
32. England, B. "Concentration of Reovirus and Adenovirus from Sewage and Effluents by Protamine Sulfate (Salmine) Treatment," *Appl. Microbiol.* **24**, 510-512 (1972).
33. Kelly, S. M. and W. W. Sanderson. "Effect of Chlorine in Water on Enteric Viruses. II. Effect of Combined Chlorine on Poliomyelitis," *Amer. J. Publ. Health* **50**, 14-19 (1953).
34. Gravelle, C. R. and T. D. Y. Chin. "Enterovirus Isolations from Sewage: A Comparison of Three Methods," *J. Infect. Diseases* **109**, 205-209 (1967).
35. Smith, J. E. and R. J. Courtney. "Concentration and Detection of Viral Particles by Passive Hemagglutination," In *Transmission of Viruses by the Water Route,* G. Berg, Ed. (New York: Wiley-Interscience, 1967), pp. 89-108.
36. Johnson, J. H., J. E. Fields and W. A. Darlington. "Removing Viruses from Water by Polyelectrolytes," *Nature (London)* **213**, 655-657 (1967).
37. Wallis, C., S. Grinstein, J. L. Melnick and J. E. Fields. "Concentration of Viruses from Sewage and Excreta on Insoluble Polyelectrolytes," *Appl. Microbiol.* **18**, 1007-1014 (1969).
38. Grinstein, S., J. E. Melnick and C. Wallis. "Virus Isolations from Sewage and from a Stream Receiving Effluent of Sewage Treatment Plants," *Bull. WHO* **42**, 291-296 (1970).

39. Lund, E. and C. E. Hedstrom. "Recovery of Viruses from a Sewage Treatment Plant," In *Transmission of Viruses by the Water Route*, G. Berg, Ed. (New York: Wiley-Interscience, 1967), pp. 371-378.
40. Grinrod, J. and D. O. Cliver. "A Polymer Two-Phase System Adapted to Virus Detection," *Arch. Ges. Virusforsch.* **31**, 365-372 (1970).
41. Chang, S. L., G. Berg, K. A. Busch, R. E. Stevenson, N. A. Clarke and P. W. Kabler. "Application of the 'Most Probable Number' Method for Estimating Concentrations of Animal Viruses by Tissue Culture Technique," *Virology* **6**, 27-42 (1958).
42. Weidenkopf, S. "Inactivation of Type I Poliomyelitis Virus with Chlorine," *Virology* **5**, 56-67 (1958).
43. Kelly, S. M. and W. W. Sanderson. "The Effect of Chlorine in Water on Enteric Viruses," *Amer. J. Publ. Health* **48**, 1323-1333 (1958).
44. Kelly, S. M. and W. W. Sanderson. "The Effect of Chlorine in Water on Enteric Viruses. II. Effect of Combined Chlorine on Poliomyelitis," *Amer. J. Publ. Health* **50**, 14-19 (1960).
45. Clarke, N., G. Berg, P. Kabler and S. Chang. "Human Enteric Viruses in Water: Source, Survival and Removability," *Adv. Water Pollution Res.* **2**, 523-541 (1968).
46. Clarke, N. and S. Chang. "Enteric Viruses in Water," *J. Amer. Water Works Assoc.* **51**, 1299-1317 (1959).
47. Kjellander, J. and E. Lund. "Sensitivity of *E. coli* and Poliovirus to Different Forms of Combined Chlorine," *J. Amer. Water Works Assoc.* **57**, 893-900 (1965).
48. Berg, G. "Virus Transmission by the Water Vehicle. III. Removal of Viruses by Water Treatment Procedures," *Health Lab. Sci.* **3**, 170 (1966).
49. Wattie, E. and C. T. Butterfield. "Relative Resistance of *E. coli* and Eberthella typhosa to Chlorine and Chloramines," *Publ. Health Rep.* **59**, 1661 (1944).
50. Butterfield, C. T. and E. Wattie. "Influence of pH and Temperature on the Survival of Coliforms and Enteric Pathogens when Exposed to Chloramine," *Publ. Health Rep.* **61**, 157-192 (1946).
51. Clarke, N. A. and P. W. Kabler. "The Inactivation of Purified Coxsackie Virus in Water by Chlorine," *Amer. J. Hyg.* **59**, 119-127 (1954).
52. Clarke, N. A., R. E. Stevenson and P. W. Kabler. "The Inactivation of Purified Type 3 Adenovirus in Water by Chlorine," *Amer. J. Hyg.* **64**, 314-319 (1956).
53. Scarpino, P. V., G. Berg, S. L. Chang, D. Dahling and M. Lucas. "A Comparative Study of the Inactivation of Viruses in Water by Chlorine," *Water Res.* **6**, 959-965 (1972).
54. Shah, P. and J. McCamish. "Relative Resistance of Poliovirus 1 and Coliphase f_2 and T_2 in Water," *Appl. Microbiol.* **24**, 658-659 (1972).
55. Dahling, D. R., P. V. Scarpino, M. Lucas, G. Berg and S. L. Chang. "Destruction of Viruses and Bacteria in Water by Chlorine," Presented at the Annual Meeting of the Amer. Soc. for Microbiology (1972).
56. Neefe, J., J. Stokes and J. Baty. "Inactivation of the Virus of Infectious Hepatitis in Drinking Water," *Amer. J. Publ. Health* **37**, 365-372 (1947).

57. Cramer, W. and C. W. Krusé. "Disinfection of f_2 Bacteria-Virus and Poliovirus by Chlorine and Iodine," Presented at the Annual Meeting of the American Society for Microbiology (1973).
58. Wyckoff, H. A. "Coliform Death Rates Resulting from Chlorination of Raw Sewage," *Sewage Works J.* **21**, 484-490 (1949).
59. McKee, J. E., J. Brokaw and R. T. McLaughlin. "Chemical and Colicidal Effects of Halogens in Sewage," *J. Water Pollution Control Fed.* **32**, 795 (1960).
60. Lothrup, T. L. and O. J. Sproul. "High-Level Inactivation of Viruses in Waste Water by Chlorination," *J. Water Pollution Control Fed.* **41**, 567-575 (1969).
61. Shuval, H., S. Cymbalista, A. Wachs, Y. Zohar and N. Goldblum. "The Inactivation of Enteroviruses in Sewage by Chlorination," In *Adv. Water Pollution Res.* Proc. 3rd Internat. Conf. Water Pollution Res., Water Pollution Control Federation, Washington, D.C. **2**, 37 (1966).
62. Marais, A., E. Nupen, G. Stander and V. Hoffman. "A Comparison of the Inactivation of *E. coli* and Poliovirus in Polluted and Unpolluted Waters by Chlorination," Presented at the Conf. of Water for Peace, Washington, D.C. (May 23-30, 1967), pp. 670-687.
63. Warriner, T. "Field Tests on Chlorination of Poliovirus in Sewage," *J. San. Eng. Div. ASCE* **93**, 51-65 (1967).
64. Burns, R. W. and O. J. Sproul. "Virucidal Effects of Chlorine in Waste Water," *J. Water Pollution Control Fed.* **39**, 1834-1849 (1967).
65. Liu, O. C., H. R. Seraichekas, E. W. Akin, D. A. Brashear, E. L. Katz and W. J. J. Hill. "Relative Resistance of Twenty Human Enteric Viruses to Free Chlorine in Potomac Water," In *Proc. 13th Water Qual. Conf.*, Virus and Water Quality: Occurrence and Control, University of Illinois, Urbana (1971), pp. 171-195.
66. England, B., R. E. Leach, B. Adams and R. Shiosaki. "Virologic Assessment of Sewage Treatment at Santee, California," In *Transmission of Viruses by the Water Route*, G. Berg, Ed. (New York: Wiley-Interscience, 1967).
67. World Health Organization. *European Standards for Drinking Water* (Geneva, Switzerland: 1970).
68. Geldreich, E. E. and B. A. Kenner. "Concepts of Fecal Streptococci in Stream Pollution," *J. Water Pollution Control Fed.* **41**, R336-R352 (1969).
69. Pavlova, M. T., E. Beauvais, F. T. Brezenski and W. Litsky. "Rapid Assessment of Water Quality by Fluorescent Antibody Identification of Fecal Streptococci," In *Adv. Water Pollution Res.*, Proc. 6th Internat. Conf. (1971), pp. 63-72.

CHAPTER 18

DISINFECTION OF WASTEWATER EFFLUENTS FOR VIRUS INACTIVATION

John T. Cookson, Jr.

University of Maryland
College Park, Maryland

C. Michael Robson

Matz, Childs & Associates
Baltimore, Maryland

INTRODUCTION

There is no doubt that secondary sewage effluents will discharge significant virus concentrations. There are over 100 viruses shed from the enteric and urinary tracts. These viruses may pass through sewage treatment plants, persist in water courses, and enter water supply facilities.

Enteric viruses have been isolated from surface water samples collected and studied throughout the world [1] (Table 18-1). In the United States water samples taken from polluted tidal rivers contained enteric viruses 27-52% of the time[2] and much of this work was done before sensitive virus detection techniques were available. Using newly developed techniques for sampling river water in the Houston (Texas) area, nearly 100% of samples collected 5 miles downstream from the nearest sewage outfall were positive for enteric viruses.[3]

Considerable data are available on viral contamination of surface waters. Berg[4] recovered 19 viral plaque forming units (PFU) per 50 gallons of Ohio River water. This constitutes a considerable hazard since these waters are used for drinking water supplies. This concentration also exceeds the level of one detectable infectious virus unit per 10 gallons as recommended for recreational waters.[5]

Table 18-1. Viral Contamination of Surface and Drinking Water

Type Water	Percent of Samples Positive for Enteric Virus (%)	Study/Reference
River water (France)	21	Coin[6]
River water (France)	9	Foliquet[7]
Drinking water (France)	8	Foliquet[7]
River water (Moscow)	34	Bagdusaryan[8]
River water (Swiss)	38	Farroh[9]
River water (Swiss)	63	Farroh[9]
Domestic water supplies (Israel)	2	Shuval[10]
Domestic water supplies (England)	56	Akin[1]
Tidal rivers (U.S.)	27-52	Metcalf[2]
Illinois River (U.S.)	27	Akin[1]

In many instances the discharge of essentially virus-free effluents is necessary for adequate water management. This need becomes more pronounced when receiving waters provide little dilution, and as total renovation becomes the goal in wastewater treatment. A comprehensive plan was adopted (Resolution No. 7-1071, February 13, 1973) by Montgomery County, Maryland for the elimination of virus hazards associated with interim sewage treatment facilities. This resolution requires that the effluent discharge shall contain no more than one detectable infectious virus unit per 10 gallons. The adoption of this standard is probably the first in the country directed at virus control, and will likely provide the impetus for further attention to virus removal. Other communities in Maryland have acted to provide for virus inactivation in new wastewater treatment facilities.

Establishing a standard for virus may raise questions when such action has not been taken by the Environmental Protection Agency, and no standards for viruses in drinking water are being proposed in the revised drinking water standards.[11] Several sensitive procedures are available for detecting viruses in water, and local authorities must not wait for others to take responsive active. This philosophy was very adequately presented by Dr. Melnick of the Department of Virology and Epidemiology, Baylor College of Medicine[5] and adopted by Montgomery County.

Viruses are pathogens; they cause disease and sometimes kill. Some viruses produce disease years after the initial infection. Regardless of one's philosophy on establishing virus standards, treatment plants can be designed and operated to effectively remove viruses. Direct testing for virus removal is not always needed when processes and operating procedures are used in which viruses are reliably removed. The design of these processes and their operating controls will be discussed in this chapter.

EFFICACY OF VIRUS REMOVAL

Water can be adequately treated so that it is always biologically safe. Biological, chemical and physical treatment procedures all remove viruses from wastewaters. The major problem is that no single process will remove viruses to the degree required for a safe effluent. Each process has a different efficiency for virus removal, and the efficiency of some depends on the degree of pretreatment provided. This is indeed true for disinfection. Disinfection is the one process capable of removing all viruses, but it is not effective on raw sewage or secondary effluents.

The application of advanced waste treatment processes make virus-free effluents possible. Enteric virus inactivation in primary treatment is little if any. Only 3% virus removal has been found in primary settling, and that was attained after 3 hours.[12] Enteric virus removal in secondary treatment processes varies from 9-96% in field studies.[12] In many cases, no virus removal was achieved with trickling filters and stabilization ponds. Viruses are removed from sewage more efficiently by activated sludge treatment than by any other biological process. However, its efficiency is far from adequate for producing a safe effluent.

There is not yet a single treatment process that will devitalize virus to produce safe water. However, a combination of proper treatment processes will provide complete virus inactivation and remove health hazards. The ability of a treatment plant to achieve this removal depends on the interaction of individual treatment systems followed by terminal disinfection.

DISINFECTION FOR VIRUS REMOVAL

One of the most important processes for virus removal is disinfection. Proper disinfection can provide an essentially virus-free effluent. However, the term "proper" refers to much more than the disinfection process itself. At present, only a combination of treatment processes designed to meet specific standards and followed by terminal disinfection will produce complete virus removal. The ability of terminal disinfection to produce safe water depends on the previous treatment processes and their performance in removing substances that interfere with virus disinfection. This aspect is most important in providing proper disinfection and therefore a virus-free effluent.

Substances interfering with the disinfection of viruses can be represented by several general parameters and two specific parameters (Table 18-2).

Table 18-2. Pretreatment Significance and Control Parameters

Interfering Substance To Be Removed	Control Parameter
Soluble Organics	BOD_5, COD or TOC
Suspended Matter	Turbidity
Hydrogen Ion Concentration	pH
Ammonia[a]	Ammonia

[a]Interference is only related to need for excessive chlorine dosage to breakpoint.

The general parameters include soluble organics and suspended matter. Soluble organic compounds can be measured as Five Day Biochemical Oxygen Demand (BOD_5), and suspended matter as turbidity. A control on BOD_5 is necessary to prevent high disinfectant demands. Soluble organic concentrations above 10 mg/l as BOD_5 make disinfection for virus inactivation difficult because of their reducing demand. Control on turbidity is needed because viruses, being very small, become enmeshed in suspended solids that impart turbidity to waters. Turbidity will enable the viruses to escape the disinfecting action. It is therefore necessary that turbidities before terminal disinfection be no more than one Jackson Turbidity Unit (JTU). Low turbidities are also necessary to assure low disinfectant demands.

The two specific parameters refer to pH and ammonia, and are significant when chlorine is used as the disinfectant. The pH of the water during disinfection with chlorine must not exceed 7.5. At higher pH values, the disinfectant HOCl ionizes to the poorly virucidal hypochlorite ion, OCl^-.[13,14]

A control on ammonia is necessary if chlorine is to be used for terminal disinfection. There are several reasons which make ammonia undesirable in effluents, including the following:

(a) Ammonia consumes dissolved oxygen in the receiving water
(b) Ammonia is toxic to aquatic life
(c) Ammonia reacts with chlorine to form chloramines
(d) Ammonia increases the chlorine demand for disinfection.

It is for the latter two reasons that ammonia concentration should be low when chlorine is used for disinfection. It has been shown that combined forms of halogens such as chloramines possess very little, if any, virucidal properties.[4,15-17] Chloramines are also toxic to fish. Since chloramines are not virucidal, chlorine must be added to achieve free chlorine residuals, referred to as breakpoint. Highly nitrified effluents contain less ammonia and organic matter (BOD), reducing the chlorine requirement for virus removal.

TREATMENT PROCEDURES FOR VIRUS-FREE EFFLUENT

Various treatment schemes can be developed that will provide a virus-free effluent.[12] Some typical schemes are provided in Figures 18-1 through 18-5. There are other alternates that can be used depending on the characteristics of the waste, but those presented are typical for most wastewaters. Wastes containing large industrial discharges may need other schemes, and other variations may occur with nitrogen removal.

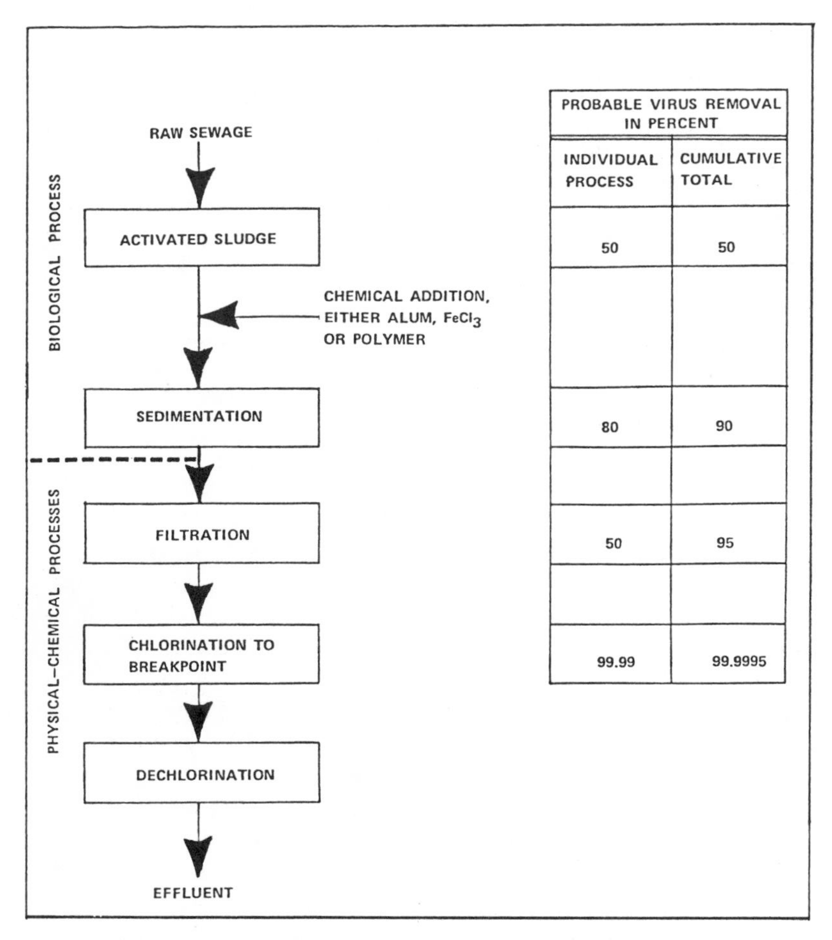

Figure 18-1. Scheme 1: Wastewater treatment for virus removal using chlorine.

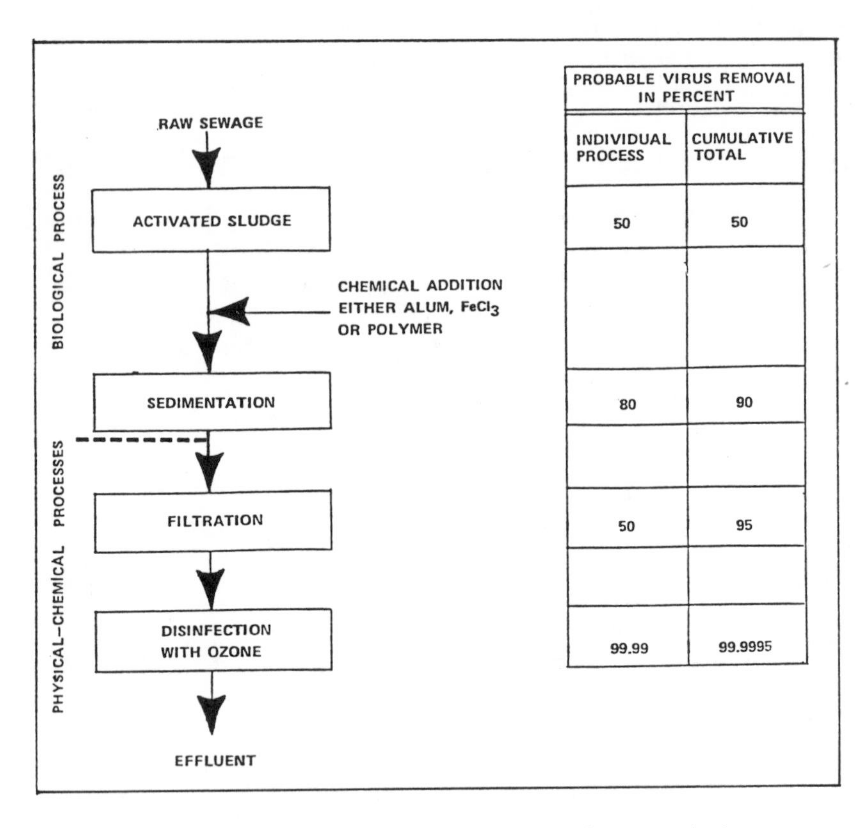

Figure 18-2. Scheme 2: Wastewater treatment for virus removal using ozone or bromine.

Although treatment schemes vary in individual processes, they all have one common feature—the use of terminal disinfection for essentially complete virus removal. As Figures 18-1 through 18-5 illustrate, individual treatment processes remove viruses to varying degrees, and thereby constitute adjunctive removal. The major importance of the individual processes toward virus elimination is the removal of those substances that interfere with terminal disinfection. Thus, they facilitate the final achievement of total removal or destruction of viruses.

Standards can be proposed to control these interfering substances, and therefore make virus removal possible by chlorine disinfection. Proposed standards are given in Table 18-3. Each parameter can be routinely monitored, and treatment systems can be easily designed and operated to conform to these standards.

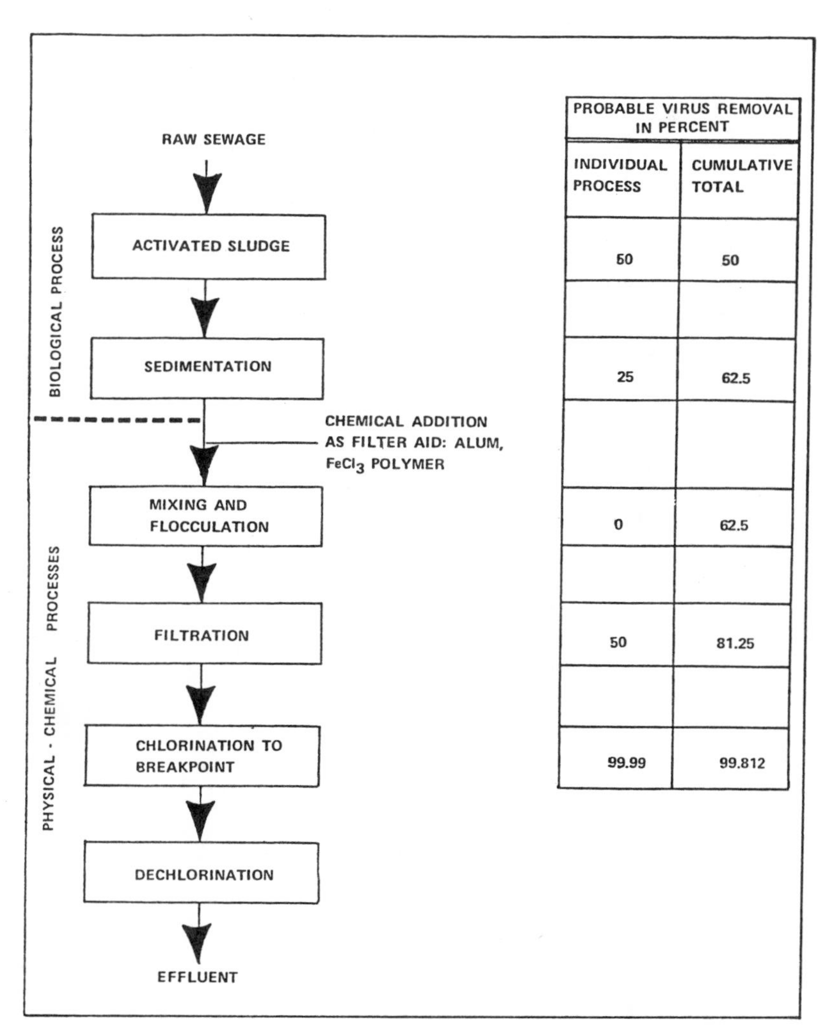

Figure 18-3. Scheme 3: Wastewater treatment for virus removal using a filter aid.

In Scheme 1 (Figure 18-1) chlorine is used for terminal disinfection. Unless nitrogen is to be removed by breakpoint chlorination, ammonia as well as turbidity must be removed to the limits specified in Table 18-3. Ammonia can be removed by designing for complete nitrification. High turbidity removal can be achieved by chemical addition to improve clarification as well as providing for phosphorus precipitation in the activated sludge clarifier. These combined processes will make breakpoint chlorination easy to achieve, providing for complete virus removal. Dechlorination with

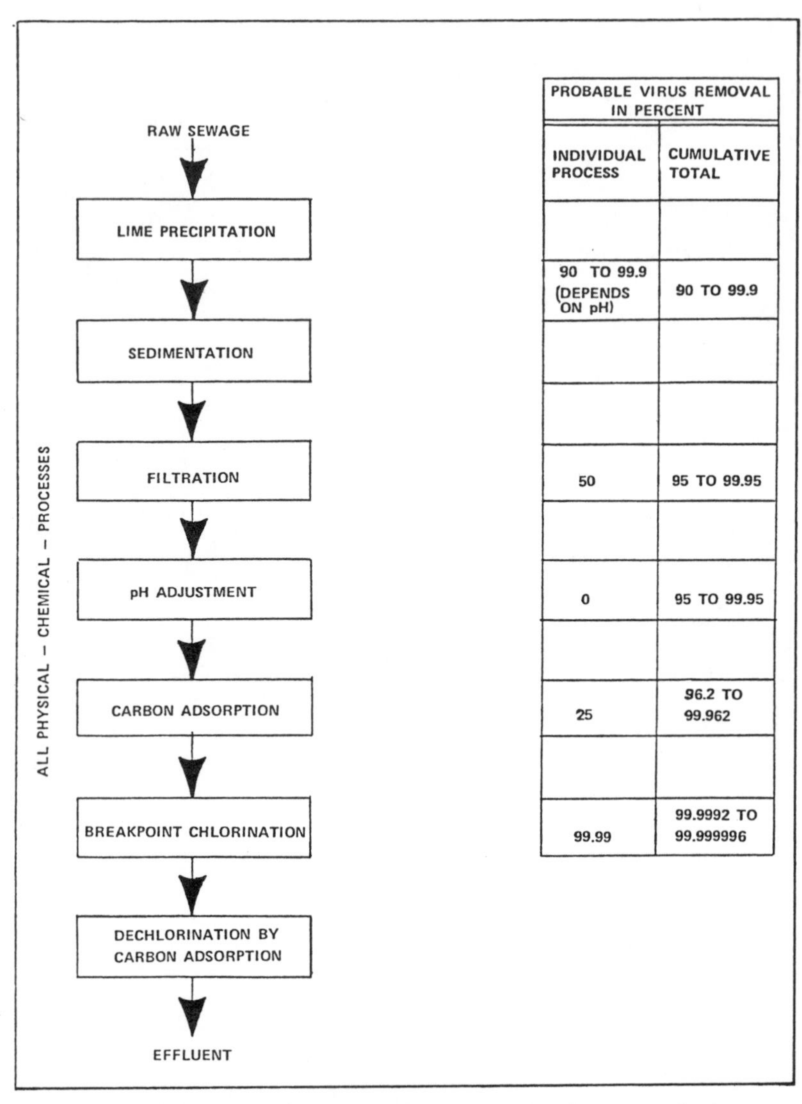

Figure 18-4. Scheme 4: Wastewater treatment for virus removal using physical-chemical processes.

chemicals, activated carbon, aeration or effluent storage may be desired before discharge.

Scheme 2 is similar to Scheme 1 except that ozone or bromine is used as the disinfectant (Figure 18-2). Therefore, ammonia removal is not needed, and the activated sludge process can be designed as a conventional

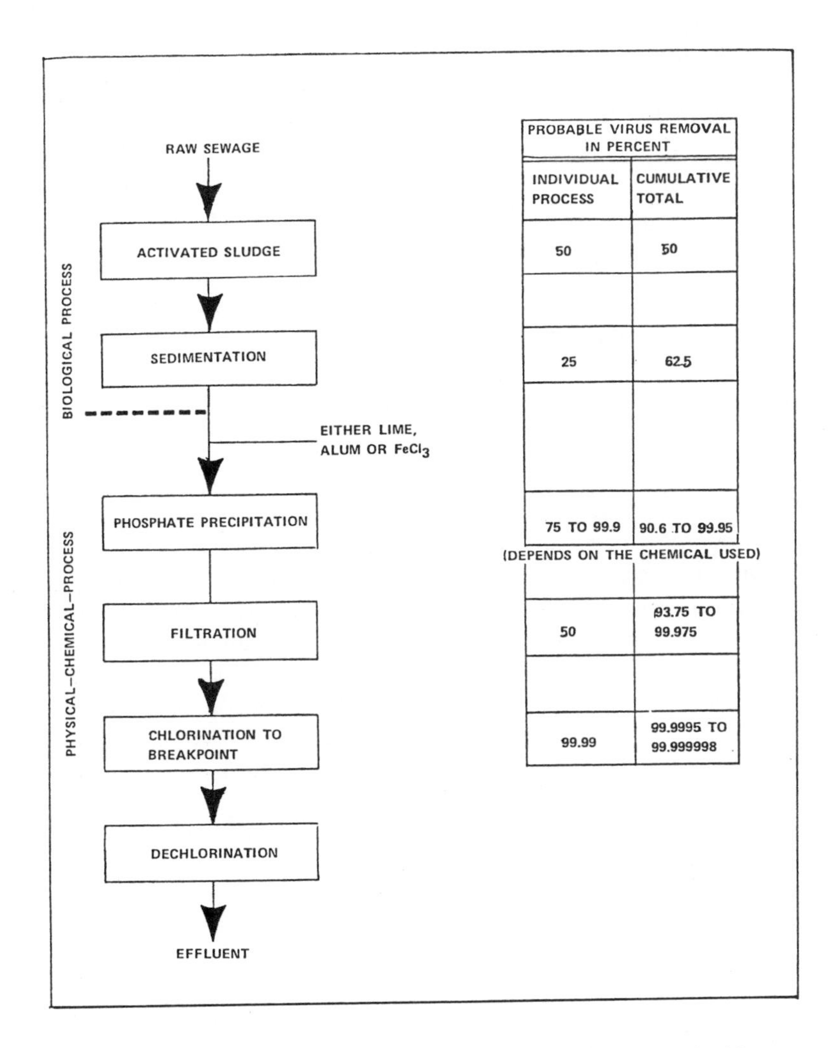

Figure 18-5. Scheme 5: Wastewater treatment for virus removal with phosphate removal.

aeration unit. As with Scheme 1, turbidity removal is achieved by chemical precipitation prior to settling and filtration. Excessive ozone concentrations will not have to be removed before discharge. Ozone is readily reduced to oxygen, which will benefit the stream. Other disinfectants not reacting with nitrogen, such as iodine, or whose reaction product is still an effective virucide, such as bromine, can also be used.

Table 18-3. Proposed Standards for Disinfection of Wastewater Effluents for Virus Inactivation

Parameter	Standard
Turbidity	Not to exceed one JTU
BOD_5	Not to exceed 10.0 mg/l
pH[a]	Not to exceed 7.5
Ammonia-N[a]	Not to exceed 1.0 mg/l

[a]These standards shall be required when chlorine is used as the disinfectant, unless breakpoint chlorination is to be used for nitrogen removal.

In Figure 18-3 phosphorus precipitation is not conducted within the activated sludge system. When this is the case, turbidity removal can be obtained by adding chemicals as a filter aid. Mixing and flocculation is provided before filtration.

Figure 18-4 presents a completely different approach. This system consists only of physical chemical processes. Turbidity removal is achieved by lime precipitation; alum or $FeCl_3$ can also be used, followed by sedimentation and filtration. The BOD_5 in this system is not removed biologically, but it is removed by adsorption on activated carbon. Ammonia is removed by breakpoint chlorination. In this scheme the dechlorination process is activated carbon which removes the free chlorine and the chlorinated compounds.

In Figure 18-5 interfering substances are removed as discussed for Figure 18-1, except that phosphate precipitation is conducted in a separate sedimentation process.

DISINFECTION TECHNOLOGY

Most current designs of wastewater treatment plant effluent disinfection systems make little attempt to optimize the disinfection process or to provide for much in terms of fail-safe operation. New effluent and design standards make greater demands on process selection and provision of standby equipment.

Recent standards for effluent disinfection have included

(a) no more than one virus PFU per 10 gallons (Montgomery County, Maryland),[12]

(b) no more than 2.2 MPN Total Coliform per 100 ml (some areas of California).[18]

These standards require that greater consideration be given to process selection than previously used.

A review of disinfection processes would include a variety of different techniques, the basic difference being the mechanism by which disinfection is achieved. Several methods are (a) the halogens: chlorine, iodine, bromine, bromine chloride; (b) ozone; (c) radiation: ultraviolet, gamma; and (d) heat.

Of these processes, the use of chlorine and ozone appear to show the greatest promise for virus removal in terms of current application knowledge. However, elemental iodine (I_2) is also a rapid virucide and may have an advantage over chlorine in effluents with pH values between 7.5 and 9.5. In this pH range, hypoiodous acid (HOI) occurs, and it is a faster virucide than I_2.[19] Bromine and bromine chloride show promise because the bromamines are good virucides compared to the chloramines. Hypobromous acid (HOBr) occurs to pH 9 so, like iodine, it is effective at higher pH than is chlorine (see Chapters 6-10). The application of three disinfection systems is presented in Figure 18-6 and below.

Breakpoint Chlorination

Breakpoint chlorination has been evaluated more extensively than the other methods of providing high levels of virus inactivation. It is defined as the addition of chlorine to water in which the chlorine residual is predominantly in the form of free chlorine, hypochlorus acid and hypochlorite, with a minimum of combined chlorine.

Chlorine in the form of hypochlorous acid (HOCl) is a rapid virucide and when it can be maintained in water, no other agent is necessary for disinfection.[19] This fact has been illustrated by many investigators. Table 18-4 provides information on the resistance of 20 enteric viruses to free chlorine.[20] Many other studies have been conducted, many of which are discussed in other chapters.

Maintenance of a high level of free residual chlorine frequently conflicts with effluent requirements due to potentially toxic effects on the aquatic life of the receiving stream.[21] Therefore, consideration of breakpoint chlorination must be accompanied by consideration of dechlorination processes.

Research[22,23] and full-scale plant studies[24] have shown increased levels of virus inactivation at reduced pH levels. Therefore, use of chlorine solution which tends to depress the pH, rather than hypochlorite solution, which tends to raise the pH, is recommended. Current studies show that depressing wastewater pH levels by acid addition may be a feasible means of enhancing chlorine virus inactivation.[24]

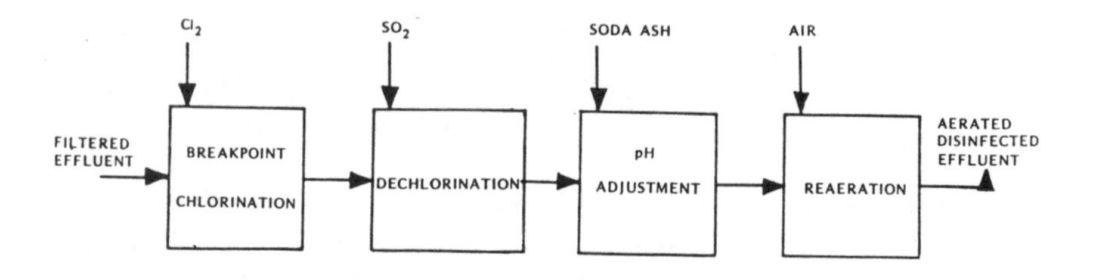

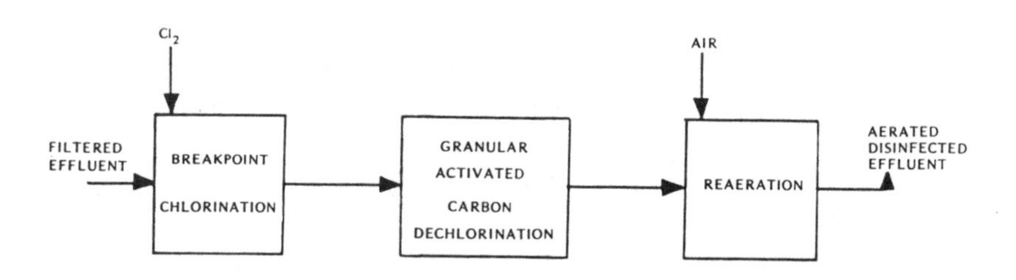

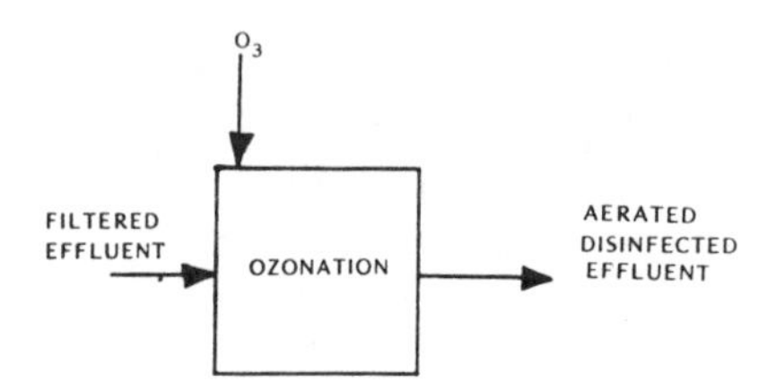

Figure 18-6. Alternate means of virus inactivation.

Maintenance of the required level of free chlorine residual necessitates the provision of a chlorination system far more complex than that installed in current wastewater treatment plants. The key to the control system is the amperometric analyzer which is currently the only practical method of measuring free residual chlorine. Operational experience with the amperometric analyzer in wastewater treatment facilities have been unremarkable to date. It can only be hoped that greater operational experience, equipment modifications, or other similar methods will improve the application

Table 18-4. Relative Resistance of 20 Human Enteric Viruses to 0.5 mg/l Free Chlorine in Potomac Water (pH 7.8 and 2°C)

Virus	Minutes[a]	Virus	Minutes[a]
Reo 1	2.7	Adeno 12	13.5
Reo 3	4.0	Echo 12	14.5
Reo 2	4.2	Polio 1	16.2
Adeno 3	4.8	Cox B3	16.2
Cox A9	6.8	Polio 3	16.7
Echo 7	7.1	Echo 29	20.0
Cox B1	8.5	Echo 1	26.1
Echo 9	12.4	Cox A5	33.5
Adeno 7a	12.5	Cox B5	39.5
Echo 11	13.4	Polio 2	40.0

[a]Minutes required for 99.99% virus inactivation based on first-order reaction.[20]

record of this piece of equipment. An amperometric membrane electrode looks promising in this respect.[25]

The recommended breakpoint chlorination system is shown in Figure 18-7 and incorporates a dual closed loop control called "cascade" loop control. The primary chlorine control is by flow measurement with subsequent secondary or "trimming" based on signals from measurement of chlorine residuals downstream of the initial chlorine mixing zone (Point A), and at the effluent of the chlorine contact chamber (Point B). The residual signal from the second residual sampler may also be used to pace the dechlorination feed system.

Dechlorination

Methods of reducing residual free chlorine levels to regulatory agency limits include the addition of reducing chemicals or activated carbon contacting. Reducing chemicals include sulfur dioxide, sodium bisulfate, sodium sulfite and sodium thiosulfate. Sulfur dioxide (SO_2) is the most commonly used, since supply and handling of the chemical is in a manner similar to chlorine. Both are supplied in pressurized liquid form and dissolved in water using the same type of equipment. Therefore, minimal addition facilities must be constructed to provide for SO_2 dechlorination, and personnel familiar with operation of a chlorination facility should have little problem operating the dechlorination facility. Figure 18-8 illustrates the system used with breakpoint chlorination plus SO_2 dechlorination. Positive control of this chemical is essential as an excess would serve as a deoxygenating

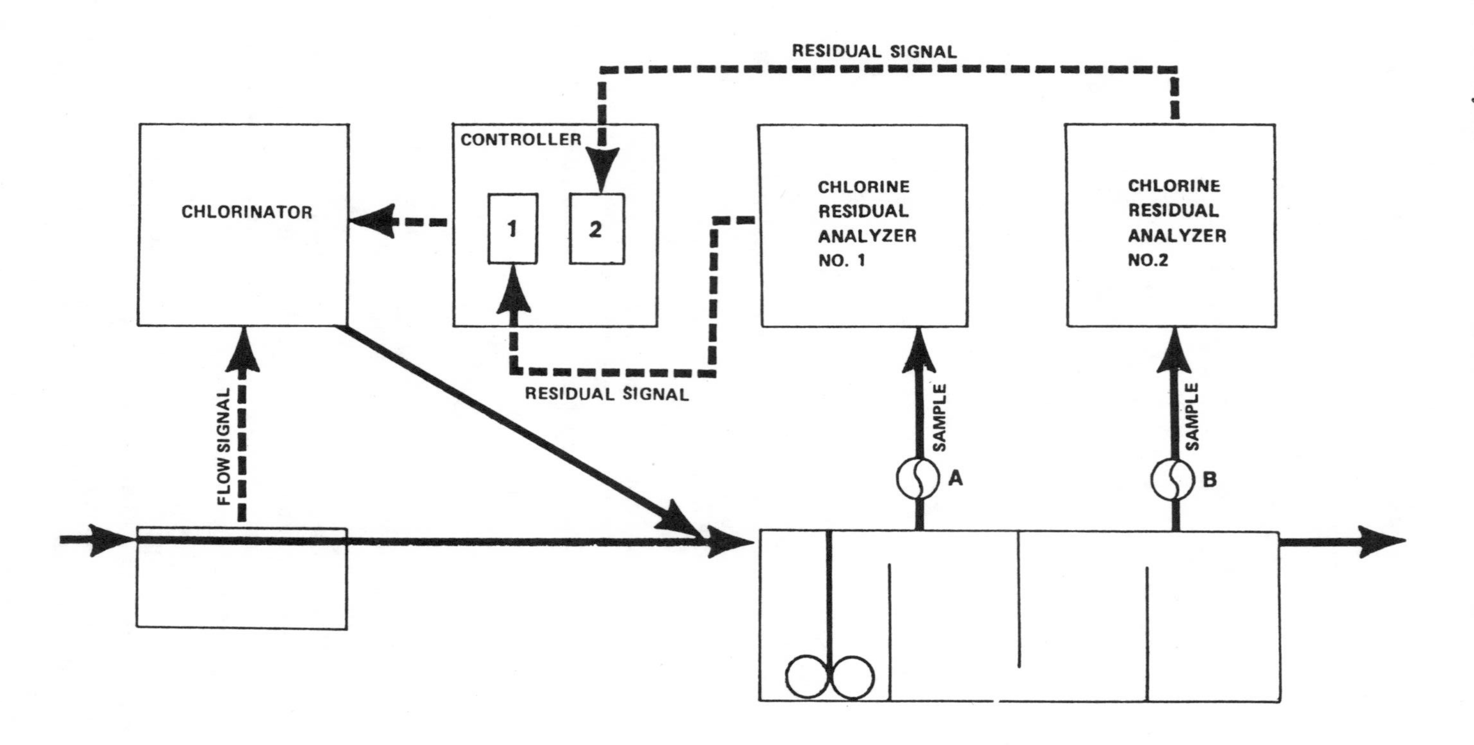

Figure 18-7. Breakpoint chlorination system.

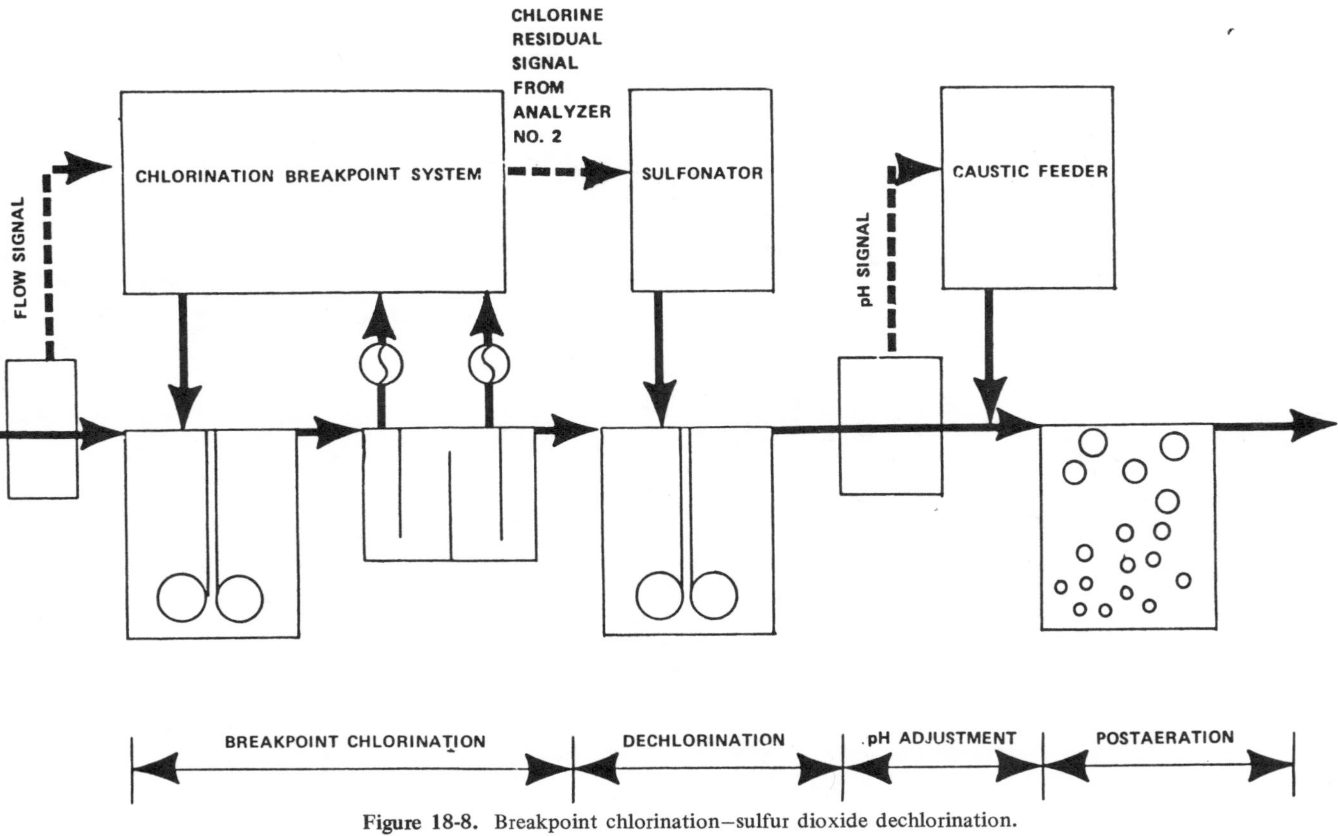

Figure 18-8. Breakpoint chlorination–sulfur dioxide dechlorination.

agent and further reduce the effluent pH. Pacing of the sulfur dioxide dissolver would be by a signal from the second residual chlorine analyzer, as there is a nominal one-to-one relationship between chlorine residual and SO_2 feed rate.

Dechlorination with activated carbon has been used in brewing and soft drink manufacturing, as well as in production of potable water for shipboard use.[26] There is little available operational data in the wastewater treatment field, but current research in physical-chemical wastewater treatment at such treatment plants as Blue Plains, Washington, D.C., Owosso, Michigan, and Pomona, California should provide more definite information. The chemical reaction is given as follows:

$$C + Cl_2 + 2H_2O \rightarrow 4\ HCl + CO_2$$

but the mechanism is not clearly understood. Chlorination-dechlorination with activated carbon is illustrated in Figure 18-9.

The effect of simultaneous adsorption of organic pollutants on the dechlorination rate and capacity of the carbon bed requires further study as the majority of the data currently available is based on dechlorination of potable water.[27] However, it is believed that an application rate of 3 gpm/sq ft is a conservative basis for design of granular activated carbon beds, using an empty bed contact time of 15-20 minutes with an influent free residual concentration of 3-4 mg/l. The life of such a bed, treating a flow of 1 mgd, would be four years based on application of potable water. The reduction of the effective carbon bed life from treating wastewater treatment plant effluent is unknown but not considered more than 20%. The rate of carbon bed dechlorination is maximum at high temperatures and low pH.

Ozonation

Ozone has long been recognized as an effective virucide.[28] It has been used extensively in Europe for potable water disinfection for more than 50 years. Its application for wastewater disinfection has, to date, been limited to pilot scale applications at research projects such as Chicago, Illinois and Louisville, Kentucky.

Ozone is produced by passing air, or pure oxygen, between electrodes separated by an insulating material. The electrodes are usually made of stainless steel or aluminum, with glass insulation. High voltage, up to 20,000 volts or more, single phase, is applied to the high tension electrode to generate ozone from dried air.

Ozone is twice as powerful an oxidizing agent as chlorine in water. Its action on viruses is often reported as instantaneous, requiring only

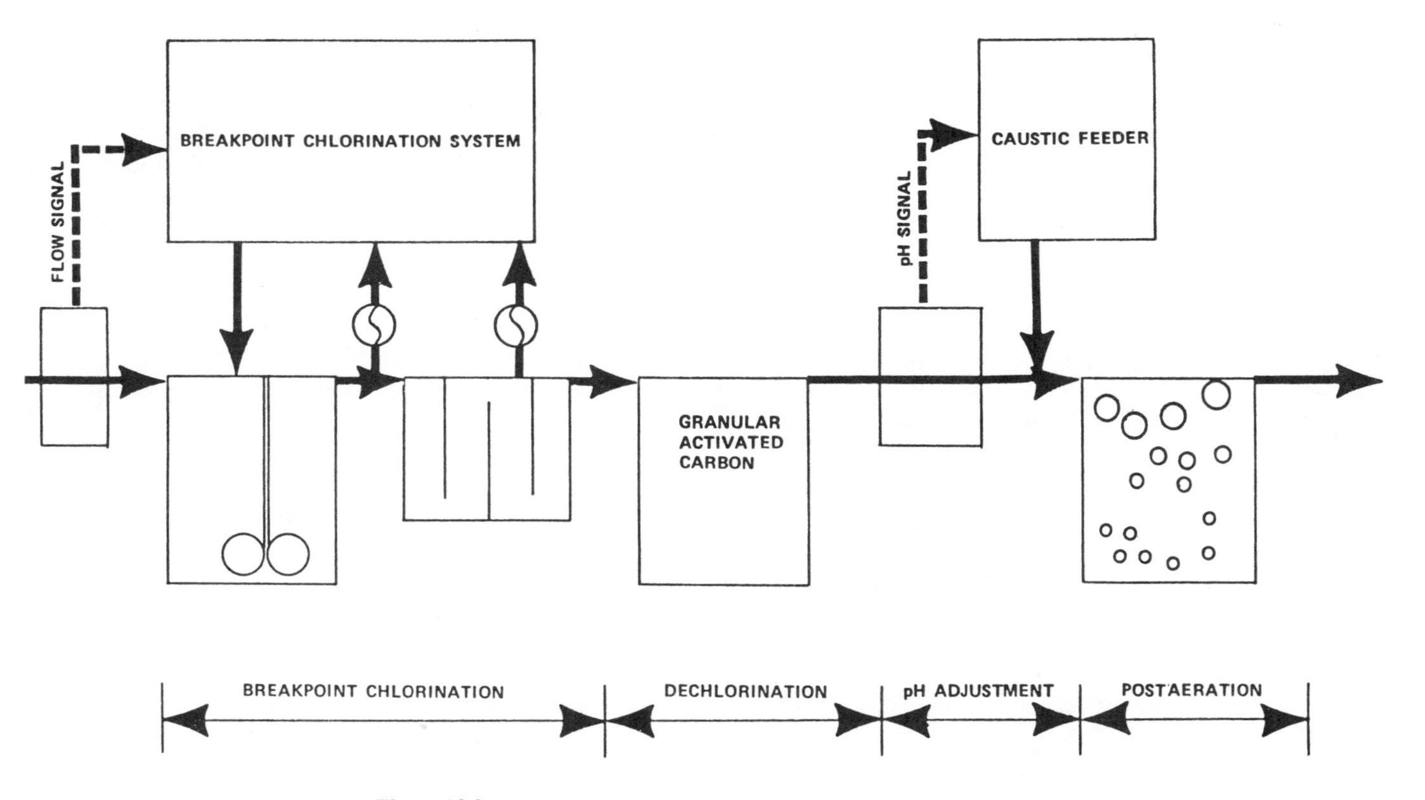

Figure 18-9. Breakpoint chlorination–activated carbon dechlorination.

instantaneous mixing for detention time. However, this would not be expected from a consideration of virus disinfection kinetics, which has been experimentally supported by Katzenelson.[29] The importance of virus inactivation kinetics as related to contact chamber design will be discussed below. The inactivation relationships for virus with ozone is poorly defined at present. This is illustrated in Table 18-5. Contact time for 99.99% inactivation varies from 2.5-64 minutes. A 99% removal appears to require at least a four-minute contact in the presence of residual ozone. Based on present data, an ozone residual of 1 mg/l at the end of a four-minute contact appears to insure 99% inactivation of Poliovirus. Other viruses, however, may may be much more resistant to ozone as with chlorine (Table 18-4). A typical ozone system for virus inactivation is shown in Figure 18-10).

Table 18-5. Inactivation of Viruses with Ozone

Virus	Time Required (min)	Ozone mg/l		% Inactivation	Reference
		Applied	Residual		
Poliovirus	4	–	0.7	99	Perlman[30]
Poliovirus	4		0.7	99.9	Perlman[30]
Poliovirus	2.5	1.27	0.23	99.99	Schaffernoth[31]
Poliovirus	2.5	1.38	0.2	99.99	Schaffernoth[31]
Poliovirus 1	8	–	0.3	99	Katzenelson[29]
Poliovirus 1	48	–	0.3	99.9	Katzenelson[29]
Poliovirus 1	8	–	0.8	99.3	Katzenelson[29]
Poliovirus 1	56	–	0.8	99.9	Katzenelson[29]
Poliovirus 1	8	–	1.5	99.9	Katzenelson[29]
Poliovirus 1	64	–	1.5	99.99	Katzenelson[29]

Ozonation may conserve plant site area as it needs only a small site area for generation equipment and the contact chamber. The system requires no on-site storage of potentially dangerous chemicals. No chemicals such as chlorides or sulfates, are added to the waste stream; in fact, foam generated by ozonation may be skimmed off, achieving additional removal of colloids. A side benefit is a plant effluent high in dissolved oxygen which eliminates the need for post aeration. The system is relatively simple to operate and, since ozone is generated on site, does not add to truck traffic on and off the site.

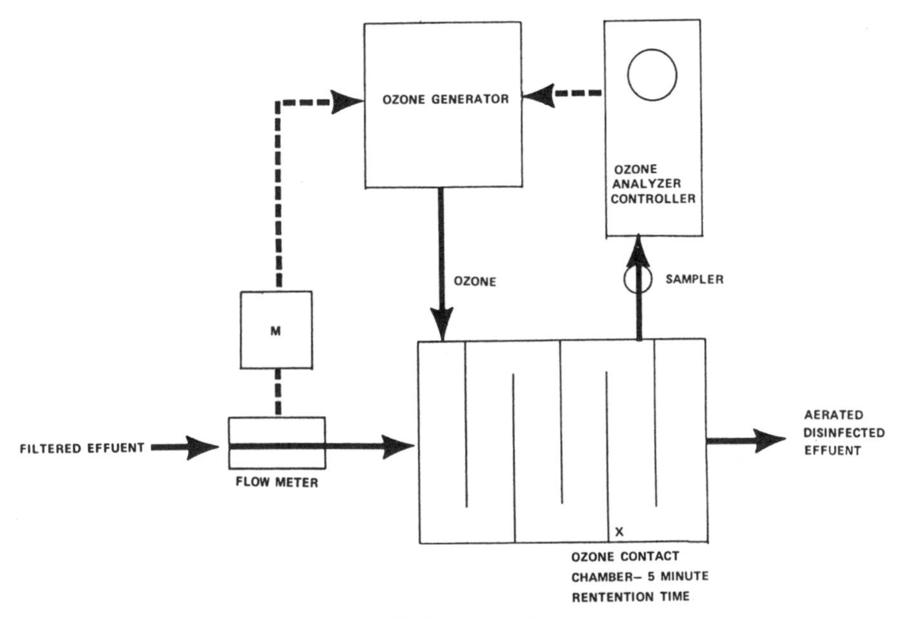

Figure 18-10. Ozonation.

The major drawbacks to greater application of ozone for wastewater disinfection are the lack of a definitive relationship on virus kill versus ozone residual and the difficulty of dosage due to its low solubility. This makes ozone difficult to control due to the low residuals found even after short contact times. It is expected that the research work currently going on in studying ozone virus inactivation will yield the required design parameters.

DESIGN OF DISINFECTION UNITS

The inactivation of virus to produce a virus-free effluent requires far greater design care than that typical of existing disinfection processes, because one must achieve 99.99% inactivation. A 90% or 99% removal is usually unsatisfactory. Removal efficiencies reported in Tables 18-4 and 18-5 cannot be achieved in the field unless special attention is given to the initial mixing of chemicals followed by a contact basin which approaches plug flow conditions. This requirement is even more significant for viruses than most bacteria, because virus inactivation does not follow a single first-order inactivation rate. Inactivation kinetics consist of at least two rate processes, as illustrated by a typical inactivation curve in Figure 18-11.

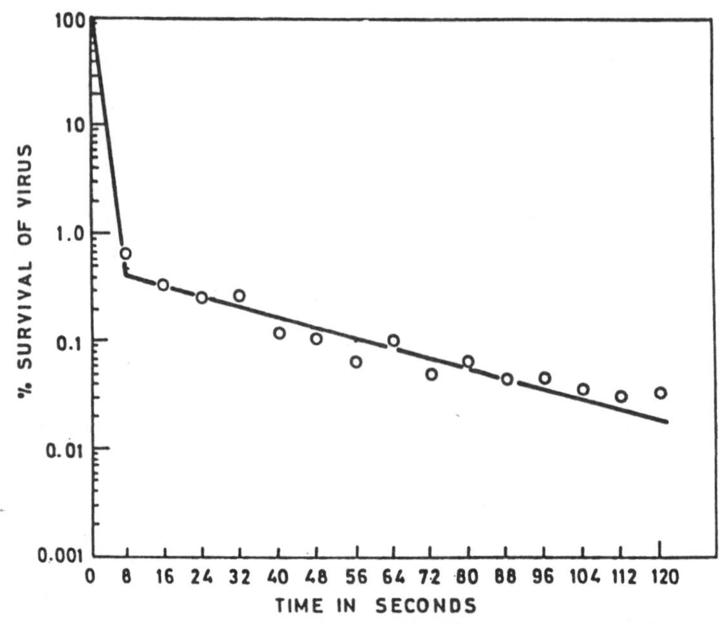

Figure 18-11. Typical inactivation kinetics of virus disinfection.[29]

Although Figure 18-11 is for ozone inactivation of Poliovirus, the same phenomenon occurs with chlorine.[29] This type of inactivation kinetics results from a mixed virus population, some more resistant than the other. The population inactivated first is believed to be single virus particles, and the more resistant population consists of virus aggregates. Thus, as Figure 18-11 illustrates, instantaneous contact periods with any disinfectant is inadequate for removals over 99%. This point is discussed in detail in Chapters 1 and 2.

The distribution of residence times for ideal plug flow, completely back-mixed flow, and for an arbitrary flow have been mathematically modeled by several investigators.[32,33] The sole use of the ratio of tank volume to flow rate for detention time means little with regard to process efficiency. For example, there exists a difference of several orders of magnitude between required tank volumes for a good versus a poorly designed contact chamber. In brief, effective virus inactivation is impossible in poorly designed contact chambers.

Mixing

Initial mixing is a major factor in optimizing virus inactivation.[34,35] Intense mixing at this point is essential to provide intimate contact of the chlorine solution and wastewater and to provide uniform distribution through the chlorine contact chamber. This concept has been verified by laboratory and field studies.[24,35,36]

Mixing at the point of chlorine solution injection may be attained in a number of ways, any of which must be designed into the system. Several means of mixing include the following:

(a) Turbulent flow in closed conduit flowing full: White[34] reported good mixing with a Reynolds number of 20,000. He suggests that a mechanical in-line mixer be used when Reynolds numbers less than 10,000 are predicted and when the pipe diameter exceeds 36 inches.

(b) Hydraulic jump: This method has been used for chemical mixing. Normal use of the hydraulic jump calls for system head losses up to 1.5 feet, which are difficult to accept in normal wastewater treatment plant design.

(c) Mechanical mixer: The use of a propellor mixer in a channel is a positive means of providing the required mixing energy.

A submerged weir is similar to a hydraulic jump with a maximum head of approximately one foot. However, the energy available for mixing is questionable for this procedure. Since a Parshall Flume is a low head loss measuring device, it is not adequate for mixing. Over-and-under baffles are better than no mixing system, but they are not considered adequate.

Contact Tank

The conventional chlorine contact tank is intended to act as a plug-flow reactor. Mixing of chlorine solution with the wastewater stream should be completed prior to entry into the contact chamber. Studies have shown that, due to design deficiencies, the actual retention time in many contact chambers does not approach the theoretical retention period, as calculated by chamber volume divided by the wastewater stream flow rate.

The optimum contact chamber configuration is a long pipeline or narrow channel. Frequently, however, an unbaffled or meagerly baffled rectangular tank is substituted for the ideal plug flow chamber. Improvement to existing facilities or design of new contact chambers should be based on approaching the plug flow configuration. Rectangular tanks should be baffled longitudinally rather than transversely into narrow channels. Fillets and vanes should be provided at the 180° turns in the channels to minimize liquid stratification. Channel velocities of 5-7 ft/min have been found suitable.

ENGINEERING A FAIL-SAFE SYSTEM

Regulatory agencies now require greater operational reliability of wastewater treatment facilities, including virus inactivation systems. Fail-safe systems will become a requirement as demands for water reclamation and reuse increase. The ability to meet continuous water quality criteria, such as virus removal, depends on the flexibility of a plant's design and the degree of operational supervision. A few mechanical failures can occur even under a consistent maintenance program. This problem is prevented by using dual or standby mechanical facilities. For example, current standards enforced in the Greater Washington, D.C. area include:

- (a) dual source of electric power
- (b) on-site standby power generator
- (c) recycling capabilities within the plant; piping facilities to permit rerouting of flows under emergency conditions
- (d) provision of backup equipment to maintain effluent quality with any treatment unit out of service, including chlorinators, residual analyzers and pumps
- (e) multiple units of smaller size versus one large unit
- (f) surge tanks to provide for flow equalization and storage potential
- (g) holding pond or basin with capacity 1.25 times average daily design flow
- (h) monitoring equipment to measure effluent characteristics such as chlorine, dissolved oxygen and pH
- (i) automatic alarm system for plant malfunction to sound in a place that has 24-hour attendance.

Economics of Terminal Disinfection

Terminal disinfection to meet virus inactivation standards requires higher capital investment and annual operating expenses. Greater capital investment is in terms of more sophisticated control systems, provision of backup equipment, as well as safety features not previously considered standard design. Operation cost increases include more chemicals and power, as well as greater manpower requirements to operate and maintain the equipment to meet stringent reliability standards.

Capital, operation and maintenance cost estimates were prepared to evaluate the economic aspects of the three disinfection systems considered. The capital cost of each system was prepared, including equipment housing, backup units and ancillary equipment considered necessary to provide a complete system to meet the standards of an operating agency. Operational and maintenance costs included power and chemicals, but did not include manpower, which is frequently more a function of agency policy than actual requirements. No major mechanical repairs or replacements were included in the cost analysis. The cost estimate was prepared for a one mgd wastewater treatment plant (Figure 18-12) given the following effluent quality prior to terminal disinfection:

Turbidity	1 JTU
Total Kjeldahl Nitrogen	2 mg/l (essentially total nitrification)

Operation costs were based on the following dosages and unit costs:

Chlorine	10 mg/l	@	\$0.08/lb
Sulfur Dioxide	5 mg/l	@	\$0.20/lb
Soda Ash	26.6 mg/l	@	\$0.04/lb
Ozone Power Requirements			11.5 kwh/lb O_3
Power Cost			\$0.015/kwh

Table 18-6 summarizes the cost estimate data showing a unit cost of 5.3, 4.2 and 5.1 cents per 1000 gallons for chlorination–SO_2 dechlorination, chlorination–activated carbon dechlorination, and ozonation, respectively. As can be seen, no small cost is involved in providing for virus inactivation systems. Approximately 10-15% of the overall plant construction cost must be allocated to terminal disinfection.

SUMMARY

The need to give greater attention to virus inactivation in water and wastewater treatment is becoming more apparent. Virus standards for potable water and treated wastewater are in existence or under active consideration. While each process in a wastewater treatment flow stream achieves certain levels of virus removal, they essentially serve only as pretreatment processes prior to the prerequisite terminal disinfection.

The only currently practicable means of terminal disinfection is breakpoint chlorination. Ozonation is becoming practicable but requires additional work on dosing and control technology. Parameters for the design of these processes for application in wastewater treatment are available, but further studies are required to optimize their application. Direct monitoring of virus inactivation is difficult and expensive; therefore, accurate monitoring of secondary standards such as free chlorine residual is extremely important. Current emphasis on the design of fail-safe features into wastewater treatment plants insures greater reliability of the processes. A review of preliminary cost estimates for the terminal disinfection processes indicates that none are inexpensive.

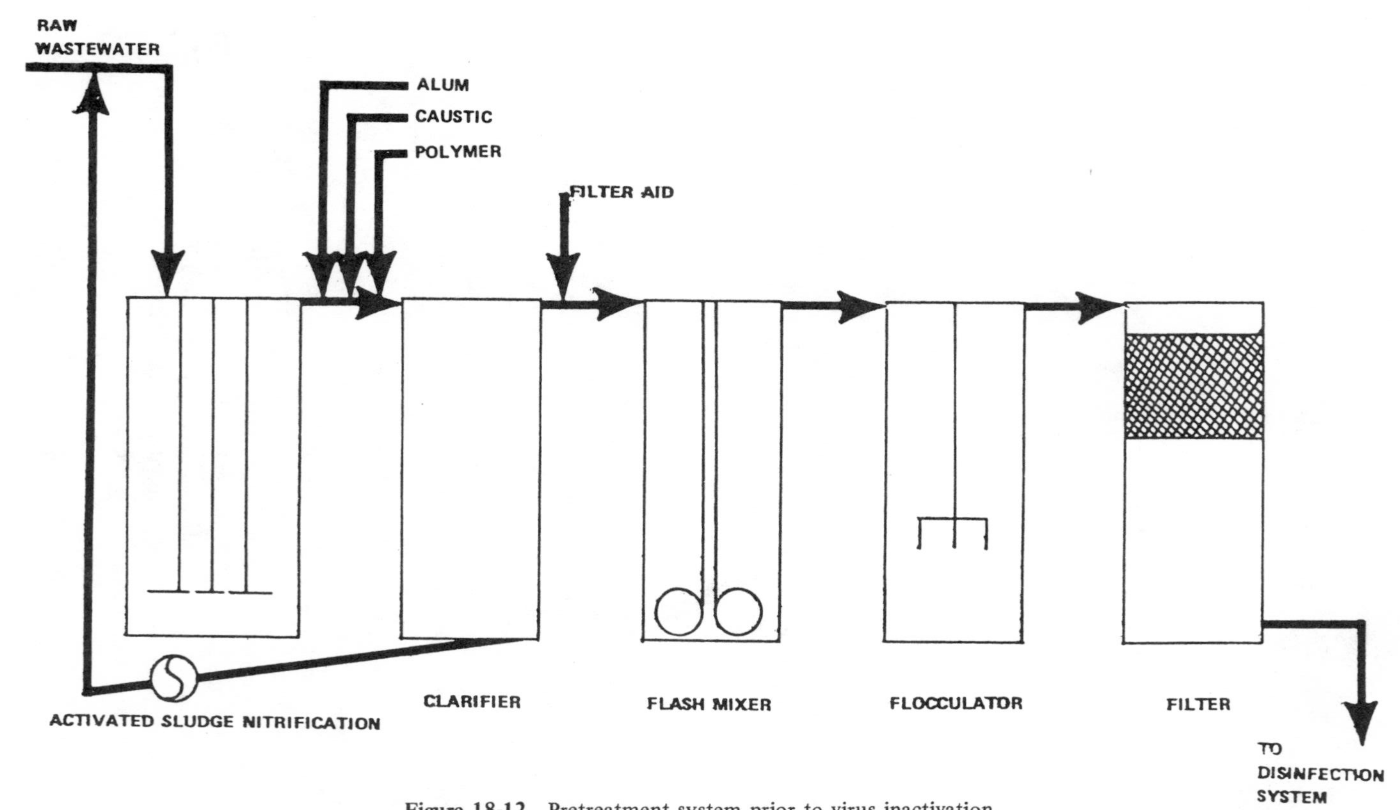

Figure 18-12. Pretreatment system prior to virus inactivation.

Table 18-6. Cost Summary of Virus Inactivation Systems

Item	Alternate Systems		
	Scheme A[a]	Scheme B[a]	Scheme C[a]
Capital			
Capital debt service (6% – 25 years)	10,550	11,010	13,190
Operation[b]			
Chlorine	2440	2440	–
Sulfur dioxide	3070	–	–
Soda ash	3130	–	–
Power	230	1710	5250
Total Annual Cost	19,420	15,160	18,440
Unit Cost (Cents/1000 gallons)	5.3	4.2	5.1

[a]Scheme A: Breakpoint chlorination - sulfur dioxide dechlorination - pH adjustment - postaeration
Scheme B: Breakpoint chlorination - granular activated carbon dechlorination - postaeration
Scheme C: Ozonation

[b]Labor or major maintenance not included.

REFERENCES

1. Akin, E. W., W. H. Benton, and W. F. Hill. "Enteric Viruses in Ground and Surface Waters: A Review of Their Occurrence and Survival," *Proc. 13th Water Qual. Conf., Virus and Water Quality: Occurrence and Control,* University of Illinois (1971).
2. Metcalf, T. G. and W. C. Stiles. "Viral Pollution of Shellfish in Estuary Waters," *J. San. Eng. Div. ASCE* **94**, 595-609 (1968).
3. Grinstein, S., J. L. Melnick, and C. Wallis. "Virus Isolations from Sewage and from Streams Receiving Effluents of Sewage Treatment Plants," *Bull. WHO* **42**, 291-296 (1970).
4. Berg, G. "An Integrated Approach to the Problem of Viruses in Water," *Proc. Natl. Specialty Conf. on Disinfection*, University of Massachusetts, Amherst (1970).
5. Melnick, J. L. "Detection of Virus Spread by the Water Route," *Proc. 13th Water Qual. Conf.*, University of Illinois (1971).
6. Coin, L., M. L. Menetrier, J. Lobonde and M. C. Hannoun. "Modern Microbiological and Virological Aspects of Water Pollution," In

Advances in Water Pollution Research (New York: Pergamon Press, 1965).

7. Foliquet, J. M., L. Schwartzbrod, and O. G. Gaudin. "La Pollution Virule Les Faux Usees, de Surface et D'alimentation," *Bull. WHO* **35**, 737 (1966).
8. Bagdusar'yan, G. A. "Investigation of Riverwater for Enteroviruses," *Hyg. San.* **33**, 134-135 (1968).
9. Farroh, K. "Research on Viruses in Water," *Rev. Immunol. Therapie Antimicrobienne* **30, 355-356 (1966).**
10. Shuval, H. I. "Detection and Control of Enteroviruses on the Water Environment," In *Developments in Water Quality Research* (Ann Arbor, Mich.: Ann Arbor Science Publishers, Inc., 1970).
11. McDermott, J. H. Director Division of Water Hygiene, Office of Water Quality, EPA. personal communication (December 1972).
12. Cookson, J. T. "The Use of Temporary Wastewater Treatment Plants: Standards and Procedures for Elimination of Health Hazards," Report prepared for Montgomery County Council, State of Maryland (December 5, 1972).
13. Mosley, J. W. "Water-Borne Infectious Hepatitis," *New Engl. J. Med.* **261**, 703, 748 (1959).
14. Neefe, J. R. and J. Stokes. "An Epidemic of Infectious Hepatitis Apparently due to a Water-Borne Agent," *J. Amer. Med. Assoc.* **128**, 1063 (1945).
15. Clark, N. A., G. Berg, P. W. Kubler and S. L. Chang. "Human Enteric Viruses in Water: Source, Survival and Removability," *Adv. Water Pollution Res.* **2**, 523-536 (1964).
16. Clarke, N. A. and S. L. Chang. "Enteric Viruses in Water," *J. Amer. Water Works Assoc.* **51**, 1299 (1959).
17. Krusé, C. W., V. P. Olivier and K. Kawata. "The Enhancement of Viral Inactivation of Halogens," *Proc. 13th Water Qual. Conf.*, University of Illinois (1971).
18. Collins, Harvey F. personal communication (July 26, 1973).
19. Berg, G. "Removal of Viruses from Water and Wastewater," *Proc. 13th Water Qual. Conf.*, University of Illinois (1971).
20. Liu, O. C., H. R. Seraichekas, E. W. Akin, D. A. Brashear, E. L. Katz and W. J. Hill. "Relative Resistance of Twenty Human Enteric Viruses to Free Chlorine in Potomac Water," *Proc. 13th Water Qual. Conf.*, University of Illinois (1971).
21. Tsai, C. F. "Effects of Chlorinated Sewage Effluents on Fish in Upper Patuxent Rivers," *Maryland Chesapeake Series,* **9**(2), 83-93 (1959).
22. Kelly, S. and W. W. Sanderson. "The Effect of Chlorine in Water on Enteric Viruses," *Amer. J. Publ. Health,* **48**(10), (October, 1958).
23. Chambers, Cecil W. "Chlorination for Control of Bacteria and Viruses in Treatment Plant Effluents," *J. Water Pollution Control Fed.* **43**(2) (1971).
24. Krusé, C. W., K. Kawata, V. P. Olivier and K. E. Longley. "Improvement in Terminal Disinfection of Sewage Effluents," *Water Sewage Works* **120**(7), 57-67 (June, 1973).
25. Edwards, J. W. and J. D. Johnson. "An Amperometric Halogen Analyzer," Preprints of Papers Presented at the 167th National Meeting, American Chemical Society, Division of Environmental Chemistry, Los Angeles, Calif, Vol. 14, No. 1 (April, 1974), pp. 169-173.

26. Magee, V. "The Application of Granular Activated Carbon for Dechlorination of Water Supplies," *Proc. Soc. Water Treat. Exam.* **5**, 17 (1956).
27. Hager, D. G. and M. E. Flentje. "Removal of Organic Contaminants by Granular-Carbon Filtration," *J. Amer. Water Works Assoc.* **57**(10) (November, 1965).
28. Grabow, W. O. K. "The Virology of Wastewater Treatment," In *Water Research*, Vol. 2 (New York: Pergamon Press, 1968), pp. 675-701.
29. Katzenelson, E., H. Schechter, N. Biederman, Y. Rotem and H. Shuval. "Ozone Inactivation of Water-borne Viruses," *Proc. 4th Scientific Conf. of the Israel Ecological Soc.*, Tel-Aviv (April 8-9, 1973).
30. Perlman, R. G. "Water Resources Engineering/Technology Briefs," *Water Eng.* **6**, 14 (1969).
31. Schaffernoth, T. J. "High Level Inactivation of Poliovirus in a Biological Treated Wastewater by Ozonation," M.S. Thesis, University of Maine, Orono (1970).
32. Collins, H. F., R. E. Selleck, and G. C. White. "Problems in Obtaining Adequate Sewage Disinfection," *J. San. Eng. Div. ASCE*, **SA5**, 549 (October, 1971).
33. Thomas, H. A. and J. E. McKee. "Longitudinal Mixing in Aeration Tanks," *Sew. Works J.* **16**, 42 (1944).
34. White, G. C. *Handbook of Chlorination.* (New York: Van Nostrand Reinhold Co., 1972).
35. Selleck, R. E., H. F. Collins and G. C. White. "Kinetics of Wastewater Disinfection in Continuous Flow Processes," presented at International Association Water Pollution Research Conference, San Francisco (July 29, 1970).
36. White, G. C. "Chlorination and Dechlorination: A Scientific and Practical Approach," *J. Amer. Water Works Assoc.* **60**, 540-561 (May, 1968).

INDEX

acid test, simple 80
activated carbon dechlorination
352
activated sludge effluent
167,175,239-241
activity coefficients 55
alkalinity 70,95,108
alpha coefficients 56
aluminum alginate filter 362
amebiasis 193,194
ammonia 59,70-72,76,122,174,
180,185-187,221,225,278,
280,283,286,305,306,309,
323,325,335,339,354,397
chlorine treatment 75,78,79
removal 352
amoebic cysts 193
amoebic dysentery 193,194
amoebic meningoencephalitis 194
amperometric analyzer 402
amperometric membrane electrode
403
amperometric titration 80,84-87,
109,314
APHA Standard Methods 80
AQUA-CHECK 97

B. coli 125,135,137
Beer's Law 105
benzyldimethylamines 257
Biochemical Oxygen Demand (BOD)
62,209,221,224,225,228-230,
269,395
biomass 52
blowing-out process 115
breakpoint
chlorination 52,59,147,278,339,
371,397,400,401,405,407,415
curve 71,75,96,311
data 306
reaction 70,128,302,309
bromamines 125,128,163,174,176,
180,181,185,189,302-304,335
half-life 122
residuals 122
bromide salts 136
bromide toxicity 138
brominations of aromatic compounds
123
bromine 63,68,79,114,145,163,179,
189,193,249,305,306,310,315,
319,320,323,396
economics 116
production 114
residual 314
total residual 314
vapor pressure-temperature curves
119
bromine chloride 113,117,130
dissociation 118
vapor pressure-temperature curves
119
bromoform 272

caliche 115
carbon adsorption 51
carbon dechlorination bed 347,348
carcinogens 51

Chemical Oxygen Demand (COD) 51,209,221,229,230,235,244
Chick's Law 1,2,11,60
chloramination 278
chloramines 64,72,77-79,82,87, 99,102,109,126,163,170-176, 181,185,277,278,302,304, 350,394
  false positive 101
  organic 180,187
  removal 352
  residual 72,181,185
  toxicity 245
chlorate ion 251,265
chlorinated cyanurates 136
chlorinated guaiacols 269
chlorinated hydrocarbons 52,233
  pesticides 54
chlorinated organic compounds 53,64,269
chlorination 4,52,63,67,74,96, 146,166,171,172,245,254, 262,269,271,279,339,352, 402,406,413
  chemistry 68
  effectiveness 4
  products 271
  simple (marginal) 75
  also see viruses; wastewater
chlorine 40,43,52,55-59,64,67,68, 119,126,129,130,137,145, 164,167,170,172,174,179-186,193,213,220,233,249, 281,301,305,306,342,348, 369,372,394
  activated carbon reactions 344
  chemistry 67
  combined 72,92,99,102,111, 182,344
  combined available residual 74
  contact tank 411
  demand 75,182
  demand-free 279,312
  disinfection 360,368
  dose 75,204
  free radical 262
  nonavailable 74
  odor 74
  residuals 91,167,233,339,340, 368
chlorine
  total available 83,85
  vapor pressure-temperature curves 119
chlorine dioxide 214,251,252,257, 264,266-268
  stability 251
chlorinium ion 251
chlorite ion 264,265
p-chloroanisole 262
chloroform 52,272
chloromelamine 136
chlorophenols 122,261
o-chlorotoluene 262
cholera 49
clumps (aggregation of organisms) 3
  composite survival 15
  destruction 12,14,15
  destruction rate constant 12,40
  geometry of formation 28
  initial concentration 12
  size 3,12,13,18,20,33,34,41,44, 46
    distribution 20,45,46
    mixed 17
  surface area 22,34
coliforms 8,127,165-171,180,184, 188,221-227,244,245,360,368, 369,376,385,386
coliphages 147
color compensation technique 101
color removal 67,94
complete kill 8,10
  doses 10
  time 10
concentration-contact time relationship 61
concentration constant, total 57
cooling water treatment 126
cross connections 194
cyanide 225
cysticidal agents 204,207
cysts 24,35,62,195,196,200
  clump size 24,25,28,30,33
  destruction 8,58
  survival 24
cytopathogenicity (CPE) 365
cytosine 271

decay or reaction time, total 8
dechlorination 75,76,182,234, 245-247,339,340,405-407, 413
  also see activated carbon dechlorination
dialysis for concentrating virus 362
1,2-diamines 265
dibromamines 164,174,188,303, 309
dichloramines 70,82,103,108, 130,280,301,335,352
1,3-dichlorodimethylhydantoin 130,136
dichloroiodate ion 131
diethylene paraphenylene diamine (DPD) methods 79,84,97, 99-102,108-111,164,170
dilution coefficient 3
dip stick test 97
discrete organisms, percentage 16
diseases
  clinical 51
  waterborne 49,59,145
disinfectants 49,50,55,58,61,92, 113,176,193,249,260,301
disinfection 7,11,50,63,68,121, 128,137,340
  action 14,136
  agents 1-3
  data 1,2,11,59,240
  dose 4
  efficiency 1,56,181,190,240, 244
  equation 4
  of halogens 145
  of viruses 147,180
  of water and wastewater 146,163,179,211,391
    organisms 4
    ozone 234,239,240
    reactions 4
    research 1
    species 4
  terminal 394,412
    interference substances 400
    units, design of 409
    also see viruses; wastewater
dissociation constant 55,56
dissolved organic carbon 51,52
dissolved solids 55,76
  total 94-96
double summation model 13
drop dilution method 81
dual-disinfectant system 63

electrophile 263
Entamoeba histolytica 8,12,193-196-206
  cysts 57,58,61
enteroviruses 35,41,365,367,371, 372,386
  inactivation by iodine 35,44,46
Environmental Protection Agency (EPA) 85,87,392
environmental stress 54
epidemiological studies 50
equilibrium constant 250,252
Escherichia coli 35-43,50,51,58, 150,151,156-160,181,223, 366,385
  cysts 61
ethylenediamine tetraacetate dihydrate (EDTA) 165

FACTS I and II 83,105-109
  evaluation 109-111
fathead minnows 127,128
fat storage of chlorinated pesticides 53
first-order kinetics 60
fish toxicity 126
fraction of inactivation 5
free available chlorine (FAC) 69,70,76,79,91
  field method 91
  residual 59,74-76
free bromine, molar absorptivity 315
free chlorine 35-39,42,43,72,79, 83,87,182,185,187,190,289, 344,370
  fraction of 55,82
  residual 278,402
f_2 virus 145,240,367

Globaline tablet 194,206
glycine addition 102,195
graphic analysis 44,46
graphic "peeling off" 17,20

half-cell potentials 253,254
haloamines 204,207
haloforms 272
halogens 113,146,159,163,170, 179,182,187,196,200, 203-207
  dose 127,152,168,175
  mixtures 113
  species 121,261
  residuals 127,169,183
  virucidal activity 145
Hammett substituent constants 257
headspace gas chromatograms 52
hepatitis 49,59,148,359,372
holding ponds 64
hydrogen peroxide 343
hydrogen sulfide 124
hypobromous acid 206,302-305, 309,311,313-317
hypochlorite ion 41,55,58,206, 253,254,258,259,263
hypochlorous acid 55,59,76, 206,258,259,266,279-283, 297,366,401
  bactericidal efficiency of 41
  dependence of concentration on pH 56
  pK value of 56
hypochlorous acidium ion 249, 262

icosahedral virions 147
inactivation rate constant k 18
inactivation time 54
indicator organisms 50
interhalogens 113,301
  absorption spectra 134
  compounds 113,117
iodide
  ion 147,156
  residues 86
  salts 138
  toxicity 138
iodination 147,150,156
iodine 68,79,114,145,146,179,193, 249,260,301,399
  economics 116
  production 115
iodine bromide 113,132
iodine chloride 113,118
iodine monochloride 129
iodometric titration 79
iron 95,107,108,124,126

k-chlorine residual relationship 41
killing time 6-8
Kjeldahl nitrogen 186

leuco-crystal violet (LCV) 82,85, 97,99-103,108-111
linear-free energy relationship 256
log diagram of HOCl-H_2O system 56,57

manganese 81,94,99,102-104,108, 110,124
mass balance 55
membrane filter absorption 361
metabolic products 52
methyl orange 83,85,98
microstraining 212,217,219,221, 222,224
millipore filter method 171
modified orthotolidine-arsenite method (MOTA) 98-103,110
monobromamine 174,175,189, 303
monochloramines 70,74,82,108, 110,111,170,174,188,253, 261,278,284,301,302,350
  interference 102,103
monochlorophenols 261
most probable number (MPN) 7, 165-167,171,182,185,373
  tests 10
multi-hit equation 12,33,41
multi-hit model 12
Multi-Poisson distribution equation 15
Multi-Poisson distribution model 11

mutagens 51

Naegleria gruberi 20-25,194
amoeba 20
cysts 26,27,30-34,45
N-bromo-N-chloro-5,5-dimethyl-hydantoin 137
N-bromo-N-chlorosulfanyl benzoic acid 137
N-chloroamino acids 99
N-chlorobenzenesulfonamide 136
N-chloro-4-toluene sulfonamide 136
neutral o-tolidine titration method 81,85
nitrification 128
nitrites 124,225
nitrogen trichloride 70,74,277
decomposition 282,290
formation 280,283
hydrolysis 294
nucleic acids 50,62,64,159
nucleophiles 263,284

odor 212,249,279
Ohio River 50,54,391
organic compounds 120
chlorinated 52
chlorine-releasing 136
ozonated 62,219
organic haloamines 204
organic molecules 50
organic nitrogen 195
orthotolidine-arsenite residuals 370
Ortho-Tolidine Arsenite (OTA) text 81,85-87
orthotolidine methods 80,86,110
Otto injection system 215
oxidation potentials 252,259,260
oxidative hydrolysis 267
oxychlorine 249
oxygen content, high dissolved 211
ozonation 217,219,229,234,239, 241,245,406,408,413,415
ozone 62,68,79,130,180,181, 211,233,396,398,408
ozone
contactor design 237
demand 62,244
disinfection 234,239,240
gas-to-liquid ratios 235
generation and injection 215
operating costs 229
oxidizing potential 214
potassium iodide reaction 214
properties 214
residuals 214
sterilization 212
ozonizers 221,222

pathogenic organisms 51,146,180
pH 56,200
effect on disinfectant chemicals 55
phenols 213,225,268,270
photoionization potentials (IP) 257
pili 159,160
plaque assay method 365
plaque-forming units (PFU) 365, 391
poised systems of constant potentials 260
Poisson probability 14
Poliovirus I 125,147,153,156,213, 260,364,367,368,370,374,377, 386,408,410
survival data 374
polychlorinated biphenyls (PCB) 269
polychlorophenols 261
polymer complexes 135
polyvinylchloride 237
powder pillows 93
prereacted experiments 195
pyrimidine 271

raw water 64
recycle 49
residual chlorine 67,71,76,78
available 74
compounds 67,68
minimum safe levels 78
also see free available chlorine
resin acids 269

retardant die-away curve 34
reverse osmosis 51
Rhine River 50,52,54
ribonucleic acids (RNA) 150,153, 271,367,368

Salmonella typhosa 220
sewage effluent 50
  secondary 62,127,200,206, 391
sewage treatment 54,217
sludge treatment 230
sodium bisulfite 341
sodium chlorite 251
sodium sulfite 341
sodium thiosulfate 342
sonication 24,25,35
Specific Lethality Coefficient 2-5,10
  amoebic cysts 6
  enteric bacteria 6
  spores 6
  viruses 6
stabilized neutral ortho-tolidine test
  (SNORT) 82,85,87,98-100, 103,108-111,182
  SNORTA 103,104,110,111
standard potential, E° 252,254, 256,257,259
steaming-out process 115
Stirling Approximation 8
stoichiometry 306,308,316,335, 336
  also see tribromamine
Streptococcus faecalis 386
sulfhydryl hypothesis 146
sulfhydryl inactivation 150
sulfur compounds, dechlorination 340
sulfur dioxide, dechlorination 403
  dissolver 406
superchlorination 75
survival curves 12,16-21,40,45,46
  aberrant 12,13,17,20,28,34
  retardant die-away 45
  shoulder formation 12
  sigmoid 45
  stepladder-shaped 17,45
survival probability of organisms 6,12
susceptibility coefficient 3
syringaldazine procedures 83,97, 99,101,104,105,107,109-111

tablet reagents 93
Taft $\Sigma\sigma^*$ values 257
taste 67,249,279
teratogens 51
tertiary amines 257,266
test kits 91
  accuracy 100
  buffer capacity 93,94
  configuration 93
  interference 94,95,99
  modified 101
  selection 96
  specificity 93
  water characteristics 93
tetraglycine hydroperiodide 138,206
thermodynamic constants 255
  also see standard potential
timing method 101
tribromine
  decomposition 301
  differential rate equation 327, 335
  half-life analysis 333
  integrated rate equation 330
  kinetics 301
  pH effect 326
  stoichiometry 308,317
trichloramine removal 352
trichlorocyanuric acid 130
triethylamine 256,267
turbidity 94,95,394
typhoid 49

ultrafiltration 362
ultraviolet absorption 303
urine 278
USPHS Drinking Water Standards 51,77

van't Hoff
  analysis 336

van't Hoff
  conditions 322
  order and rate determinations 317,319
  plots 323-329
Venturi throat 215,225
viral components 147
viral contamination 391
viral nucleic acid 54
viral protein 150
viruses 41,44,51,59,62,64,145, 146,157,220,245,260,261, 359,392
  chlorine disinfection 366-369, 371
  concentration methods 360-363
  Coxackie 377
  disinfection 370
  Echovirus 382
  inactivation 391,408,410
    systems 415
  removal 393
  titration procedures 365
  also see enteroviruses; f_2 virus; Poliovirus
wastewater 52,53,269,278
  bromination chemistry 180
  chlorination 277
  disinfection 119,120,129, 163,170,233,301,402,411
  effluents 391
  samples 372,392
  treatment 51,63,339,340,369, 392,395-400,406,413
    tertiary 359,370
water
  disinfection 12,67
  drinking 55,136,212,213, 216,272,277,392
    USPHS standards 51,77
  natural 94,95,99,103
  organically contaminated 95,96
  storage 63
  synthetic 99
  treatment 63,91,212,277
waterborne epidemics 359